Humanizing Cities Through Car-Free City Development and Transformation

Rahma M. Doheim
Assiut University, Egypt & University of Business and Technology, Saudi Arabia

Alshimaa Aboelmakarem Farag
Zagazig University, Egypt

Ehab Kamel
University of Central Lancashire, UK

A volume in the Advances in Mechatronics and Mechanical Engineering (AMME) Book Series

IGI Global
PUBLISHER of TIMELY KNOWLEDGE

Published in the United States of America by
IGI Global
Engineering Science Reference (an imprint of IGI Global)
701 E. Chocolate Avenue
Hershey PA, USA 17033
Tel: 717-533-8845
Fax: 717-533-8661
E-mail: cust@igi-global.com
Web site: http://www.igi-global.com

Library of Congress Cataloging-in-Publication Data

Names: Doheim, Rahma M., 1976- editor. | Farag, Alshimaa Aboelmakarem, 1979- editor. | Kamel, Ehab, 1975- editor.
Title: Humanizing cities through car-free city development and transformation / Rahma M. Doheim, Alshimaa Aboelmakarem Farag, and Ehab Kamel, editors.
Description: Hershey, PA : Engineering Science Reference, [2020] | Includes bibliographical references and index. | Summary: "This book explores the concepts of car-free cities and city humanization as possible solutions to reduce the deteriorating effect on the environment and the community. The publication discusses the urban initiative to implement pedestrianization and humanization of cities and public spaces to promote the concept of car-free living"-- Provided by publisher.
Identifiers: LCCN 2019054309 (print) | LCCN 2019054310 (ebook) | ISBN 9781799835073 (hardcover) | ISBN 9781799835080 (paperback) | ISBN 9781799835097 (ebook)
Subjects: LCSH: City planning--Environmental aspects. | City planning--Social aspects. | Urban transportation--Planning. | Land use, Urban. | Sustainable urban development.
Classification: LCC HT166 .H87 2020 (print) | LCC HT166 (ebook) | DDC 307.1/216--dc23
LC record available at https://lccn.loc.gov/2019054309
LC ebook record available at https://lccn.loc.gov/2019054310

This book is published in the IGI Global book series Advances in Mechatronics and Mechanical Engineering (AMME) (ISSN: 2328-8205; eISSN: 2328-823X)

British Cataloguing in Publication Data
A Cataloguing in Publication record for this book is available from the British Library.

For electronic access to this publication, please contact: eresources@igi-global.com.

Advances in Mechatronics and Mechanical Engineering (AMME) Book Series

ISSN:2328-8205
EISSN:2328-823X

Editor-in-Chief: J. Paulo Davim, University of Aveiro, Portugal

MISSION

With its aid in the creation of smartphones, cars, medical imaging devices, and manufacturing tools, the mechatronics engineering field is in high demand. Mechatronics aims to combine the principles of mechanical, computer, and electrical engineering together to bridge the gap of communication between the different disciplines.

The **Advances in Mechatronics and Mechanical Engineering (AMME) Book Series** provides innovative research and practical developments in the field of mechatronics and mechanical engineering. This series covers a wide variety of application areas in electrical engineering, mechanical engineering, computer and software engineering; essential for academics, practitioners, researchers, and industry leaders.

COVERAGE

- Nanomaterials and nanomanufacturing
- Bioengineering Materials
- Tribology and surface engineering
- Vibration and acoustics
- Manufacturing Methodologies
- Biologically Inspired Robotics
- Mechanisms and machines
- Micro and nanomechanics
- Computer-Based Manufacturing
- Control Methodologies

The Advances in Mechatronics and Mechanical Engineering (AMME) Book Series (ISSN 2328-8205) is published by IGI Global, 701 E. Chocolate Avenue, Hershey, PA 17033-1240, USA, www.igi-global.com. This series is composed of titles available for purchase individually; each title is edited to be contextually exclusive from any other title within the series. For pricing and ordering information please visit http://www.igi-global.com/book-series/advances-mechatronics-mechanical-engineering/73808. Postmaster: Send all address changes to above address.

Titles in this Series

For a list of additional titles in this series, please visit:
http://www.igi-global.com/book-series/advances-mechatronics-mechanical-engineering/73808

Modeling and Optimization of Solar Thermal Systems Emerging Research and Opportunities
Jagadish (National Institute of Technology Raipur, India) and Agnimitra Biswas (National Institute of Technology Silcha, India)
Engineering Science Reference • © 2020 • 170pp • H/C (ISBN: 9781799835233) • US $185.00

Airline Green Operations Strategies Emerging Research and Opprtunities
Yazan Khalid Abed-Allah Migdadi (Qatar University, Qatar)
Engineering Science Reference • © 2020 • 215pp • H/C (ISBN: 9781799842552) • US $185.00

Recent Technologies for Enhancing Performance and Reducing Emissions in Diesel Engines
J. Sadhik Basha (International Maritime College Oman, Sohar, Oman) and R.B. Anand (National Institute of Technology, Tiruchirappalli, India)
Engineering Science Reference • © 2020 • 298pp • H/C (ISBN: 9781799825395) • US $215.00

Practical Approach to Substrate Integrated Waveguide (SIW) Diplexer Emerging Research and Opportunities
Augustine Onyenwe Nwajana (University of East London, UK) and Kenneth Siok Kiam Yeo (Universiti Teknologi Brunei, Brunei)
Engineering Science Reference • © 2020 • 171pp • H/C (ISBN: 9781799820840) • US $195.00

Diverse Applications of Organic-Inorganic Nanocomposites Emerging Research and Opportunities
Gabriele Clarizia (Institute on Membrane Technology, National Research Council, Italy) and Paola Bernardo (Institute on Membrane Technology, National Research Council, Italy)
Engineering Science Reference • © 2020 • 237pp • H/C (ISBN: 9781799815303) • US $195.00

For an entire list of titles in this series, please visit:
http://www.igi-global.com/book-series/advances-mechatronics-mechanical-engineering/73808

IGI Global
PUBLISHER of TIMELY KNOWLEDGE

701 East Chocolate Avenue, Hershey, PA 17033, USA
Tel: 717-533-8845 x100 • Fax: 717-533-8661
E-Mail: cust@igi-global.com • www.igi-global.com

Editorial Advisory Board

Table of Contents

Detailed Table of Contents

Creating a street where people have the opportunity to interact with each other and socialize rather than suffer loneliness used to be a concern for many scholars over the past decades when automobiles dominated the street scene. This chapter highlights considerable contributions to recognize the role that streets play in the life of a community. It draws out essential requirements for restoring the social role of the street as a place. International practices emerging across the world are outlined in this chapter to draw on the lessons learned from their fresh approaches in putting people first when considering street designs. The chapter concludes with design recommendations that act as the guiding principle to integrate street function and user needs in a way that motivates positive opportunities to create streets as places.

This chapter describes how greenways can be a constituent of sustainable urban mobility (SUM) systems that reduce automobile dependence while simultaneously having positive environmental and social co-benefits. It begins by providing a brief background on the harmful effects of automobile dependency. A chronological review of the evolution of greenways as a typology, divided into five generations starting from pre-1900 until today, demonstrates how various economic, political, environmental, and social factors have shaped blue-green corridors in different cities, mainly in English-speaking countries. The discussion then focuses on the integration process between greenways and SUM planning, as well as highlighting some of the planning challenges and opportunities of (re)developing greenways to support as non-motorized transport corridors. By critically analyzing the evolution of greenways in relation to urban mobility and their integration process, this chapter supports green space, transport, and design professionals to work toward a shared vision of sustainable cities.

Streets, plazas, and parks are important components of a city that play a key role in affording socio-cultural, political, and economic activities for the benefit of society. The physical nature of these urban spaces facilitates sharing of resources, infrastructures, good, services, experiences, and capabilities. Recent socio-economic and technological changes have resulted in a new generation of city design and planning paradigms shifting the way that urban public and semi-public forms and spaces are designed, managed, and used. This chapter addresses the foundational changes brought by smart, or autonomous (AV), vehicles; smart city technologies; and the business models and associated technologies of sharing. The primary goal is to examine how these three socio-economic and technological changes may influence the use of current and future urban public space. It further informs designers on how urban spaces can provide opportunities to create new relationships of use and engaging public experiences through technology.

This explanatory research investigates the impact of configurations in the urban fabric on walkability regarding connectivity and global integration. This study aims to examine this effect by comparing street networks in two residential neighbourhoods in Heliopolis and New Cairo cities. The research methods use DepthMapX and Walk Score to provide relative indicators about distinctive features that enhance the walkability in the case studies. The results show significant differences between the walkability patterns and pedestrians gate counts in two case studies. In study areas, the pedestrian gate count, connectivity, and global integration give signs about people interactions during the day time. The findings figured out that the spatial configurations of street networks—that create urban fabric—play a vital role in enhancing the walkability. The possibility of simulation, to a great extent, can investigate other cases in future research with real situations on the ground. The results confirm a worthy choice for the space syntax technique as a predictor in analysing walkability.

As various cities around the world are implementing car-free policies, the need to understand it from a dynamic point of view becomes more pronounced. In effect, by invoking the organic nature of urbanism, a complexity involving the growth of cities in relation to their environment and human society emerges. Seeking to contribute to an understanding of the production processes of space in the contemporary street, the discussion of the planning future of cities, and perspectives on urban transformation, this chapter aims to build an understanding of the production of spaces for public life in Indonesian cities from the perspective of planning, production, and culture in car-free (day) movement in Bandung. This chapter contributes to the process of spatial production in car-free (day) and implies a reflective paradigm of practice and its potential to illustrate in planning the street transformation-productions of public spaces within the current process of globalization in car-free scheme.

This chapter discusses the challenges and debates related to the concept of shared space street design via demonstrating the case of the ongoing Masterplan development

Section 2
Macro Perspective

but rather a crucial need that should be a global trend. The biggest challenge in this transforming process is to minimize the dependency on private cars. This chapter reviews thoroughly some global practices of inspiring models of transforming into car-free cities around the world. This review aims to identify the success measures for the transformation of a car-free city through investigating the challenges that affected the adoption of the transformation process. This would potentially guide governments and policymakers to select the approach that copes effectively with the cultural, social, geographical, and economic characteristics of their countries.

Foreword

Since the industrial revolution, cities have been and continue to be the birthplace of innovation and embrace economic growth. The Gross Domestic Product (GDP) of some cities exceeds that of entire nations. For example, Guangdong, China, is wealthier than Russia or Mexico. In Latin America, São Paulo State is wealthier than Argentina, Uruguay, Paraguay, and Bolivia combined. Nevertheless, cities are among the reasons for the contemporary environmental crisis. They emit up to 70 percent of greenhouse gases that contribute to climate change. Many practitioners and scholars, such as Jen Gehl, attempted to transform the contemporary city to be car-free. They interrogate the use of public spaces, including streets, roads, and avenues, aiming to enhance the quality of urban life and elaborate schemes for cities that are friendly to pedestrians and cyclists.

In the post-WWII period, many scholars called for a gentle, kinder city with public places for its residents. Jane Jacobs wrote, "Not TV or illegal drugs but the automobile has been the chief destroyer of American communities." In 1968, Henri Lefebvre introduced the concept of the right to the city, where he argued that the city is not just a place for human coexistence and interaction but as a space that shapes the residents' lives. In 2008, David Harvey extended the concept to not only accessing what exists per se but the right to change it.

The quest for the humane city continues. Humanizing Cities Through Car-Free City Development and Transformation that Doheim, Aboelmakarem, and Kamel edited is another step in that direction. The book contains of ten chapters that present experiences from different cities and schools in both developed and developing countries. The chapters offer conceptual and methodological views for both practitioners and those in academia to consider. Delving through the chapters of the book, the reader could deduce lessons learned and best practices that could apply to his/her city.

To the city roads, sidewalks and avenues are like arteries and veins in the human body. They connect the different parts of the city as well as their hinterlands. In Chapter 1, Ahmed argues for transforming the street from a just space used as a path into a place. She resonates with Jacobs' argument. Streets are public space for people to interact and engage. Ahmed concludes the first chapter with design recommendations to transform streets from just paths into livable public spaces.

In Chapter 2, Zawawi, Porter, and Ives examine the role of greenways in reducing automobile dependence, thus generating environmental and social benefits that contribute to the sustainability of a city. Following the review of the development of the concept of greenways, the authors proceed to investigate means to integrate greenways and sustainable urban mobility planning, highlighting challenges and opportunities of (re)building greenways.

Technological advances often impacted the city and society at large. The emergence of sharing economy practices, such as Uber and Lyft, and attempts to transform cities from traditional into "smart" through adopting technologies, such as Big Data and the Internet of Things, are changing planning practice and city administration. Pasalar, Hallowell, and Lu examine how new social, economic, and technological transforms are influencing the use of public spaces. They conclude Chapter 3 by highlighting how technological advances can enhance the use of public spaces.

Chapter 4 provides the reader with research techniques to predict the impact of urban fabric on walkability. Abedo, Salheen, and Elshater examine how configurations in the urban fabric affect walkability. They compare street networks in two residential neighborhoods in Cairo using DepthMapX and Walk Score to provide relative indicators about distinctive features that enhance the walkability in the case studies. The urban fabric, which results from the spatial configurations of street networks, plays a critical role in improving walkability and pedestrian interaction with the environment.

Chapters 5-7 present the reader with experiences and framework for car-free cities. Prasetyo offers an understanding of the processes of producing spaces for public life using Car-Free day in Bandung, Indonesia. Car-free, a placemaking modality, enables the people to capture back the public space. Chapter 6 is another case study from Preston, England, presenting the applicability of the shared space design concept. Turner, Kamel, and Ramadan list the challenges of applying the concept and stakeholders' reactions. Turner, Kamel, and Ramadan conclude that updated and detailed legislative directives in the UK, coupled with an appreciation of the cultural context of the conventional pedestrian and vehicular behaviors, are crucial for the successful execution of shared space projects. In Chapter 7, Friman and Olsson offer the Three-Dimensional Balanced Transport Intervention Ladder as a framework for car-free cities. Their argument echoes that of Jacobs, Lefebvre, and Gehl – the automobile destroys the city.

The second section of the book, i.e., Chapters 8-10, examine the urban topology of car-free cities, the impact of highways, and measures for successfully transforming cities into car-free, respectively. Kone presents in Chapter 8 the different topological parameters of car-free cities. She offers a conceptual framework for a car-free urban environment. In Chapter 9, Dyson questions how much land the British devote to highways? He uses an analytical tool to estimate the land-take of highways by applying mapping techniques to produce a plan or percentage figure. It makes a comparative analysis across several urban centers. The outcome might associate with different measures of civic success, such as urban happiness, pedestrian safety, or retail occupancy. In Chapter 10, Doheim, Aboelmakarem, and Badawi conclude the section and the book with recommendations to transform the city into car-free. To reach these recommendations, the authors reviewed cases from around the world, inferring best practices and lessons learned.

As an educator and practitioner, I am pleased to have read the book and foreword it. I am sure the planning community will welcome this addition to the recent literature on the city.

Ahmed O. El-Kholei
Arabian Gulf University,Manama, Bahrain

Preface

The concept of this book started with exchanged discussions on the need for academics and specialists to put their efforts together to develop alternatives for bringing back urban spaces to people. No doubt, we now live in a world, where there has always been a conflict between urban development and human needs.

Cities have been developed in a way to serve industrial development that had a massive impact on the design of transportation and connection in-between cities to provide a supporting infrastructure for the transportation of industrial materials and goods. On a city scale, this urban trend, along with the rapid urbanization, has resulted in sprawling cities that require a robust transport system to connect the city parts. Additionally, the single-use neighborhoods, big blocks, and segregated pathway network make the private cars the most appropriate transport option for such urban patterns. The cars facilitated the way people travel, eliminating the constraints of distance while offering personal, fast, and convenient ways to commuting. Thus, private cars became the primary means of travel in contemporary cities, and excessive use has led to reshaping cities to accommodate cars rather than people. This has led to deteriorating the natural and built environment that resulted in health problems and social distancing in urban areas. Additionally, the crowded and occupied streets lost essential aspects of urban design for people, such as comfort, enjoyment, and social engagement.

The question we raise here is whether humankind still has, and deserves, to live in the same settings, especially that such settings have proved to threaten the quality of life for both current and future generations.

Car-free approaches were found to be newly experimented in different cities and neighborhoods across the world, as a solution to improve the quality of life for people, reduce traffic and to minimize carbon footprint, and help to reduce the global threat of climate change. It is people's right to get livable cities, where they have a human, safe, enjoyable, accessible, clean, and healthy environment that is mainly designed for them rather than cars to live and work in. Yet, relatively very few academic articles are published on this topic to discuss it academically. As a result, the three editors of this book combined their efforts to co-edit this book to introduce and provide a comprehensive overview discourse of the concepts of city humanization through a car-free city implementation as a possible solution to reduce the deteriorating effect on the community and the environment.

It is worth mentioning that during the final stages of this book editing, the world had witnessed a challenge that the planet Earth experiences for the first time, a total global lockdown due to the pandemic of COVID-19 that is expected to have a significant impact on many aspects of life. Although the fact that people around the world suffered and struggled with the lockdown from many aspects, especially on the mobility level, surprisingly, the COVID-19 global lockdown had a significant and undeniable impact on Earth recovery and air purification due to the almost complete ban of all motorized transport modes. No doubt that after people have experienced better quality of air, breath in cleaner air, enjoyed the clear sky, they will not give it up and maybe willing to make compromises and spend efforts to maintain it. Therefore, while governments put human health and wellbeing as a priority, efforts from authorities and communities should be compiled to take advantage of the current global situation to push governments to adopt car-free city policies and accelerate decision-making with regard to reducing the use of motorized vehicles. This global lockdown experience has highlighted more than ever the importance of humanizing our cities for a better quality of life. Hence, this book provides a discourse on adopting car-free strategies as one of the many approaches that lead to a human-centered city.

AIMS AND OBJECTIVES

The high car reliance in contemporary urban contexts poses a health crisis due to the vehicles' toxic emissions on air quality, and it is responsible for killing more than one million people around the globe every year in traffic accidents. Additionally, the city's residents are deprived of enjoying public spaces, and social inclusion, as a result of taking over the spaces for the benefit of car drivers. Therefore, there is an essential need for reducing the dominance of car use in our cities and transforming them into greener, healthier, and more humanized places to live.

This book aims to promote the car-free city urban design and planning as a potential solution towards humanizing public spaces, which can provide an efficient approach for bringing back urban spaces to their owner, who is the community. The book discusses the initiative of car-free cities as one of the urban initiatives that advocate reducing the dependence on cars and leave streets for pedestrians where they can experience a healthier, safer, and happier life.

The question the book raises here is whether it is possible to bring cities to a human scale, restore the social relationships, and enhance the quality of air. The key objective of this book is to explain the potentials, challenges, and strategies for implementing car-free urban spaces, and how this can promote a healthier city lifestyle, and community wellbeing in general. Due to the current shortage of such information among existing scholarly publications, this book targets professionals, urban planners, and policymakers to initiate a collective willingness to shift the city format, and influence transport policies.

TARGET AUDIENCE

The target audience of this book is professionals and researchers working in the field of Architecture and Urban Planning and Design, particularly in the field of Urban Mobility. Also, the book provides perceptions, insights, and support to policymakers, mobility planning agencies and authorities, and environmental management, urban developers, educators and advanced-level students (Masters/Ph.D.), as well as technology developers, and innovator communities with regards to the challenges and opportunities existing in the shifting of the existing cities to car-free cities.

The book could be recommended to universities' library collection, as it could be a reference for some educational courses, especially that the book topic tackles critical issues in the field of urban design and planning.

BOOK CHAPTERS

The book discusses various concepts and strategies that revolve around adopting car-free concepts and strategies as an approach towards a more humanized city design. The book chapters offer separate approaches that tackle the concept from both micro and macro perspectives; the first seven chapters explore various micro-design tactics, with the focus on street, quarter or neighborhood scale, while the last three chapters demonstrate a more macro-design perspective that relates to city design and urban planning. The authors participating in this book are coming from different countries with various backgrounds, and different experiences and perspectives. This diversity of perspectives enriches the book with different approaches at different levels, emphasizing the importance of the car-free city initiative, and city humanization.

In this book, the first chapter (Contemplating the Street as a Place: An Approach to Restoration), the author highlights the importance of humanizing the street by designing it for people rather than designing for vehicles, where the people have the opportunity to interact, and socialize rather than suffer loneliness. This chapter calls for restoring the social role that the street plays in the life of the community. The chapter highlights the essential requirements to restore the street as a place.

International practices emerging across the world are outlined in this chapter to draw on the lessons learned from their fresh approach in putting people first as a start point when considering the street design. By the end of this chapter, the chapter concludes design recommendations that could guide the urban designers to integrate street function and user needs in a way that motivates positive opportunities to create streets as places.

In the following chapter (Greenways and Sustainable Urban Mobility Systems), the authors call for a shared vision of sustainable cities. They discuss the issue of reducing car dependency by integrating greenways with sustainable urban mobility systems, while simultaneously having positive environmental and social co-benefits. One of the strengths of this chapter is the chronological critical review and analysis of the evolution of the greenway concerning urban mobility, in which this analytical review is conducted along five generations starting from pre-1900 until today, demonstrates how various economic, political, environmental, and social factors have shaped blue-green corridors in different cities. This chapter highlights some of the planning challenges and opportunities of (re)building greenways to support non-motorized transport corridors.

In the third chapter (Smart Cars, Smart Cities, and Smart Sharing: The Changing Nature of Urban Public Spaces), in a micro-scale level of streets, plazas, and parks, the authors discuss the vital role of the urban spaces in affording socio-cultural, political, and economic activities for the benefit of society. In this chapter, the readers can understand how the recent socio-economic and technological changes have resulted in a new generation of city design and planning paradigms shifting the way that urban public and semi-public forms and spaces are designed, managed, and used. This chapter addresses the foundational changes brought by smart or autonomous (AV), vehicles; smart city technologies; and the business models and associated technologies of sharing. The primary goal of the chapter is to examine how these three socio-economic and technological changes may influence the use of current and future urban public space. It further informs designers on how urban spaces can provide opportunities to create new relationships of use and engaging public experiences through technology.

The key finding of Chapter 4 (The Down of Walkability) is understanding how the spatial configurations of street networks—that create urban fabric—play a vital role in enhancing walkability and pedestrian interaction with the environment. In this chapter, the authors used DepthMapX, the spatial analysis software, to investigate the impact of spatial configurations in the urban fabric on walkability, by comparing street networks in two residential neighborhoods in Heliopolis and New Cairo Cities, Egypt. They also used the pedestrian gate count method to understand people's interactions during the daytime. This chapter provides relative indicators about distinctive features that enhance walkability and confirms the validity of the space syntax technique as a predictor in analyzing the walking environment.

In Chapter 5 (The Car Free [Day] Movement: Transformation of Space and Place in Bandung, Indonesia), the author discusses the implementation of the policy of Car-free day from a dynamic point of view. The author criticizes the process of spatial production of space for public life in Bandung city, Indonesia, from the perspective of planning, production, and culture. This discussion contributes to an understanding of the production processes of space in the contemporary street, the urban planning future of cities, and perspectives on urban transformation. This chapter implies a reflective paradigm of practice and its potential to illustrate in planning in the street transformation-productions of public spaces within the current process of globalization in the Car-free scheme.

In Chapter 6 (Confronting the Conundrum of Shared Space Street Design), the authors examine the design, development, and delivery of the University of Central Lancashire's Masterplan Programme of Capital Projects in the City of Preston, England and the impact that shared space streetscape concept design has had on the historical dominance of motor vehicles within a strategic city center interchange. The authors investigate the origins of the shared space design concept, the challenges posed in the context of an urban regeneration scheme, and the reactions of the communities and stakeholders to those opposing this specific type of development. The chapter concludes by emphasizing how updated, and detailed formal legislative guidance in the United Kingdom (UK) and a better appreciation of cultural context from the perspective of traditional pedestrian and vehicular behaviors is paramount to the successful implementation of shared space projects in the future.

In Chapter 7 (Roads to Car-Free Cities: Introducing the Three-Dimensional Balanced Transport Intervention Ladder), the authors discuss the cities interventions to reduce car use in relation to people's autonomy (freedom of choice and self-determination). The discussion is extended to include potential effects on travelers' perceptions of accessibility, satisfaction, and wellbeing after (voluntary or less voluntary) travel mode changes. The authors of this chapter introduce a "Three-dimensional balanced intervention ladder" as a framework that can be used to describe autonomy in relation to reduced car use, balancing this against perceived accessibility and people's wellbeing. The chapter ends with a discussion on the implications for research and policymaking.

With a macro-design perspective, Chapter 8 (Urban Topology of Car-Free Cities), focuses on different topological parameters of car-free cities. The aim is to understand the topological difference between an automobile city and a car-free city. The chapter also looks into Masdar city, UAE, as an example of a city employed to create an urban car-free environment. The chapter introduces a conceptual framework of a car-free urban environment in a sustainable approach.

In Chapter 9 (Highway Robbery: Measuring the Land Use Priorities of a British Town), the author raises critical concerns regarding the urban planning of our cities and how much land we devote to highways, road surfaces, verges, embankments, cuttings, car parks, and other related infrastructure by-products. This chapter introduces an analytical tool to estimate the land-take of highways using mapping techniques to produce a plan or percentage figure and makes a comparative analysis across other urban centers.

In the last chapter of this book (Success Measures for Transforming Into Car-Free Cities: Recommendations for Implementation), the authors review some global practices of inspiring models of transforming into car-free cities around the world thoroughly. This review identifies the success measures for the transformation of a car-free city through investigating the challenges that affected the adoption of the transformation process in different cities around the world. The success measures would potentially guide governments and policymakers to select the approach that copes effectively with the cultural, social, geographical, and economic characteristics of their countries. The strength of this chapter lies in the recommendation for the successful implementation of the car-free practice, where the authors specify the three pillars that are essential for successful transformation into car-free cities: planning, policies, and people. Those three pillars should be balanced and incorporated together for a successful car-free implementation process.

KEY CONCEPTS

In this section, the editors clarify the key concepts and terms that have been significantly used in this book to reveal the unique meanings beyond those terms for the audience. Those terms are used in this book and are essentially related to the topic of this book:

- **Accessibility:** The quality of being easily reached, entered, or used by people who have a disability.
- **Autonomous Cars:** The self-driving car is a vehicle that is capable of sensing its environment and moving safely with little or no human input.
- **Autonomy:** Is defined as a travel license given by the parents referring to the child's likelihood to travel without adult supervision.
- **Car-Free (Zone):** It is an urban area where the people rely primarily on public transport, walking, or cycling for transport.
- **Car-Free (Day):** It is an event that is held in some cities to encourage people to give up their cars for one day, which is probably held on the first Sunday of the month.

- **Commuting:** A term used to refer to periodically recurring travel between locations.
- **Complete Streets:** Streets that are designed and operated to enable safe access for all users, including pedestrians, bicyclists, motorists, and transit riders of all ages and abilities.
- **DepthMapX:** Is a multi-platform software platform to perform a set of spatial network analyses designed to understand social processes within the built environment.
- **Humanized City:** It is a concept used to seek to improve a city's living conditions for its citizens by creating more pedestrian-friendly streets and sidewalks, rather than creating spaces for vehicle accommodation.
- **Manual for Streets:** A manual that contains guidelines and principles to help designers to design, construct, adapt, and maintain new and existing residential and non-residential streets.
- **Mobility:** It refers to the people's ability to access and move, from place to place, by walking, cycling, or public transport.
- **New Urbanism:** A planning and design approach that emphasizes the function and ethics of (re)building urban areas to be walkable, diverse, connected, mixed-use, aesthetically pleasing, and compact while protecting natural environments.
- **Non-Motorized Transport:** It refers to human-powered modes such as walking, bicycling, skates, rickshaws, skateboards, scooters, and wheelchairs.
- **Public Participation:** A term that refers to the inclusion of the public in the decision-making process.
- **Quality of Life:** It is a term used to refer to individuals' wellbeing, including different life aspects from physical health, family, education, employment, wealth, safety, security to freedom, religious beliefs, and the environment.
- **Shared Space:** An urban design tactic where streets are equally used by vehicles, cyclists and pedestrians without hierarchical priorities, with less/no reliance on conventional traffic signals.
- **Smart Cities:** It refers to the cities that are designed to implement new Information and Communication Technology applications, to improve the quality of life of citizens.
- **Smart Mobility:** It refers to using the available technologies to improve the coordination and integration of different transportation modes to provide users with a more efficient, clean, and equitable transport network.
- **Space Syntax:** Is a science-based, human-focused approach that investigates relationships between spatial layout and a range of social, economic, and environmental phenomena.

- **Sustainable Urban Mobility:** It refers to afford the travel needs of the current generation with minimal environmental impacts, and without affecting the ability of future generations to meet their own.
- **Transportation Infrastructure:** The fundamental transportation facilities and systems that are serving a country, city, or area.
- **Transport Modes:** They are different means of transportation by which passengers move from place to place.
- **Travel Behavior:** It refers to the people's decision to arrange their travel, regarding travel mode choice, route choice, departure time choice, and destination choice.
- **Urban Compactness:** The term is to refer to the dense arrangement of urban features in urban areas.
- **Walk Score:** Is a method to measure the walkability of any urban area.

FUTURE DIRECTION

This book has raised many questions and highlighted several concerns that could open new potentials for researchers and academics to investigate new solutions for the current challenges. Additionally, the current situation of the COVID-19 global lockdown that the world witnesses has raised more challenges orienting researchers in the field to investigate the potential future of the cities.

Although the fact that people around the world suffered and struggled with the lockdown from many aspects, especially on the mobility level, surprisingly, the global lockdown had a significant and undeniable impact on Earth recovery and air purification due to the almost complete ban of all motorized transport modes. And, within the lockdown period, governments developed different smart alternatives to handle people's needs and deal with mobility challenges through the virtual world (e.g., distance learning, work from home, delivery services, home-medical services, digital governmental services). This made us realize that all what was hard, resistible, and impossible regarding the implementation of car-free before the global lockdown, is now easy, acceptable, and possible. Therefore, we should think of a car-free city differently and have new approaches to free streets and cities from cars. This would be the starting point for a new era of future car-free cities that are healthier and more humanized.

However, there is a universal concern now regarding the expected change of travel attitude, in which people are expected to use more private modes than before the pandemic. People would be more concerned with infection and prefer to avoid any kind of social inclusion in mass transportation modes. Such public concerns may reform how people perceive conventional social distances, and can, accordingly, transform urban design measures in general, and public transportation, in particular. Therefore, researchers should investigate safer, healthier, and more sustainable mobility alternatives. Another challenge that urban planners and designers will face is to study the balance between the virtual and physical public spaces due to spatial distance, as the expected reformed 'new normal', and accordingly how this will affect urban streets. Smart solutions are highly required to motivate people to socialize again in public spaces overcoming their fear of the threats of another pandemic.

There is a need for research and scientific discussions that invest in, and investigate the current situation of the forced lockdown and find possible alternatives. No doubt that the future will bring new challenges for researchers that would need smart investigations and resilient solutions.

Acknowledgment

The Editors would like to express their gratitude to all who contributed to the development and delivery of this book. In the process of editing this book, we enjoyed and benefited from reading the work submitted, and we would like to deeply thank all the chapters' authors who enthusiastically contributed their knowledge and experience to enrich this scholarly publication, meeting usually tight deadlines. Equally, the Editors show their deep appreciation to the panel of reviewers for offering their valuable comments, which contributed to improving the criticality of the book content. In addition, we would like to expand our deepest gratitude to Professor Ahmed O. El-Kholei, who, in his forward, put our publication into the wider context of current academic discourses. At last, we cannot forget to thank our families, who work on providing a supportive environment to work in.

Section 1
Micro Perspective

Chapter 1
Contemplating the Street as a Place:
An Approach to Restoration

Nora Osama Ahmed
Housing and Building National Research Center (HBRC), Egypt

ABSTRACT

Creating a street where people have the opportunity to interact with each other and socialize rather than suffer loneliness used to be a concern for many scholars over the past decades when automobiles dominated the street scene. This chapter highlights considerable contributions to recognize the role that streets play in the life of a community. It draws out essential requirements for restoring the social role of the street as a place. International practices emerging across the world are outlined in this chapter to draw on the lessons learned from their fresh approaches in putting people first when considering street designs. The chapter concludes with design recommendations that act as the guiding principle to integrate street function and user needs in a way that motivates positive opportunities to create streets as places.

INTRODUCTION

The official definition of urban design is *the art of making places for people*, so each component of the city should be designed as a place for people. Urban design is about creating and retaining areas that operate effectively for all the individuals who are components of that location and are appealing to them. While' Ease of Movement' may seem to be the one goal most strongly linked to what transportation experts do. To date, however, transportation engineers, designers, and planners

DOI: 10.4018/978-1-7998-3507-3.ch001

have concentrated primarily on developing roads that are a major part of our cities ' public space as a vehicle only, regarded as traffic pipelines, as a way of reaching and moving cars safely and efficiently, and overlooking streets as public spaces.

In reality, this one-minded focus regarded the environment from behind a window and saw the street as an issue for vehicles to be solved. On streets designed for fast-moving cars with their broad lanes, traffic signals, and stripping the noise, exhaust fumes, and the danger of injury forces pedestrians and bicyclists to retreat, and the outcome is the most known city today: sprawling, traffic flows, and the danger of injury. This car-centered planning has failed to take into account that the street can be a place for personal motor vehicle motion and storage the same as being a place for the public where a wide variety of uses is taking place.

The main objective of this study is to address the problem of neglecting the variety of street functions, which is one of the factors affecting transforming into car-free cities and humanizing cities.

The study calls for changing the way that streets are thought about and designed. It takes into the consideration that understanding how streets are utilized, and what the space allotted is made for, is determined by the design and functions that had chosen for the streets as well as the way of integration between street function and user needs. That is why treating the street just as a 'corridor' devoted for the moving cars and disregarding its role as a 'space' where much urban life occurs restricted its contribution to better communities with distinctive character and understanding the social and economic benefits that can result from strengthening their role as multiuse places.

THE STREET AS A PLACE - A HISTORICAL PERSPECTIVE

In a historical perspective, the city used to be a dense, traditional city, spatially well-deðned public space with urban social functions where everyone moved by foot these principles of the Middle Ages predominated until the ðrst half of the 20th century when cities faced severe health challenges and the spread of diseases as cholera. The modernism movement appeared and considered the medieval qualities as the problem proposing urban building solutions to help offset these challenges. This perspective was supported by a series of essays titled 'Towards a New Architecture' published by Le Corbusier in 1923 that advocated rational modern buildings with straight, large buildings, highways, and large green areas and functional cities. Although its good intentions to create better conditions for people and lessen the severity of diseases through abandoning overpopulated dense cities and building a modern one that permits light and the follow of air, modernism was criticized for concentrating on the form more than life; prioritizing buildings over surroundings. (Gehl & Svarre, 2013)

In the years between the two world wars, the principles of modernism gradually prevailed until it displayed its strength in the 1960s when the automobiles became an essential element of everyday life and the street scene. This modern movement was synchronized with the rise of functionalism, which brought social suffering upon many urban societies and created additional social distress. Consequently, a new counterattack began against the social inadequacy of the modernism planning ideology. It took place at the last meeting of the Congré d'Architecture Moderne (CIAM) in 1956, where Alison and Peter Smithson, among others, presented their doubt towards the city's sterility as envisioned by the modernists. The aim of the oppositionists was, according to Moughtin (2003), *the rehabilitation of the street as a legitimate element of civic design.*

Many architects and sociologists supported this counteroffensive and tried to restore the city generally and the street specifically as a true social arena. Amongst Jacobs (1993) saying *If we can develop and design streets so that they are wonderful, fulfilling places to be – community-building places, attractive for all people – then we will have successfully designed about one-third of the city directly and will have had an immense impact on the rest.* According to Gehl (2010), Jane Jacobs was the first strong voice to request for radical reform in the development of cities, for almost 50 years, Jacobs raised the discussion of 'death and life defiantly' in cities in her seminal book The Death and Life of Great American Cities in 1961. She pointed out that the dramatic increase in car traffic and urban planning of modernism, which divides the city's uses and stresses the importance and significance of self-employed individual buildings will end the urban environment and city life and contributes to the absence of people in desolate cities.

The theories of Jacobs were developed further by Jan Gehl in his influential book Life between buildings, first published in 1971, documented the social inadequacy of the modernistic public space, and called to think about people and not just design. In doing so, he focused his studies on the space between buildings and developed methods for studying the interaction between public space and public life to quantify which kind of conditions induced people to move and work.

In the same way, William H. Whyte who had experienced great success as a sociologist in his book 'City- Rediscovering the Center', which was first, published in 1988; appraised the social use of urban streets and give guidelines how to design and manage urban spaces to affect social use positively. Whyte's main message is to emphasize that the fascinating things about the street are the interchanges between people that take place in it, Whyte (1988) said *what attracts people most, it would appear, is other people.*

Around 1970, studying public life and its interaction with public space has gained momentum and has witnessed the emergence of many researchers who contributed greatly to the ðeld. Although their signiðcant work was a rebound of Jane Jacobs's argument of the important social dimensions of the street as space, each researcher addressed the issue through a variety of lenses. The exclusive focus on streets as traffic space rather than places for people was a concern of Donald Appleyard and Allan Jacobs. The former, in his book Livable Streets (1981), empirically proved the social harm of traffic on the residents and helped to create a conversation n about designing new kinds of light or no traffic streets. The latter, in his book Great Streets (1993), collected numerous examples from streets around the world, illustrating those who work well and others who are less successful.

On the other hand, human needs were considered by Christopher Alexander, who values in his book A Pattern Language (1977), people's needs, and their ability to translate it into a design than do architects and planners, and he demonstrates essential qualities to permit this. Similarly, Clare Cooper Marcus highlighted the scarcity of considering people in the residential planning in her book Housing as if People Mattered (1986), and set out the concepts to be taken into account in the design of outdoor space in residential areas, with a particular focus on the needs of children. In addition, Peter Bosselmann, in his book Sun, Wind, and Comfort (1984), illustrated the contrast between two perspectives; professionals versus users in experiencing a particular environment; the first are stationary experiences in public space whereas the second is a moving experience over time.

Accordingly, it could be argued that the years 1960-1985 had witnessed the emergence of the basic books in the ðeld of studying street as a place, which formed the cornerstone for research and enabled to convert the knowledge and the approach into practice started from the mid-1980s to the turn of the century. There was a shift back to the city within these 15 years (i.e., 1985-2000) and the ideas of the urban environment as diversity, prioritization of cars for pedestrians, and the general focus on public space conditions launched in the 1960s and 1970s were supported.

Planners and politicians continue the accumulated movement of Jane Jacobs and others that helped form this study area, in which nonprofit organizations were surfaced to put theories and ideas into practice. For example, but not limited to the Project for Public Spaces (PPS) founded by Fred Kent, who assisted William H. Whyte on The Street Life Project in the 1970s and provides practical recommendations for cities on how to change conditions with particular emphasis on citizen participation. Similarly, Walkable and Livable Communities (WALC) Institute, co-founded by Dan Burden, who sought to create a holistic vision for healthy communities that are pedestrian and bicycle-friendly.

Besides, city officials contributed to this battle for restoring the street from the car dominance such as the former mayor of Bogota, Enrique Peñalosa, who established a new bus-based transit system, bike paths and pedestrian streets with sidewalks that have been highly successful, and transformed a degraded city center street into a dynamic public pedestrian space.

The application of these methods and principles could be considered as a response to the critique formulated against the suggestions of this people-centered perspective and specifically Gehl's urban design recommendations to design for the human scale rather than the building scale. The criticism is directed at the different challenges that face its implementation side; firstly, the comprehensiveness of the process of urban planning incorporates different stakeholders with different interests and wills, so agreeing on one opinion could not be guaranteed. Secondly, the complexity of urban development policies, including building codes, land use plans, design standards, etc. makes the implementation of Gehl's principles more difficult the larger the city because it is better realized on a smaller scale that goes beyond a government's ability to handle the detail. Lastly, the market requirements characterized by certain budget and investment interests make building an iconic visual building that attracts investments a priority even if it has a negative influence on over its role in a particular space (Teipelke, 2016).

It is worth mentioning that there were earlier attempts to challenge modernist approaches to design and urban planning started by Alison and Peter Smithson, who, as stated by Heathcote (2017), introduced the concept of 'streets in the sky' to housing as an attempt to humanize modernism. Smithson explored the idea in numerous urban projects and characterized by numerous sustainability potentials that are positive as offering a calm, safe network of footpath passing through gardens and public places, between sculptural concrete towers that provide spectacular, picturesque views over the city. However, the idea is criticized for being non-public realms, which disguise as social spaces; it is an exclusive private space rather than an inclusive public one. Elevated walkways were modernist compensation for the lost qualities in older urban environments, but it lacks a variety of social issues. Firstly, it is designed for pedestrian movement and does not offer any attractions to motivate pedestrians to remain in the walkway as a place used to be poorly maintained and ill-lit, which reduced its richness and uses. Secondly, it lacked the natural surveillance that came from the continuous flow of users. Lastly, the policy of traffic segregation has not appeared to deliver the safety benefits anticipated, especially in the evening and at night, on the contrary, it created social segregation and lifelessness. Even though these social concerns some cities adopt this kind of project as a kind of rejuvenation, as the city of Seoul in reusing its rail tracks as a sky garden in Figure 1.

Although these efforts have been going on for decades, the pressure from cars and functionalist city planning is increasing in many cities that motivate current thinkers as Eric Mumford provides an overview of how key design figures responded

Figure 1. Seoullo 7017, Seoul sky garden, South Korea (Google Maps)

with design proposals and projects to changing social, technical, and economic conditions. Whereas, Anthony Vidler in his book 'Scenes of the Street', underlines the importance of paying attention and observing the psychological pressures and challenges of people due to new construction or renovation projects, just as the thought of how a particular building or public square looks.

It became a necessity to abandon the old way of thinking that governed the street design and adopt a new paradigm to gives rise to streets that unify a community rather than dividing it. In this context the study determined two fundamental requirements that should be taken into consideration and had another look at in re-establishing the streets ' social function to restore the street; the first is to redefine the street as a place rather than a traffic route and the second is to reconsider the classification system of the street.

FUNDAMENTAL REQUIREMENTS TO RE-ESTABLISH THE STREETS ' SOCIAL FUNCTION

Redefining the Street as a Place Rather Than a Traffic Route

A distinction should be made between a street and a road as a prerequisite in reclaiming the street, as in Figure 2. The classic study of Duany and Plater-Zyberk (1999), pointed out different points that distinguish a 'Street' from a 'Road' stating that a street is a slow-moving local way, ideal for general and central areas. It has façades for buildings with a higher density, including banks, stores, apartment buildings, and row housing. It is urban, with its high curbs, open drains, wide sidewalks, a parallel car park, and trees in the different planting areas. Nonetheless, the character can differ somewhat reacting to commercial or residential uses. On the contrary, a road is a slow-moving regional way that is ideal for edge and rural areas. It has frontage to buildings and houses with low density, and instead of sidewalks, it may have clustered plantations and trails. By controlling those elements, the level of the rural or rustic character of a road may be modified.

Figure 2. A distinction between a road and a street, Portland, Oregon, USA (Google Maps)

There is no standard definition of streets; different explanations emerged according to different perspectives as the descriptive definition conducted by Hess and Milroy (2006) stating that a street is descriptively a length, space, not a line from A to B. It refers to the main elements of that area, including the lane, with pavement for vehicles separated into streets with roadways and probably medians. There are side walking and plantings in between the street and the private area, some of them but not all the right-of-way (ROWs). The backyard area often referred to as the front yard, is the land between the ROW and the buildings that line the street. While outside the ROW, the setback area is noticeable and technically part of the roadway.

Although this definition well explained the street, it solely focused on its physical features neglecting its social one, and undoubtedly, the streets are much more than the tarmac between the pavements, though that is where attention has been placed. Many writers who, in their interpretation of the street, found the street to be a place rather than a road and viewed it as a quality place in themselves have long argued this. This was indicated in the definition of Rapoport (1991), which said that streets are more or less small, linear spaces lined by settlement buildings that are used for traffic and sometimes-other activities. His main concern was the use of streets by pedestrians who need to treat the street as a place, which was asserted by the correlation illustrated by Norberg – Schulz (1980) between the street as a place and creating a sense of place in the street.

The street is also a path in terms of Lynch (1960), illustrating that the street is a path activated by several nodes where other paths intersect or where activities increase to such an extent that place and rest compete for domination with path and movement role. However, Moughtin (1991) said that it is more than just a path. It is a set of integrated places to stay in and not just to pass around. His thought came from his investigation about the European city, where he found that that the city has two main functions: it is both a path and a place; it is a means of access and movement but also an arena for social expression,

Consequently, two main concepts dominate the demarcation between the street as a place rather than a traffic route that are; streets are characterized by being places for social integration in addition to its multi-functional role compared to the mono-functional view of roads.

The Street as a Place for Social Integration

There is a social logic in city streets, as Marshall (2009) wrote that the way cities are organized with public streets, and private plots and buildings enable the right combination of security, anonymity, personal control of one's environment, and so on. This contributes to making this space perfect for people who do not know each other and essentially allows the life of a city as a place for strangers.

This had been confirmed long years ago by many pioneers who defended the social usage of the street amongst Jacobs (1961), who concentrated on the space of the street as a practical venue of daily life, and its role as a spatial container of social activities. Jacobs struggled to make modern planners and architects, in particular, comprehend that there are additional factors than the quality of the physical built environment that has an impact on the urban quality of life. Thereby, in the way places are designed car is no longer king. Instead, streets should be designed primarily as social spaces – places where people come first, and this needs to create places where people would like to stay.

Jan Gehl went on to develop Jacobs' perspective by illustrating how the physical environment could be improved to become more socially attractive. Gehl (2011) divided human activities into three categories: necessary activities such as going to school and work, optional activities related to enjoying life, and finally, social activities, including various kinds of community activities as just seeing and hearing other people. His study concluded that only optional activities were dependent on the physical environment of good quality that generates social activities. Therefore, Gehl's study sought to rehabilitate the urban space to become attractive for optional use; consequently, it regains the urban social scene as an alternative to the social distress of the dormitory suburb.

On behalf of Allan Jacobs, Streets are places that ought to be able to endure people of many different social backgrounds, which are consistent with Whyte's (1980) opinion that the social life of a place flourishes when streets and sidewalks function as public spaces. It is important to build and maintain the streets in a way that attracts people and promotes street development to consider others taking into account that creating a place is only emphasized by street life and outdoor activities and cannot be achieved by a street landscape project alone. From the previous, it is concluded that the street actually has represented a social arena for millennia and still possesses this quality if they are designed accordingly.

The Street Multi-functional Role in the Built Environment

Whereas a word such as 'road' suggests movement from one place to another, the word 'street' suggests an area for public use but not exclusively devoted to circulation. The street is a multi-functional space by definition that provides enclosure, activity, and movement; a variety of functions characterizes it. The Architecture and the Built Environment [CABE] (2002) summarized them in five principle functions for the street that are: a place for human interaction and sociability, movement for vehicles and pedestrians circulation, access to buildings, parking as storage space, and a route for utilities.

Among these main functions, there are two key functions that designers should pay attention to and understand that they should be treated differently, as stated by the Project for Public Spaces [PPS] (2008); they are the street as a conduit for motor vehicles and the street as a place. The former is analyzed and designed as a two-dimensional space (length and proportions), whereas in the latter function, a sense of enclosure is added to its analysis and design. This third dimension provides the function for the street as a place or exterior room in the city it must possess similar qualities of the enclosure as the public square, so instead of blank walls in the wide fast-moving traffic street, there are shop windows displaying merchandise, people-sized doors, and places to sit in a slow comfortably scaled street.

Reconsidering the Classification System for the Street

The multiple uses of streets, rather than mainly vehicle movements are inevitable to create street as a place, precisely the balance between their functions as a link and a place. The two should be taken together with their relative importance, according to the role of the street within a network. It is only by considering both aspects that the right balance will be achieved. It is often acceptable to concentrate exclusively on one to avoid the other, often on streets carrying heavy traffic volumes, such as high streets.

There are in principle, innumerable ways of describing and potentially classifying streets. However, these different ways of classifying streets used to focus on one function of the street; as a link and denied the street's function as a place (ARTISTS Consortium, 2004).

The study prepared by Marshall et al. (2004) criticized the street classification system for being functional; focusing on one dimension on classifying the street whether link status or place status, it criticized neglecting the multifunctional nature for the street. It illustrated that in treating the street as a link, the classification of the street is strongly correlated to traffic and transport-related criteria and is typically manifested in a road hierarchical classification system; classified into arterial roads, district distributor, local distributor, access roads and shared surface roads; the conventional road design hierarchies. Whereas in treating the street as a place, the classification is related to geographical scale concerning frequency and type of use, the ranking is typically done by relating to the geographical scale of their catchment areas. For example, different urban centers may be distinguished by their degree of specialization – a city center usually has more specialist shops than a district center; a city government has more specialist functions than a local community center. Although paying attention to the place function positively affects the quality of the street, the study argued that place designations are no more or less descriptive than the standard designations of road roles since the role of place in the traditional classification is no direct equivalent.

A New Framework for the Hierarchy of the Streets

This shortcoming in the classification system was clarified by ARTISTS Consortium (2004), illustrating that this conventional classification for the street is often a means to organize types of streets into hierarchies that are more often of the type of road hierarchies than street hierarchies. The focus on motorized traffic is striking, and the place status is, in principle, independent of the link function of the street. One could have 'city status' shops on a street that performed a local arterial role, or local shops on a primary (city status) arterial; the 'functional' specialization of the place status can nevertheless be expressed in geographical terms since it relates to the catchment area of shops or coverage of services.

Thus, both classifications are a descriptive hierarchy as termed by CABE (2002), and it is difficult to find systems that allow for both; a high traffic function and a high urban function. At a closer analysis, it becomes evident that this traffic function is not the same as traffic flow but rather a kind of network function. A new framework to categorize the street has become essential that would be applied through a new classification system for the street and a new hierarchy for the street user.

A new Classification System for the Street: Moving Away From the one-dimensional System of Classification Towards the two-dimensional System of Classification.

The fundamental basis for a new classification system, according to ARTISTS Consortium (2004), is to look at link function and urban place function as two independent variables. By doing so, the system would allow streets to have any combination of these two variables and still be included in the classification system. The basic elements of a new classification system are:

- Link function; defined as the significance of a street as a link in the overall road network
- Urban place function; defined as the significance of a street as an urban place relative to the overall urban system, town center, district center, local center, specific shopping street, etc.

In combining these two elements, the link function and the urban role will be considered in the broader city context. This means that both the link function and urban place function will depend not only on the immediate attributes on the micro-level (including physical form and use) but also in the broader street and urban system considered as a whole. It also ensures that when designing the street in the local area, the specific space and time demand will be reconciled concurrently with the overall functioning of the city (the pedestrians will cross this; for one stream of traffic to take a right or left through another path; and for street trading, etc.). With this approach, any urban street is recognized and has a location in the classification system. Accordingly, rather than the linear ranking typical of conventional practice, it should be a two-dimensional classification framework that comprises independent variables instead of one being the inverse of the other. Each street segment in Figure

Figure 3. Classifying each street section (locale) according to two criteria: link status and place status (Marshall et al., 2004; ARTISTS Consortium, 2004)

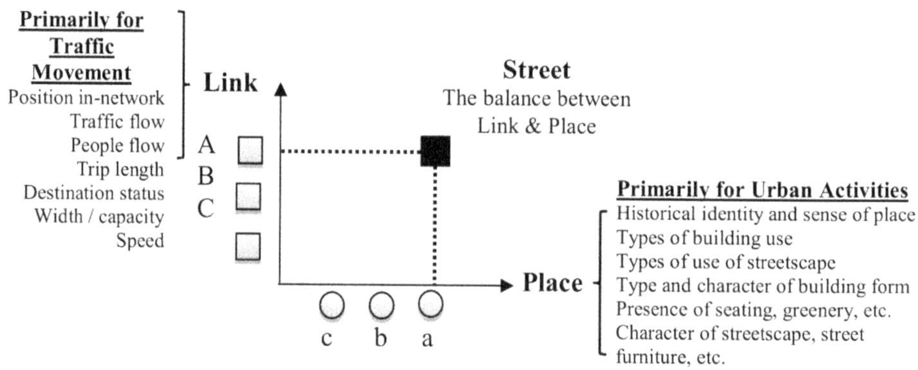

3 is classified according to its link status and place status in order on a rated scale; in both cases, based on geographical importance, both axes have the same scale.

The EU project ARTISTS with the support of the study of the 'Manual for streets' developed this two-dimensional hierarchy as a way of informing street design to balance between a link and a place.

ARTISTS

The Arterial Street towards Sustainability (ARTISTS) project was an EU-sponsored initiative in 2004. The ARTISTS category has two axes, link status, and place status; each axis is placed on an equal geographic scale. This recommends a practical classification system that will identify all street users and uses rather than vehicle movements. Link status is the relative significance of a street section as a link in the network, based on its geographical scale; routes of national significance have more status than routes of city significance, and so on; routes of local significance have the least status. Whereas place status is the relative significance of a street section as an urban place, based on its geographical significance and specialization of its land uses, places of national significance have more importance than places of city significance, and so on. The ARTISTS project represented street types as 'cells' or 'small scale systems' in a 'periodic table' of street types in Figure 4. The number of types recognized (related to the number of levels recognized) and their labels would be tailored to each city's context of the application, and a generic notation is used to demonstrate the two-dimensional basis of the classification. However, in practice, each institution or language would use its appropriate labels (Aurbach, 2009).

Figure 4. ARTISTS street classification matrix (Marshall et al., 2004)

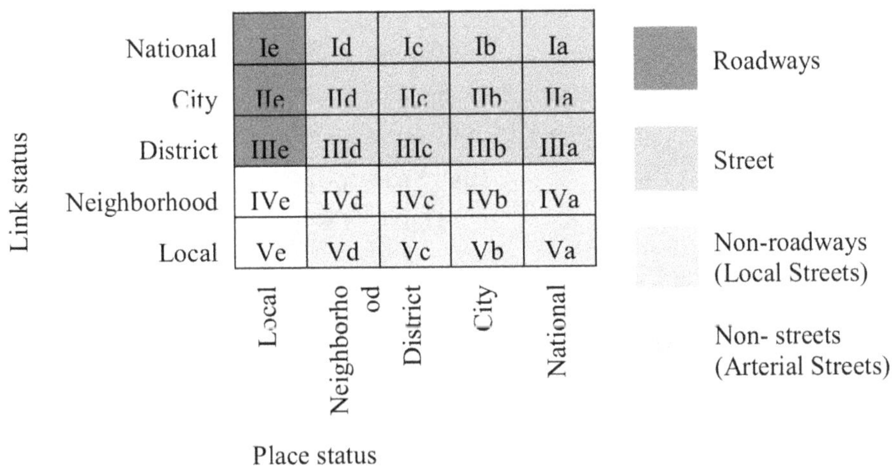

Manual for Streets

The study of the 'Manual for Streets' (MfS) prepared by the Department for Transport and released in 2007, the UK, provides technical guidance for the design of livable, walkable streets such as residential and mixed-use streets as it acknowledges the role that streets play in the community's life, in particular, the positive incentives they can offer for social interaction (Aurbach, 2009).

It builds on the original principles of balancing movement and place status at any particular location, as its key recommendation is that, in determining the character of the streets, among its five functions, place, and movement are the most important. The study examined common street types in different contexts to demonstrate how context and user needs inform a balanced approach to design; as a result, setting the relative importance for place and movement functions of certain streets or roads will help inform design choices.

The Manual expressed this way of looking at streets as a two-dimensional hierarchy termed place and movement matrix in Figure 5. The axes are defined in terms of place status; the relative significance of a thoroughfare 'in human terms,' considering especially local distinctiveness, visual consistency and ability to facilitate social interaction in addition to the status of movement; a combination of traffic efficiency (volume) and geographic size of the destinations served. It recognizes that while some streets are more important than others in terms of traffic flow are; some are also more important in terms of their place function and worth being held to a different standard.

Figure 5. Mfs place and movement matrix, which illustrates typical road and street types in the place and movement hierarchy (Department for Transport, 2007)

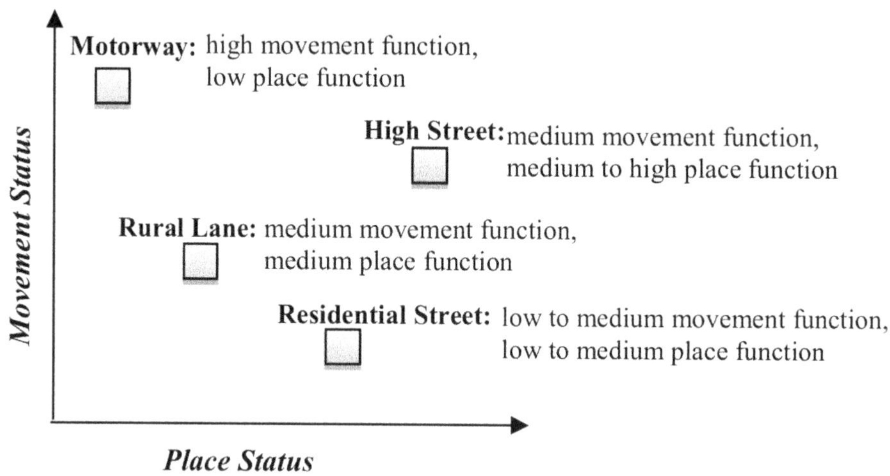

A New Hierarchy for Street User: Moving Away From a Descriptive Hierarchy Towards User Hierarchy.

In the completion of the new approach of street classification that considers the 'place' function of streets, this approach also helps designers to break from earlier approaches to hierarchy, where street layouts are based solely on road traffic criteria, and where the automobile is the pioneer towards a new approach to identifying street hierarchies that position pedestrians and cyclists at the center of the hierarchy of street users.

The two studies ARTISTS project and Manual for Streets asserted this new hierarchy for the street user by proposing a new hierarchal approach to categorize the user of the street in Figure 6 that begins with the proposition that people, not vehicles, should be put first in consideration of street design. Both studies recommend the flexible application of a user hierarchy, in which pedestrians are considered first, then cyclists, public transport users, service vehicles, and lastly, other motor traffic.

Pedestrians are taken into account first, and others should be treated accordingly. It makes the street work in a balanced manner for all of its users. There might be cases where there is no higher-level mode-for instance; buses may not need to be placed on a short, narrow street where car access is required.

Figure 6. Illustrate the new user hierarchy of street design (Department for Transport, 2007)

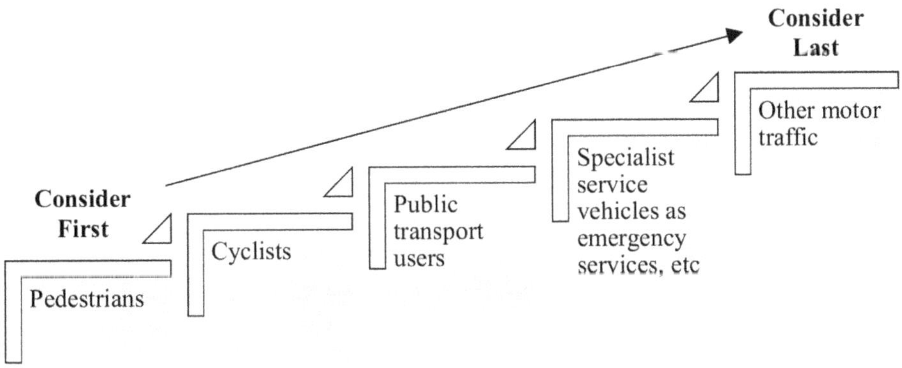

Meeting All Users' Street Needs

People are the most important elements that make a street a place; *places need to offer variety to their users. They need to be unique and different from one another – each rooted in their own particular historical, geographical, physical, or cultural context* (Tibbalds, 1992), however, the conventional approach in treating street as a movement corridor used to design the street according to the car user only and neglect meeting the needs of all other street users.

Streets should work better for everyone; compromising is inevitable between all street users. It should be designed and managed to ensure that they are safe for all; children and young people, cyclists, drivers and passengers, older people, impairment, motorcyclists, pedestrians, public transport users, specialist service vehicles -emergency vehicles, other motor traffic-private, and commercial motor vehicles.

In doing so, it is essential to figure out the need of each user and desire, the study prepared by Marshall et al. (2004) classified users into 'through users' and 'locale users' in Table 1 which contributed in identifying users' direct demands; things they want to do.

INTERNATIONAL PRACTICES THAT SUCCEEDED IN EMPHASIZING THE PLACE FUNCTION OF THEIR STREETS

Many countries from all over the world had begun to focus on the street as an important urban planning element and started to fight back since 1985 in the sense of freeing their cities from traffic dominance and returning it to the people. This chapter highlights three cities that restored their streets by satisfying necessities defined in this study; firstly, they applied two simultaneous actions, on the one hand, design the street as a continuous artery, facilitating efficient 'people movement' along the street. On the other hand, design the street as some locales made up of the buildings, the spaces between the buildings, and the related activities. Secondly, in reestablishing their street to its former position before being occupied by parked cars, these selected cities thought again about their street classification system and adopted a new classification system for the street that moved away from a one-dimensional system of classification towards the two-dimensional system of classification. Besides, it applied a new hierarchy for the street user that abandons descriptive

Table 1. A primary division between 'through users' and 'locale users' (Marshall et al., 2004)

	Through Users	Locale (Place) Users
Definition	A 'through user' is someone wishing simply to pass through the street section as quickly and safely as possible.	A 'locale user' is someone wishing to make use of the attributes of the particular street area (locale) as a place – whether in terms of a market place, a play area, or social interaction.
Subdivision	Through users may be further subdivided by mode of movement, such as pedestrians, cyclists, car users, bus passengers, goods vehicle users, and so on. While there is some competition for the use of the available street width, such as between general-purpose lanes, bus lanes, cycle lanes, and footways, the modes themselves are effectively 'parallel alternatives', in that one typically choose one mode or the other, to serve the same end.	Locale users may be further subdivided into users of different land uses or users engaged in different activities. Unlike different modes of movement, these 'locale uses' tend to be quite diverse and not interchangeable, for the user's point of view, in the sense that one may expect to do any or all of them, without one substituting for the other.
The trade-off of Street-space	The through user just needs to get from A to B. Therefore, the trade-off between rights of way used for different kinds of vehicle users and pedestrians moving along the street is a relatively self-contained transport problem of assessing the different flows and capacities, the proportions of people desiring (or able) to use the different modes for different purposes, and their different consequences for sustainability.	A 'locale user' of street-space needs to use that particular area to fulfill their activity purpose. Thus, from the supply point of view, the local authority may have to prioritize or substitute one use for another, where there is insufficient space to provide for all.

hierarchy to implement user hierarchy, taking into account the inclusiveness in satisfying uses' needs.

Cities replaced the presumption that the car driver has the right of way and not even to expect to have priority over any part of the street with redistributing a street space in favor of non-motorized transportation; allowing people to walk or relax, children to play, bicycles and cars to pass, all in the same space at the same time.

'Haarlemmerdijk' Street, Amsterdam, the Netherlands

It is considered as the best busy shopping street in Amsterdam. It reduces traffic speed through; a narrow street with narrow traffic lanes, parallel car parking, bicycle parking, nearly widened sidewalks. The street in Figure 7 is shared with the bikers and allows access to homes, so the street user immediately understands and feels it is a 30 km/h local street.

Street Classification

ARTISTS Street Classification: Vc - Local link, District place

Place and Movement Matrix for street Classification: Medium to Low movement function and High place function

User Hierarchy

Walking, biking, and motorized vehicle modes, respectively, are given the highest priority in the street user hierarchy.

Street Space Configuration

Design Features as a Link

The street balances the needs of pedestrians with cars and bikes and designed as a one-way street and two-way bike path, marking road bike path for cyclist movement, widened sidewalks, and parking orientation usage act as a buffer between the bike lane and the pedestrian space and a calming traffic measure. Besides, there are a white marking pavers for car services with a marking sign to create safer environments for bicycle users and pedestrians, a lack of signage and markings, and a narrow traffic lane.

Figure 7. Design features & configuration of 'Haarlemmerdijk' street, Amsterdam, The Netherlands (schlijper.nl)

Design Features as a Place

The street uses different elements that are chosen to add vitality to the street, as the widened sidewalks that include cafés, newsstands, and other inviting activities. The landscape design features as garbage bins, an effective alternative grate design, outdoor benches, night lighting. In addition to the street red brick pavements, the outdoor cafes, and the active frontage along the street increase the levels of interest and activity in Haarlemmedijk Street and ensure a vibrant street and urban community with a variety of activities and functions.

'Strædet' Street, Copenhagen, Denmark

It is the first designed shared street in Copenhagen. In the center of Copenhagen near the famous pedestrian street Strøget in which the street in Figure 8 has been rebuilt over some years running through a well-preserved historic district. The purpose of the renovation was to promote the street as a shopping street and to preserve its distinctive architectural character while keeping the necessary traffic in the street.

Figure 8. Design features & configuration of 'Strædet' street, Copenhagen, Denmark (flickr.com)

Street Classification

ARTISTS Street Classification: Vd - Local link, Neighborhood place

Place and Movement Matrix for street Classification: Low movement function and High place function.

User Hierarchy

It is a so-called Shared Area, where cars, bicycles, and mopeds can travel at reduced speeds (normally below 15 km/h). The driving traffic (i.e., cars and bicycles) has to adjust to the characteristics of the pedestrians, so priority is given to pedestrians then bikes and vehicles, and no access to buses.

Street Space Configuration

Design Features as a Link

It is designed as a slowly mixed street that carries a low volume of traffic and designed to slow down traffic, so bicyclists feel secure, but instead of dividing the street, the signs and the street equipment mark the traffic status of the area. It is the steles, benches, and parking spaces that are speed reducing. Its shared street design permits one-way travel for cars, but bicycles can travel, either way, allow two-way bicycle traffic via a contraflow lane. This will make Strædet an important cycle link. A typical sidewalk paving, with cobblestones, characterizes the street and concrete that takes up its usual space and dimension but does not have a curb to differentiate it from the rest of the street section, as a kind of calming traffic measure. Car parking places are marked with white pavers, so drivers know where permitted, and the use of the bike parking at the intersection serves as a means of calming the traffic.

Design Features as a Place

It is designed as a pleasant street to be on, especially for cyclists and pedestrians. It is possible to sit down, eat, drink, and enjoy life, especially during the summer months when the outdoor service at several cafés spill out the street and function as speed reducers. The street's patio café culture and active frontages make the street lively from morning until past midnight and become a popular public space to walk along or to stop at one of the many patios. In addition to the variety in the used materials between silver markings and cobblestone pavement at the intersection and the fountain square at the end of Strædet St. in the center of Strøget added value to the street.

'Swanston' Street, Melbourne, Victoria, Australia

It is Melbourne's main thoroughfare, was renewed as Swanston Street Walk among the different renovations that Melbourne's downtown went through to become a truly livable urban center, and transformed from city's primary artery for north-south car traffic to a pedestrian priority street opened only to pedestrians, public trams, and a few vehicles. Swanston Street Walk-in Figure 9 has retained the familiar shape of a car-bearing road, with distinct traffic lanes, curbs, and sidewalks.

Street Classification

ARTISTS Street Classification: IIb - City link, City place

Figure 9. Design features & configuration of 'Swanston' Street, Melbourne, Victoria, Australia (flickr.com)

Place and Movement Matrix for street Classification: Medium movement function and High place function.

User Hierarchy

The Street currently functions as the tram, bicycle, and pedestrian spine of Melbourne, which balances the needs of all of these users besides the needs of service vehicles and Taxis.

Street Space Configuration

Design Features as a Link

It is designed as a one-way street, curb extensions, widened footpaths, very little on-street parking, a lack of signage and markings, narrow traffic lanes. As a Traditional Street divided into an asphalt roadway and sidewalk, it provides the required facilities for pedestrians, cyclists, tram users, and motorists. Footpaths have been widened for train stops and include cafés, newsstands, and other inviting activities, an on-street bike path with Line marking for cyclist movement and bike hire locations, a tram stop has a raised platform that allows for access to trams for people of all abilities. Additionally, each tram stop will feature bluestone paving and new street furniture. The tram stop area also contains banded paving to signify priority cyclist movements.

Design Features as a Place

The new Swanston Street Walk had become an eminently usable public space through the installation of trees, major public artworks, and a consistent and elegant suite of street furniture – newsstands, drinking fountains, information pillars, new lighting, and self-cleaning public toilets; in which it improved its aesthetic appeal. Incorporation of small-scale street vendors such as the Newsstand and magazine kiosk, the fruit vending stall and flower stall, all address practical retailing requirements but also stimulate use and street activity levels. Other items such as the news pillar, retractable curbside cafe canopy, and information hub make a strong impact on the personal scale, vitality, and color of the streetscape. The street is also characterized by its active edges, which ensured that ground-floor facades appeal to pedestrians and contribute good lighting and levels of interest and activity.

DESIGN RECOMMENDATIONS

The struggle between designing the street according to either its link function or its place function used to be the dilemma in deciding about the street design. Neglecting one of the two dimensions produces a pattern of streets that are out of their context contradict rather than support. An equilibrium between these two main functions in Figure 10 should have strived in designing any street apart from being with high movement function and low place function.

It should be noted that the selection of which elements to be used is based on the categorization of the street, so if it is with a low link and high place, more elements of place will be used and vice versa. The point is to keep in mind the two dimensions in classifying any street, and as indicated by Marshall et al. (2004) in designing the street, the challenge is to match form and regulation to support the use and mediate between competing activities and afford them an appropriate share of space and time.

This study argues that in designing the street in terms of its link function, it should comprise design elements that assist proper movement for accommodating all modes (i.e., motor vehicle traffic, bicycle, and pedestrian traffic). Whereas in designing the street in terms of its place function, it should comprise the actual elements that make up a street for people; walking, talking, watching, sitting, and waiting.

Addressing the Link Function

In designing the street space for through use, it has to be concerned about the different modes of movement not only vehicles so pedestrian, cyclists, and transit riders should all treated equally and reflects the basic facilities that Through Users would need in the street to get from somewhere to somewhere else, such as:

Figure 10. The process of designing a street as a place (the researcher)

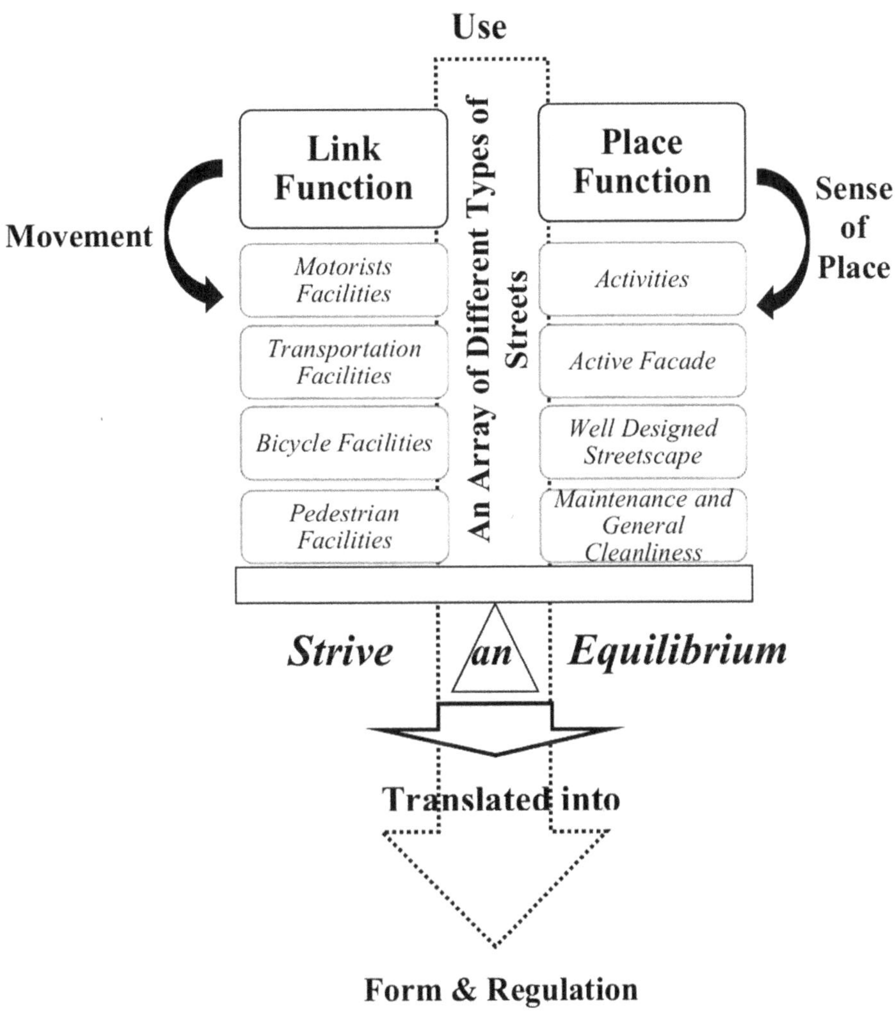

Motorists Facilities

A motorist needs on-street parking that should be designed into streets rather than being an unplanned occurrence, and it should be metered to manage to park. The speed should be controlled with well-designed traffic calming elements that are positioned to form part of the overall design and worked on enhancing safety, same as improving the quality of the built environment, unlike traditional technical techniques, and loading and unloading of goods should be regulated.

Pedestrian Facilities

A pedestrian requires a sidewalk that is wide enough for two-way foot traffic with convenient flooring with no obstructions well illuminated on both sides, a signalized crosswalk of marked zebra with special visibility light, and short block lengths that encourage walking and slow drivers, and provide a comfortable space for people to leisurely walk and experience the surroundings.

Bicycle Facilities

A cyclist needs safe, comfortable bike lanes that permit car traffic but give bicyclists strict right-of-way priority over the entire breadth of the street, special traffic lights at intersections, usually with priority signaling for bikes, and bicycle parking, storage, changing facilities, and lockers.

Transportation (Buses or Tramways) Facilities

Public transportation should be efficient, clean, inviting, and safe, characterized by bus schedules and equipped with comfortable bus stops.

Addressing the Urban Place Function

In designing the street space for locale/place use, it has to be concerned about users of different land uses or users engaged in different activities and reflects the basic facilities that locale/place users would need to stay in the street and enjoy it. The use of different design features that create an environment-friendly place for people on foot and a sense of place and improve the quality of the street environment such as:

Activities

Ensure the presence of the number of activities in Figure 11 that not only brings pedestrian movement to the street but also allows a diverse range of people to voluntarily engage in a wide variety of stationary activities as to stop, rest or do anything other than just keep going.

Figure 11. People are engaged in stationary activities in 'Strædet' Street, Copenhagen, Denmark (Google Maps)

Well-designed Streetscape

Trees and plants should form an intrinsic part of the street design through paying attention that vegetation does not obstruct pedestrian or driver sightlines, select the right flora that provides shade, comfort, and beauty for pedestrians, and maintenance.

Street furniture should be integral to the overall street design in Figure 12 and contribute to creating comfort, safety, and a sense of place using generous pavements with tree planting, various pavement textures, and physical structures.

Figure 12. Well-designed streetscape in 'Swanston' Street, Melbourne, Victoria, Australia (pps.org)

Street lighting should be appropriate to the domestic setting illuminates not only the roadbed, but also the pedestrian areas through positioning lamp columns in a way that define space, enclosures, gateways, sculpture, and variations in levels.

Finishing materials of the street should consider the ease of maintenance and replacement, safe for purpose, durable, appropriate to the local character, a variety of surface treatments, colored and textured surfaces.

Parking lot places should consider diversity as the provision of motorcycle and cycle parking, user convenience, and safety and security concerns by adding walkways, shading, and features that create natural surveillance within parking facilities.

Drainage and rainwater run-off systems should be designed within the street in a way that does not impede the movement of pedestrians, cyclists, and wheelchair users.

Maintenance and General Cleanliness

Sufficient space should be given for periodic and emergency maintenance to keep the street clean and in good repair, taking into account that the street is designed with easy to maintain materials.

Active Façades

Ensuring that building frontages are 'active' by paying attention to detail in the front elevation such as maximizing windows and doors, balconies, canopies and bay windows enliven the frontage, transparent windows enable communication between inside and outside, 'Spill out' space adds vitality to the public realm in Figure 13.

Figure 13. Outdoor cafes act as a 'spill out' space add vitality to the 'Haarlemmerdijk' street, Amsterdam, The Netherlands (schlijper.nl)

SOLUTIONS AND RECOMMENDATIONS

A change in the mindset about designing streets has become essential; redefining the street as a place rather than a traffic route as well as reconsidering the classification system of the street should be the fundamental stone of any approach to restoring the street as a place. Most importantly, to shift from the 'one size fits all' conventional approach in designing the street and classifying its types by focusing only on the street movement or place function and move towards a people-oriented approach with a variety of street types to the design and management of streets. This should be followed by moving away from the descriptive hierarchy that is oriented only to car users towards a new hierarchy giving priority for the user by their use of the street as a place.

Moreover, it is essential to give up the mono responsibility towards streets, since roads authorities have traditionally carried out street classification. In this new classification system, it is expected that the designation of link status would ultimately remain the responsibility of the roads authority, while the judgment of the place status of a street is most likely to be appropriately carried out by the urban planning authority, which normally is charged with making decisions of urban status and land use. However, the overall classification process of judging link and place status of a street section should ideally involve all stakeholders' views and the engagement of any professionals (i.e., planners, the public, etc.) with a stake in the role of the streets in the discussion.

FUTURE RESEARCH DIRECTIONS

This study opens the possibilities for a variety of street space configurations that need further research in treating the street as a general category, which involves a set of sub-classes where Individual Street is formed within it. Besides, there is a need to strengthen the professional background for the selection of solutions for such a typology, which evolves different types of streets that should be treated differently to act as a potential solution towards humanizing public spaces.

CONCLUSION

This study criticized the post-World War II global approach in making roads wider, straighter and faster, which resulted in a decrease in the quality of the street and, consequently, the quality of the city as it suffered from increasing car traffic, which goes together with exhaust gas pollution and noise. The study adopted a view of a point that streets as one of an essential element in an urban set-up can significantly make foremost contributions to the improvement of the quality of urban life if paid attention to their significant social role in shaping the community. In this context, two fundamental requirements were outlined that should be restored in the street, to return to the social role of streets as places, redefining the street as a place rather than a traffic route, and reconsidering the classification system of the street.

In understanding the street as a place, the study discussed the social use of the street, illustrating that many authors highlighted the role of streets in contribution to the quality of public life and emphasized how streets and sidewalks can be captured for social purposes and the need to be designed to support these activities. A distinction between roads and streets was illustrated as well to clarify that if road connects places, then a street connects people, so both should be treated differently. The second fundamental requirement to design the street as a place is reconsidering the classification system of the street. This requires a new framework to define street hierarchies, where people come first, based on the expected presence of pedestrians and bicyclists, and their significance in terms of both place and movement, which connect the 'meaning' of a roadway with its appropriate design and use. Besides, this new framework is completed by classifying street users into users and locale (place) users to meet all user's street needs, and it asserted that the street must satisfy the needs of both users to create this balance. A review of international practices was examined, proving that each studied country had helped their communities to get the most out of their street as an 'exchange space' rather than just 'movement space'. These cases recognized the social importance of streets and demonstrated the potential for more effective use of urban street space where distributing street space among all users is their key to creating and sustaining livable communities.

Finally, the study concluded with an array of design recommendations that address the wide range of different types of streets and support the application of the two-dimensional perspective that enables movements and other activities to be accommodated in the street space.

REFERENCES

ARTISTS Consortium. (2004). *Arterial streets towards sustainability. A Guide for European Cities*. Lund University.

Aurbach, L. (2009). *Towards a functional classification replacement*. PedShed.net.

CABE. (2002). Paving the way: How we achieve clean, safe, and attractive streets. CABE (Commission for Architecture and the Built Environment) Space Publications.

CABE. (2008). Civilized streets. CABE (Commission for Architecture and the Built Environment) Space Publications.

Department for Transport. (2007). *Manual for Streets 1*. Thomas Telford Publishing.

Duany, A., & Plater-Zyberk, E. (1999). *The lexicon of the new urbanism*. Duany Plater-Zyberk & Company.

Gehl, J. (2010). *Cities for people*. Island Press.

Gehl, J. (2011). *Life between buildings: using public space* (6th ed.). Island Press.

Gehl, J., & Svarre, B. (2013). *How to study public life*. Island Press.

Heathcote, E. (2017, May 5). *How architects have tried to create 'streets in the sky*. Retrieved from https://www.ft.com/content/8f2bab62-2b32-11e7-bc4b-5528796fe35c

Hess, P. M., & Milroy, B. M. (2006). *Making Toronto's streets*. The Centre for Urban Health Initiatives, University of Toronto.

Jacobs, A. B. (1993). *Great streets*. MIT Press.

Jacobs, J. (1961). *The death and life of Great American cities*. Vintage Books.

Lynch, K. (1960). *The image of the city*. MIT Press.

Marshall, S. (2009). *Cities design and evolution*. Routledge.

Marshall, S., Jones, P., Svensson, Å., Hydén, C., Draskoczy, M., Papaioannou, P., Thomsen, J., & Boujenko, N. (2004). *ARTISTS - Arterial streets for people: guidance for planners and decision-makers when reconstructing arterial streets*. European Commission Fifth Framework Program.

Moughtin, C. (1991). The European city street. Part 1: Paths and places. *The Town Planning Review*, *62*(1), 51–77.

Moughtin, C. (2003). *Urban design: street and square*. Architectural Press.

Norberg-Schulz, C. (1980). Genius loci: towards a phenomenology of architecture. Rizzoli, International Publications, Inc.

Project for Public Spaces, Inc. (2008). *Streets as places: using streets to rebuild communities*. Author.

Rapoport, A. (1991). *Pedestrian street use: culture and perception in public streets for public use*. Columbia University Press.

Teipelke, R. (2016, April 13). *An Implementation Critique of Jan Gehl's "Cities for People"*. Retrieved from https://placemanagementandbranding.wordpress.com/2016/04/13/an-implementation-critique-of-jan-gehls-cities-for-people/

Tibbalds, F. (1992). *Place matter most in making people-friendly towns: improving the public environment in towns and cities*. Taylor & Francis.

Whyte, W. H. (1980). *The Social Life of small urban spaces*. The Conservation Foundation.

ADDITIONAL READING

Cullinane, K., & Cosgrove, T. (2010). *Smarter Travel in Six European Cities: A Precedent Study*. University of Limerick.

Glaser, M. (Ed.). (2012). *The city at eye level: Lessons for street plinths*. Eburon Uitgeverij BV.

Jacobs, A., & Appleyard, D. (1987). Toward an urban design manifesto. *Journal of the American Planning Association*, *53*(1), 112–120.

Kott, J. (2016). Streets of yesterday, today and tomorrow. *World Transport Policy and Practice*, *21*(4), 42–51.

Marcus, C., & Francis, C. (1990). *People Places: Design Guidelines for Urban Open Space*. Wiley.

Marshall, S. (2004). *Building on Buchanan: evolving road hierarchy for today's streets-oriented design agenda*. In *Proceedings of the European Transport Conference*.

Moudon, A. V. (1987). *Public streets for public use*. Van Nostrand Reinhold.

Southworth, M., & Ben-Joseph, E. (2003). *Streets and the Shaping of Cities and Towns*. Island Press.

KEY TERMS AND DEFINITIONS

Link Function: The role of the street as a 'movement space'.

Locale (Place) User: Any user who makes uses of the features of the particular street area (locale) as a place whether to engage in different activities or use different land uses.

One-Dimensional System of Classification: Linear thinking about the street and classifying its types by focusing only on the street either as a link or as a place function.

Place Function: The role of the street as an 'exchange space'.

Through User: Any mode of movement, whether pedestrians, cyclists, car users, bus passengers, goods vehicle users, etc. who need to get from A to B as quickly and securely as possible.

Two-Dimensional System of Classification: Non-linear thinking about the street and classifying its types by bringing together its two main roles as a link and a place.

User Hierarchy: A people-oriented approach to organize users of the street into hierarchies according to their priority in the street.

Chapter 2
Greenways and Sustainable Urban Mobility Systems

Abdulrahman A. Zawawi
https://orcid.org/0000-0002-6754-8711
King Abdulaziz University, Saudi Arabia & University of Nottingham, UK

Nicole Porter
https://orcid.org/0000-0001-7734-0052
University of Nottingham, UK

Christopher D. Ives
https://orcid.org/0000-0003-0438-7225
University of Nottingham, UK

ABSTRACT

This chapter describes how greenways can be a constituent of sustainable urban mobility (SUM) systems that reduce automobile dependence while simultaneously having positive environmental and social co-benefits. It begins by providing a brief background on the harmful effects of automobile dependency. A chronological review of the evolution of greenways as a typology, divided into five generations starting from pre-1900 until today, demonstrates how various economic, political, environmental, and social factors have shaped blue-green corridors in different cities, mainly in English-speaking countries. The discussion then focuses on the integration process between greenways and SUM planning, as well as highlighting some of the planning challenges and opportunities of (re)developing greenways to support as non-motorized transport corridors. By critically analyzing the evolution of greenways in relation to urban mobility and their integration process, this chapter supports green space, transport, and design professionals to work toward a shared vision of sustainable cities.

DOI: 10.4018/978-1-7998-3507-3.ch002

INTRODUCTION

Automobile dependency in cities negatively impacts people and the environment directly (i.e., air and noise pollution), indirectly (i.e., cardiovascular and respiratory disease), and accumulatively (i.e., climate change). Newman and Kenworthy (1999) define automobile dependence, as "a situation in which a city develops on the assumption that automobile use will predominate so that it is given priority in infrastructure and the form of urban development." (p.60). Sustainable urban mobility (SUM), by contrast, rebalances the priorities of transportation modes in cities. Transitioning towards SUM necessitates limiting the usage of the private automobile and its associated infrastructure, promoting public and non-motorized transport modes, reducing the need to travel, shortening distances, as well as using renewable energy sources (such as biofuels, solar, etc.) to power vehicles (United Nations Advisory Committee of Local Authorities; UNACLA Secretariat, 2013). SUM encourages an ideological shift from the connotation of urban mobility planning as being a mere means of transport provision, to a "focus on the human right to equitable access to opportunity." (UN-Habitat 2013, p.1). Five essential strategies to reduce automobile dependence include prioritization of public transit and non-motorized modes of mobility over car-use, implementing traffic-calming schemes, integration between land-use and transportation planning, application of growth management approaches to delimit urban sprawl, and introducing automobile usage taxations (Newman & Kenworthy, 1999).

This chapter focuses on how the integration of land-use and transportation planning can reduce automobile dependence, specifically through incorporating greenways as a constituent of SUM systems, particularly non-motorized transport (NMT) infrastructure. Sustainable urban mobility planning (SUMP) of cities found in Southern, Central-eastern, and Eastern Europe is not yet implemented due to location-based urban challenges, differences in culture and planning practices, and lack awareness of its potential (CIVITAS, 2019). The same holds in many cities of the Middle East, Africa, and parts of Asia where car-based planning is the norm, and its outcomes are affecting the realization of creating sustainable cities (Dora and Hosking and Mudu & Fletcher, 2011). By highlighting the integration process of greenways and SUMP, as well as the challenges and opportunities, this chapter support automobile-dependent cities to adopt and promote SUMP. At the same time, reduce the gap between the treatment of green infrastructure (one aspect of land-use planning) or, more specifically, greenways and transportation planning as separate issues, like in many developing nations.

The chapter is divided into four sections. First, the authors of this chapter recap the multiple negative impacts of car dependence and identify green infrastructure approaches, and the provision of greenways in particular, as a means of addressing this problem. In the second section, an updated comprehensive chronological review of greenways' evolution worldwide is presented. Moreover, section 2 describes how the realization of greenways as mobility corridors was influenced by several global events, phenomena, and planning initiatives. Relevant literature for these two sections was identified via searching an initial search of the Scopus database using strategically chosen keywords (i.e., greenways, SUMP, and NMT). Additional key literature was identified by snowballing from reference lists and through personal communication with urban landscape planning experts. A qualitative review of this literature was then conducted (Bryman, 2016), with key themes identified and distilled.

Analysis and interpretations of Section 2 findings are displayed in Section 3 by explaining the integration process between greenways and SUMP, in particular, NMT or 'soft mobility' - a human-powered mode that includes walking, cycling, roller skate, and skateboards (La Rocca, 2010). In addition, Section 3 describes how the integration of both can aid in overcoming the automobile dependency of many cities Worldwide. This chapter is concluded, in Section 4, by the reflection on some of the challenges, and opportunities, found in past projects and academic studies of planning urban greenways as non-motorized transport corridors.

SECTION 1: AUTOMOBILE DEPENDENCY AND THE POTENTIAL OF GREEN INFRASTRUCTURE

"the right to have access to every building in the city by private motorcar in an age when everyone possesses such a vehicle is actually the right to destroy the city." (Lewis Mumford, 1963)

The incremental dependency of the automobile has dramatically fragmented the landscapes of post-industrial cities (the second half of the 20th Century) as a primary mode of transport (Carmona, Tirsdell, Heath, & Oc, 2010, p.29; Speck, 2012). This is most clearly seen in the case of leapfrog urbanization, exemplified in North America, whereby population growth was accommodated by low-density greenfield development outside city boundaries. Enabled by immense vehicle ownership, car-oriented transportation networks, and cheap and easily attainable oil, this pattern of development contributed to the loss of the traditional density of cities and the rise of urban sprawl. An extensive body of scholarly work published by both academic, governmental and non-governmental bodies described the detrimental effects of this process on public health, environment, social structure and cohesion, and on non-motorized modes of transit (Newman & Kenworthy, 1999, 2015; Carmona, Tirsdell, Heath, & Oc, 2010). However, some scholars predict that cars will remain a 'fixture of our communities' regardless of future technological breakthroughs, and complete abandonment will cause more harm than good (Carmona, Tirsdell, Heath, & Oc, 2010, p.109; Crevero and Guerra and Al 2017; Speck, 2012). Therefore, efforts to overcome automobile dependency must work in tandem with novel planning approaches that sustain the integrity of ecosystems while simultaneously encouraging smart integrated development.

Green infrastructure (GI) offers one such integrated approach with several advantages over conventional engineering solutions. For instance, having a multifunctional transport corridor can increase land productivity, reduce expenses (implementation or operation or maintenance), and minimize resource expenditure (i.e., conserve open spaces and reduce urban sprawl). GI is generally understood as a network of hubs and links (Austin, 2014), with the links that connect larger tracts of green space known as greenways (Benedict & McMohan, 2002). There are several definitions for greenways; the following by the European Greenways Association highlights movement as a defining characteristic:

"Communication routes reserved exclusively for non-motorized journeys, developed in an integrated manner which enhances both the environment and quality of life of the surrounding area." (EGWA 2000, para. 1)

Non-motorized or 'active' transport networks (i.e., walking, cycling, and scooters) are one of the integral components of sustainable mobility systems due to their low cost, being almost pollution-free, and widely documented social and health benefits (Crevero and Guerra and Al, 2017). Investment in the active transport infrastructure enables the use of public transit, which can facilitate transport for the first and last mile of travel journeys. There is, therefore, a huge potential for greenways as a linear form of GI to help cities overcome automobile dependence. Greenways enable cycling and walking. Cycling reduces traffic congestion, carbon footprint, as well as increasing physical activity, social capital, saves time traveling to work or recreation, easily parked, and cheaper than purchasing a car (Taylor, Giang, Chau, & Aumann, 2017). According to Speck (2012), walkability is "both an end and a means, as well as a measure [...] walkability is perhaps most useful as it contributes to urban vitality and most meaningful as an indicator of that vitality" (p.4). However, before greenways as a solution to urban mobility can be further established, it is necessary to explore their historical design and use.

SECTION 2: THE EVOLUTION OF GREENWAYS

This section identifies the forms and functions, especially mobility functions, of greenways worldwide as evidenced through a chronological review of their planning and design in different contexts. The chronological review is divided into five periods, based on a synthesis of scholarly work that has defined generations of greenway/ infrastructure (Benedict and McMahon 2006; Fábos 2004; Searns 1995; Kullmann 2013; Pourjafar and Moradi 2015). The five periods are:

1. First Generation: the origin of the greenway concept (pre-1900).
2. Second Generation: Landscape Architects and Greenway Planning (1900-1950).
3. Third Generation: Greenways in the wake of the Environmental Movement and Ecological Planning (1950 – 1987).
4. Fourth Generation: Greenways beyond the attenuation of urbanization ills (1987 – 2006).
5. Fifth Generation: Greenways as infrastructure (2006 – present).

Several general observations about this chronology can be made. First, this survey of greenways' evolution is based predominantly on literature originating in the US, Canada, UK, Europe, Australia, and some parts of Asia, reflecting the dominance of literature based in these contexts (see Figure 1).

Second, none of the periods presented is discrete, since there is not a clear moment where one theme stops and another emerges. It is more accurate to suggest that greenway trends, approaches, and technologies have been successfully integrated into existing and newly built projects over time.

Third, issues related to urbanization and population growth are a continuing struggle to this day in many parts of the world. Even though each generation had unique events that affected the evolution of greenways, rapid urbanization and population growth remained constant challenges of greenways.

Lastly, mobility is only one of many functions greenways are designed for. Thus, greenways' evolution must be understood as a function of different scales, goals, landscape contexts, and planning strategies (Ahren, 1995). In this way, we shall holistically look at greenways in this section, while highlighting mobility as one important function throughout.

First Generation: The Origin of the Greenway Concept (pre-1900).

Greenways before the 20th century have been referred to as "Ancestral Greenways" (Ahern 2003; Imam 2006; Searns 1995). Ancestral greenways are a broad concept, including the formal axes and boulevards of Europe and other parts of the world reaching back to the ceremonial avenues in Ancient Egypt, the walkways found in highly geometrical Islamic gardens, the and religious ways of Rome (Turner, 2006). The axes of Versailles and the boulevards of Champs Elysees in Paris dating

Figure 1. Scopus analysis of search results (a total of 505) featuring the term 'greenway' from 1984 until 2019 (darker shades represent more research publications).

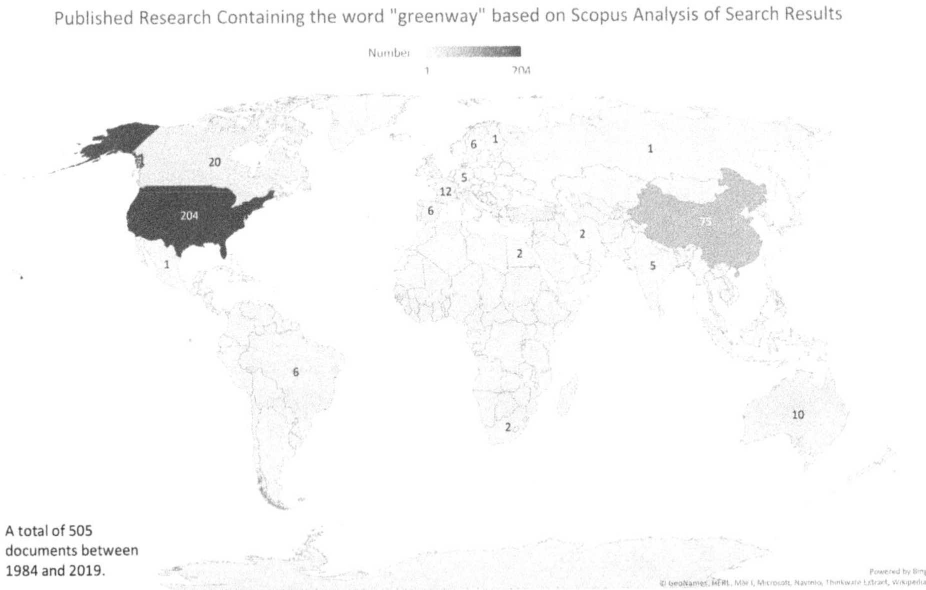

circa the 1700s are examples of what later evolved into the concept of large scale urban greenways (See Figure 2). These ancestral greenways displayed grandeur and connected access ways that manifested both beauty and power, as well as unique scenic values.

During the eighteenth century, greenways and waterways influenced the economic and spatial development of the urban landscape in many European cities. Both in Paris, France, and Campine, Belgium, power provision for mills and workshops, as well as transport were dependent on urban canals (Nijhuis & Jauslin, 2015). In the Netherlands, during the same period, paths and lanes were utilitarian features of the urban landscape (Loos & Vliet 2014). Furthermore, plants were selected based on their usability and growth rate. For instance, in the Netherlands, plants that were planted at both sides of the paths were used to produce wooden shoes. Other important events include the industrial revolutions that affected the pace and shape of development patterns and urbanization. Greenways and waterways during the eighteenth and in nineteenth-century were utilitarian features that affected the economic and spatial development of cities.

In the US, the term "parkways" was introduced by the famously recognized father of the greenway movement, Frederick Law Olmsted (Walmsley, 1995). Olmsted's book "Public Parks and the Enlargement of Towns" published in 1870, viewed the

Figure 2. Champs Elysees 1860s in Paris, France (Public Domain).

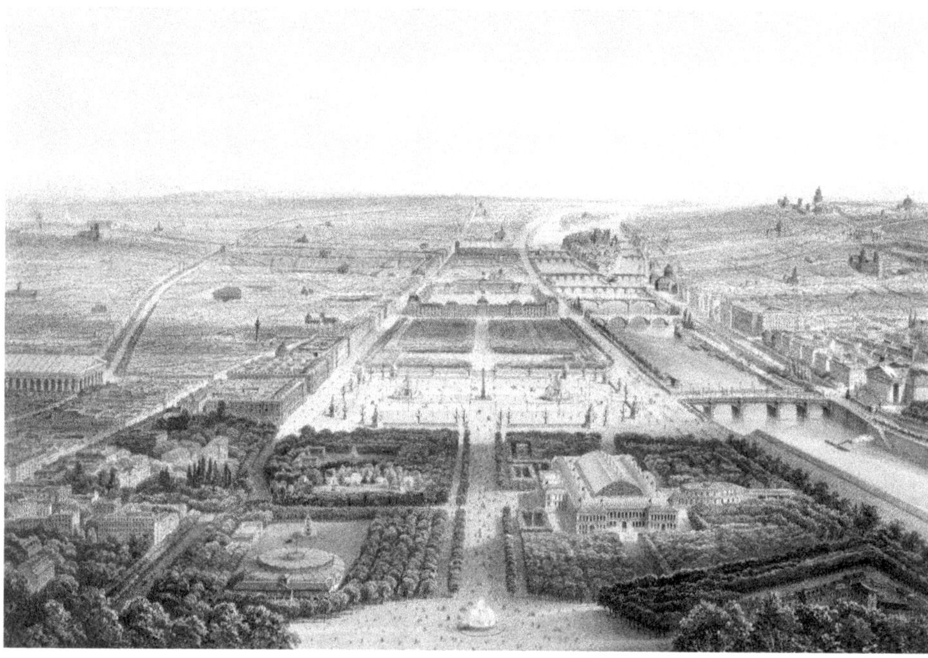

planning concept of parkways as a way to shape urban growth and introduce nature to cities. Olmsted's goal was to permeate and structure urban fabric via parkways and park systems so that no part of town is more than a few minutes away from any of these spaces. One of his greatest achievements (in addition to Central Park of New York) was the Emerald Necklace, Boston, US (1867), an integrated parkway system that not only preserved natural patches and corridors but also connected with distinctive cultural and historic locations in the city (Walmsley, 1995). Olmsted, along with the work of Peter Lenne, George E. Kessler, Horace W.S. Cleveland, Charles Eliot, and John C. Loudon (designer of the first British public parks) during the 19th century were amongst the primary influencers of what became known as greenway planning. One of the notable publications was John C. Loudon's book *Hints for Breathing Places for Metropolis*. He envisioned urban growth being shaped by green belts and open spaces "breathing zones" for London in 1829 (Turner, 1984, 75).

In summary, before the invention of the car, ancestral greenways conserved existing landscape features like river floodplains by the planner's proactive actions before the introduction of new developments in the city (Kullmann, 2013). These existing natural patterns cut into the urban fabric in the form of axes, boulevards, and parkways. The axes and boulevards of Europe and the parkways of the US were multifunctional corridors that provided many social, economic, environmental, and social benefits to societies of the 19th century.

Second Generation: Landscape Architects and Greenway Planning (1900-1950).

Following the significant influence of 19th-century greenway planning, the early 20th century witnessed several extensions and applications of this concept, particularly in Europe and the US. Key projects and policy changes included:

- The 40-Mile Loop in 1903, Portland, the US planned by Olmsted, continued to develop by his two sons to become 140 miles (230 km) long that connects more than thirty city parks. (Little, 1995).
- The Blue Ridge Parkway, the longest linear parkway system (750 km) that links 29 Virginia and North Carolina counties in the US (Fabos, 2004; Searns, 1995).
- The establishment of the UK National Parks and Access to the Countryside Act in 1949, which created a framework for the creation of national parks and maintenance of natural reserves (Turner, 2006).
- The open space plan for the Commonwealth of Massachusetts 1928 by Charles Eliot II (Imam, 2006; Searns, 1995).
- San Antonio's Paseo Del Rio (The River Walk) in 1930-41 (Searns 1995; Imam, 2006).
- John H. Forshaw and Leslie P. Abercrombie London's plan of 1943–1944 viewed "parkways" and "park systems" as one of four key aspects in the planning of London county (Turner, 2006).

These projects emphasize the importance of comprehensive greenway planning that includes coordination from a national, regional, and local scales.

At the turn of the century, Ebenezer Howard's Garden City concept, bringing nature to cities and use of greenbelts as a method to control urban expansions, as well as a protection of the rural landscape and creation of a buffer between communities, influenced a generation of planners (Zube, 1995). In 1909 Letchworth was built, Howards' first Garden City, which aimed to achieve a vision of combining the virtues of both town and country living. The success of creating a social, community, and economic model, where the captured annual value of the land is spent on building and maintaining the garden city (Howard, 1902), encouraged the creation of additional new towns in the UK, such as Hampstead Garden Suburb and Wythenshawe. Later, the British New Town Act in 1946 generated additional multiple satellite towns such as Runcorn, Ipswich, and many more. The influence of Ebenezer Howard's Garden

City concept was not limited to the UK's new towns, but also the American Greenbelt Towns in 1920-30s (Walmsley 1995). In short, greenways of this generation were greatly influenced by the English Garden City Movement in 1898.

At the same time, this generation introduced pedestrian-oriented greenways that follow human-made flood channels and other land-uses (ex. Commercial, and recreation) in direct response to the rise of the automobile (Searns, 1995). With the increased rate of car ownership, as well velocity and size, city plans introduced by Clarence Perry and Le Corbusier and many more (influenced by Ludwig Hilberseimer 1885-1968) separated pedestrian and vehicular movement by the introduction of hierarchical transportation systems that corresponded to the traffic volume and function (Carmona, Tirsdell, Heath, & Oc, 2010). An example of such separation, parkways such as the Bronx River was the first parkway explicitly designed for the recreational use of automobiles (Benedict & McMahon, 2006). Another example is the traffic-free superblock Radburn (1929-1931) by Henry Wright and Clarence Stein (proponents of the Garden City Movement) in New Jersey. Radburn's development pattern had automobiles as the second mode of transport and open spaces as the community structure that enabled the movement of pedestrians and bicyclists to several public amenities (Girling & Kellett 2005). The separation of pedestrians and automobiles, as well as the departure from the traditional towns, undermined streets as the center of public activities and made pedestrian-oriented parkway systems as the cornerstone for community structures (see Figure 3) (Walmsley, 1995).

The degradation of the quality of life and the loss of indigenous natural and cultural landscapes were key concerns of planners and architects, including Benton MacKaye (originator of the Appalachian Trail), Patrick Geddes, Lewis Mumford, and Stuart Chase. Such landscape changes stemmed from urban sprawl stirred by the development of the automobile industry and infrastructure in metropolitan regions early 20th century (Bower, 1963). Thus, the advancement and rise of comprehensive parkway systems in cities between the late 19th century and 1933 (a combination of the Romantic Park and City Beautiful Movements) was a response to socio-economic and political factors which included the steadily in-migration to urban areas (Walmsley, 1995).

This generation acknowledged links between ecology and planning (Mossop, 2006) as manifested in the work of Patrick Geddes, Benton MacKaye, Aldo Leopold, and Lewis Mumford describing the city as an amalgamation of human and natural processes. This approach is manifested in Walter Burley Griffin's 1911 design for Canberra, Australia, where significant natural environments affected the placement and form of key structures and major axes (Mossop, 2006).

At the same time, however, there was a rise in standardization and normative thinking, which led to standardized design solutions in the 20th century (Mossop, 2006; Velde & Wit, 2015). An example of such thinking is Clarence Perry's 1929

Figure 3. Segregated pedestrian and bicyclists' corridors from roads in Radburn, New Jersey. (Photo Credit: Flickr user: Design for Health, web-link: https://www.flickr.com/photos/designforhealth/6974616023).

Neighborhood Unit, which had a standardized categorization of open spaces in both scale and functions (Girling & Kellett 2005).

In summary, this generation is characterized as a period that witnessed suburbanization enabled by the massive production and ownership of automobiles. This phenomenon (in addition to events such as World Wars, economic depressions, and more) considerably affected human settlements and ecosystems around the world, and by extension, the form and function of greenways in urban areas. To accommodate such growth, the separation between pedestrian and vehicular movement was a major trend in city planning. However, concerns about the quality of life, as well as the natural environments, are led by landscape architects, among other allied professions, to the advancement of comprehensive greenway planning at various scales.

Third Generation: Greenways in the Wake of the Environmental Movement and Ecological Planning (1950 – 1987).

This generation is referred to as the environmental decades due to the prevalence of post-WWII environment negligence and responses to it. During that time, the rate of suburban development outpaced the ability of most cities in the US to plan for park systems and protect ecologically vital land (Girling & Kellett 2005). In response, Rachel Carson's *Silent Spring* in 1962 emphasized the role of ethics in science and advocated for earth stewardship to protect the endangered and fragile natural environments (Carson, 2000).

During this generation, multiple world-renowned scholars in the US and UK promoted greenways as a response to environmental problems. Those scholars include, but not limited to, Philip H. Lewis Jr., Ian McHarg, Ervin Zube, William H. Whyte, Bill Flourney, and Julius Fabos. Their innovative ecological planning and design methodologies had greenways as a part of their models (Ahern, 1995). It is important to realize that during the first three generations, there was a dominant perception of "Man vs. Nature" (Mossop, 2006): nature being inherently good, while cities being inherently bad. Many of the works that followed McHargian ideals were anti-urban and assumed that the right design outcomes followed the right process. However, in 1984, two critical publications, 'Michael Hough's City Form and Natural Process' and Anne Whiston Spirn's 'The Granite Garden' developed the discourse of integrating ecological planning with urban development in cities, thereby starting to break down the dichotomy between humans and nature.

With regard to greenways, McHarg's "Valley Floor," Lewis's "Environmental Corridors", and Julius Fabos's Metropolitan Landscape Planning Model (METLAND), are prominent examples that preserve the ecological structure of natural systems to better inform decisions regarding urban development and economic growth (Fabos, 2004). Nonetheless, McHarg's work being an influential and significant contribution to the field of ecological planning and design, many authors criticized his exclusion of social and ecological systems in cities, and the viewpoint that has led to the creation of "well-landscaped, automobile-dependent suburbs", which many view as unsustainable (Farr, 2008). In short, environmental planning models of this generation focused on identifying ecologically sensitive corridors (greenways) to guide urban development as a response to the decline of the biotic and abiotic ecosystems.

Since most new urban development occurred on urban fringes during the second half of the 20th century in the form of new towns, such as Reston, Virginia and Columbia, Maryland (Girling & Kellett, 2005), a considerable number of planning initiatives, such as (Planned Unit Developments [PUDs], Planned Residential Developments [PRDs] and Planned Mixed-Use Developments) were introduced to counteract urban sprawl, and enable the preservation of natural landscape features.

While these initiatives were facilitated by the works of Clarence Stein, Henry Wright, and others in the Greenbelt towns, they were still mainly low density and remained automobile-dependent (Beatley, 2000, Table 2.1, p.30, as cited in Walmsley, 2006). Greenways were considered in these planning initiatives as the underlying structure that guided development. However, the shortcomings of their urban development plans, being automobile-centric, contributed to the degradation of many ecosystems, and with time, it exacerbated the social, environmental, and economic impacts (Carmona, Tirsdell, Heath, & Oc, 2010). These defects caused the emergence of new planning initiatives (discussed in the fourth generation section) that also had greenway planning as a foundational layer to development at multiple scale levels, but simultaneously rewrote urban codes to overcome the shortcomings of urban planning of their predecessors (Walmsley, 2006).

Another iconic scholar of this generation was William H. Whyte who is acknowledged as the first to use the term "Greenways" in his 1959 monograph titled "Securing Open Space for Urban America" published by the Urban Land Institute (Bueno, Tsihrintzis & Alvarez, 1995; Fabos, 2004). In the 1960s, the term was used to describe linear parks in several parts of the US, and by the mid-1970s it was first applied in large built projects such as the Platte River Greenway, Denver, Colorado (Searns,1995).

Most of the greenways created by Olmsted and other scholars were transportation routes that accommodated horses and carriages in the 19th century. By the second half of the 20th century, these greenways or parkways were also occupied by automobiles that gradually made it difficult for pedestrians, bicyclists, and equestrians to use due to the safety, as well as noise and air pollution (Searns, 1995). As a result, there was an increased demand to create separated trail-oriented (or urban trails or the European's off-street bikeway concept) that are separated from automobiles.

Due to the growing interest in the Platte River Greenway model (trails along natural corridors), and abandonment of the several city infrastructures in the 1960s and 1970s (referred to as Urban Decay due to suburbanization post-WWII in the US, UK, and parts of Europe) (Walmsley 1995), created a greenway movement known as "Rails-To-Trails" or "tow-path movement" (Kullmann, 2013). This movement generated thousands of greenways in the US that connected towns to the countrysides. Furthermore, this movement was even supported by President John F. Kennedy in the 1960s due to the growing public awareness about the environmental issues, problems of automobiles, public health, and physical inactivity (Searns 1995). Furthermore, with this growing popularity, as well as the invention of the mountain bike in the early 1980s, the design standards of trails (width, surface materials, and more) were changed. For instance, in the US, the recommended width was increased from 2.4m to 3m with some trails up to 4.87m (Searns 1995). Surface materials were also expanded to include compacted soils and crushed stone surfaces. In short, the

Rails-To-Trails movement was a significant milestone in the evolution of greenways that influenced people's movements, perceptions, and interactions with nature.

In summary, even though greenways were, in the wake of the environmental movement, the underlying structure of several ecological planning models and urban planning initiatives of new towns, the automobile-centric development models remained, and so did their deleterious effects. Thus, some greenways were created based on planning efforts, while others were a result of regenerating disused city infrastructures, which later became a movement referred to as 'rails to trails.'

Fourth Generation: Greenways Beyond the Attenuation of Urbanization Ills (1987 – 2006)

The fourth generation of the greenway movement (1987-2006) is marked by the US president Ronald Regan's Commission on American Outdoors Report in 1987, which is recognized as a significant landmark by scholars of the greenway movement (Ahern 2003; Bueno, Tsihrintzisb & Alvareza, 1995; Fabos, 1995; 2004; Salici, 2013; Sharma 2015; Turner, 2006; Walmsley, 2006). Robert Searns (1995) identified this generation as the 'multi-objective greenways' where "ecological linkage, water quality, recreation, and education are woven into a multifarious set of performance objectives." (Kullmann, 2013, p. 41).

It can be inferred from the past greenway generations that they were adaptive responses to the issues that gradually manifested from urbanization, automobile dependency, air and water pollution, lack of public open spaces, accessibility to nature, and many more. Even though the greenways of this generation served the same causes, the principles of earth stewardship and sustainability are integral components that expanded the purpose of greenways of this generation (Searns, 1995, p.72). Rio Grande Valley State Park (Paseo del Bosque Bike Trail) in Albuquerque, NM, The Boulder Creek Greenway in Boulder, CO, and The Santa Monica Mountains Corridor Project outside Los Angeles, CA are examples of this new iteration of the greenway concept.

The purpose of greenways was expanded to treat human interventions that deteriorated many environmentally sensitive ecosystems as human settlements and technologies (including the car) accelerated. At the policy level, by the late 1990s in the US. Environmental Protection Agency (APA) enforced urban stormwater management programs that resulted in more cities incorporating greenways in planning strategies (Girling & Kellett 2005). Ahern (1995), as well as Searns (1995), presented several arguments in support of greenways, especially in terms of the ecological services they provide, as well as environmental protection. However, it is important to note that the direct relationship between ecosystem services and human-made systems should be based on "an ecological model of interdependency, instead of an industrial model of segregation." (Brown & Stigge 2014, p.7). In short, the increased integration of greenways with multi-sectoral planning and political strategies evolved its potential to serve the cause of achieving sustainable urban development and environmental protection goals.

The major difference in the planning initiatives of this generation (such as Green Infrastructure Planning, Smart Growth, New Urbanism, Smart Conservation, Traditional Neighborhood Development, and Transit-Oriented Development) compared to the previous ones is its integration of their processes with sustainable urban development strategies. The planning process of these initiatives borrowed concepts from previous generations, for instance, Ian McHarg and Phil Lewis's approach in the 1960s to preserve environmentally sensitive spaces and corridors, as well as Clarence Perry's 1929 Neighborhood Unit model (Girling & Kellett, 2005). Even though these initiatives redeveloped the concept of a city and neighborhood planning, as well as aided in the rewriting of urban codes, they are all "different aspects of the greenway movement, *expressing* its many possibilities, *enriching* its original concepts, *enlarging* its credibility—if need be—and *emphasizing* its importance for and relevance to current issues of sustainability and 'green' planning and design." (Walmsley, 2006, p.252).

Many scholars during this generation developed greenway planning models that preserve environmentally sensitive areas (ESA) while guiding urban development. Examples of those models (1) Charles Flink and Robert Searns (1993) (2) Ndubisi and DeMeo and Ditto (1995). Both models integrated cultural aspects in the greenway planning process, and such distinction is significant because early planning models of the 19th century and throughout the 20th century overlooked the interrelationship between nature and culture (Velde & Wit, 2015). Even though the conservation schemes shifted from protecting fragmented patches to ensuring overall ecological integrity, it is important to understand that scale:

"is responsible for the dynamic relationship between landscape configuration and function (Carlile et al.,1989), so that neither corridors nor large reserves will work as a stand-alone solution to our wildlife protection problems; an integrative solution based on functional scales of operation may represent a more balanced approach." (Linehan, Grossa & Finnb 1995, p.181)

In densely populated urban areas, greenway planning is a restorative process that involves an ecological restoration of existing infrastructures and spaces for the provision of ecological, social, and economic benefits such as storm water management (Austin, 2014). In short, greenway planning is a sensible and systematically structured approach that integrates both smart conservation and smart growth principles.

Since in the US. 80% of the total population lived in edge cities (Walmsley, 2006); any enhancements towards creating sustainable alternatives were encouraged. Planning approaches such as New Urbanism, TND, TOD, transit villages, pedestrian pockets, as well as 'Livable Neighborhood' in Australia shared similar values and principles. However, they differed primarily in "the conception of the pedestrian shed: the location of its centroid, and its extent." (Davis, Duany & Plater-Zyberk, 2002, p.3.2). Even though New Urbanist towns (such as Orenco Station in Hillsboro) had a better sense of community and public transit use, they were built next to existing automobile-dependent areas with poor non-motorized transport infrastructure. As a result, New Urbanism is "less effective as the building blocks of a whole and livable city or region." (Girling & Kellett, 2005, p.137). In short, New Urbanism, as well as other compact and transit-oriented development models, were movements that mainly focused on supplanting suburban sprawl with the provision of sustainable and transect-based development.

As for greenways, they are part of New Urbanism's open space typologies that traverse from urban core zones to rural zones. Each zone has a specific open space form and function. On the one hand, Green belts were perceived as a corridor that separates urbanized areas and may contain both recreational and agricultural uses. On the other hand, greenways were not only described as non-motorized transport routes and corridors that connect countryside to urban parks within neighborhoods but also characterized as a landscaping pattern that:

"should be appropriate to the location: naturalistic within the countryside, and formal within the neighborhoods. A greenway should follow a natural trajectory which is transformed to its purpose. Typically, these are riverfronts (riverwalk) or disused tracks (the rails-to-trails)." (Davis, Duany & Plater-Zyberk, 2002, p.5.1)

The underlying assumption of the New Urbanism's transect-based codes is that natural (environmentally sensitive areas) and agricultural open spaces are preferably separated from compact development patterns because they compromise their protection and assume that low-density development will mitigate loss and fragmentation of natural systems (Girling & Kellett 2005). Thus, New Urbanism's greenways are multi-purpose linear parks that are based on an opportunistic basis to establish ecological, cultural, economic, social, and recreational functions (See Figure. 4), while leaving 'species-rich spaces', as well as natural and agricultural grounds separated from development as an act of preservation.

In summary, greenways' functions in this generation expanded beyond mitigating the deleterious effects of urbanization. Stewardship of both lands and resources played a pivotal role in its planning and management process. This new iteration of greenways was accompanied by several sustainable urban developments and political strategies that addressed the shortcomings of past planning initiatives, notably automobile dependency. These planning and political strategies had the multifunctional greenways continue to serve human needs while sustaining ecosystems on a wide scale by implementing perseveration codes and regulations.

Fifth Generation: Greenways as Infrastructure (2006 – Present)

Greenways of this generation are characterized as integral components of the landscape, (in)forming the infrastructure of urban regions. Since the turn of the 21st century, the landscape is increasingly recognized as a model and a building block for urbanism and infrastructure (Corner, 2006; Waldheim, 2006, 2016; Mostafavi & Najle, 2003), incrementally identified as a performative medium for urbanization and the restoration of the post-industrial sites and structures. The landscape is seen to provide an armature for the regeneration of the gradually deteriorating Fordist economy via integrating ecological performance and design culture (Waldheim, 2016). In this time, streets, constituting more than 80% of public places in large urban regions (National Association of City Transportation Official, 2016), are increasingly recognized as landscape spaces, with Walmsley noting how they can form "additional links" that complete the green infrastructure systems (Walmsley 1995). With this in mind, in densely populated urban regions, where limited original ecosystems may remain, the substitution of a "potential replacement ecosystem" is an option to consider, especially if it would link or support the remaining natural spaces and corridors (Clewell, Rieger & Munro, 2005). Practices to spatially and materially transform conventional single-purpose grey networks into ecological links include the use of light-colored porous pavements, continuous tree trenches, storm water management, rich tree canopies, and ground plantings (Brown, 2014). In short, this era focuses on replacing existing and operational mono-functional structures and streets with parks and ecological connections as primary building blocks of urbanism to achieve this dual ambition of both continuous and integrated green infrastructure.

Ever since the turn of the 21st century, the use of landscape as infrastructure led to the emergence of several discourses of urbanisms. Among those are landscape urbanism (Waldheim, 2006, 2016; Mostafavi and Najle, 2003), ecological urbanism (Mostafavi & Doherthy, 2010, 2015), infrastructural ecologies (Brown 2013, 2014; Bélanger 2016; Hung et al., 2013; Brown and Stigge, 2017), flowscape (Farini and Nijhuis 2013; Nijhuis, Jauslin & Hoeven, 2015), and many more. Each of these approaches to urbanism provides a unique perspective with regards to the shape of the city and performance in ecological, social, and economic terms. These concepts advocate moving away from the conventional infrastructural silos for the provision of public services, especially since they proved to be inefficient as a system

Figure 4. A plan is illustrating both natural landscapes and parks of Villebois, Wilsonville, Oregon, US. A 30-acre greenway in this example acts as the thread that attempts to connect the ecological structure while wrapping around the Village Center, providing multiple functions for the community (i.e., bicycle trails, gathering spaces, community pool, children playgrounds, and more). (Google Earth September 17, 2019).

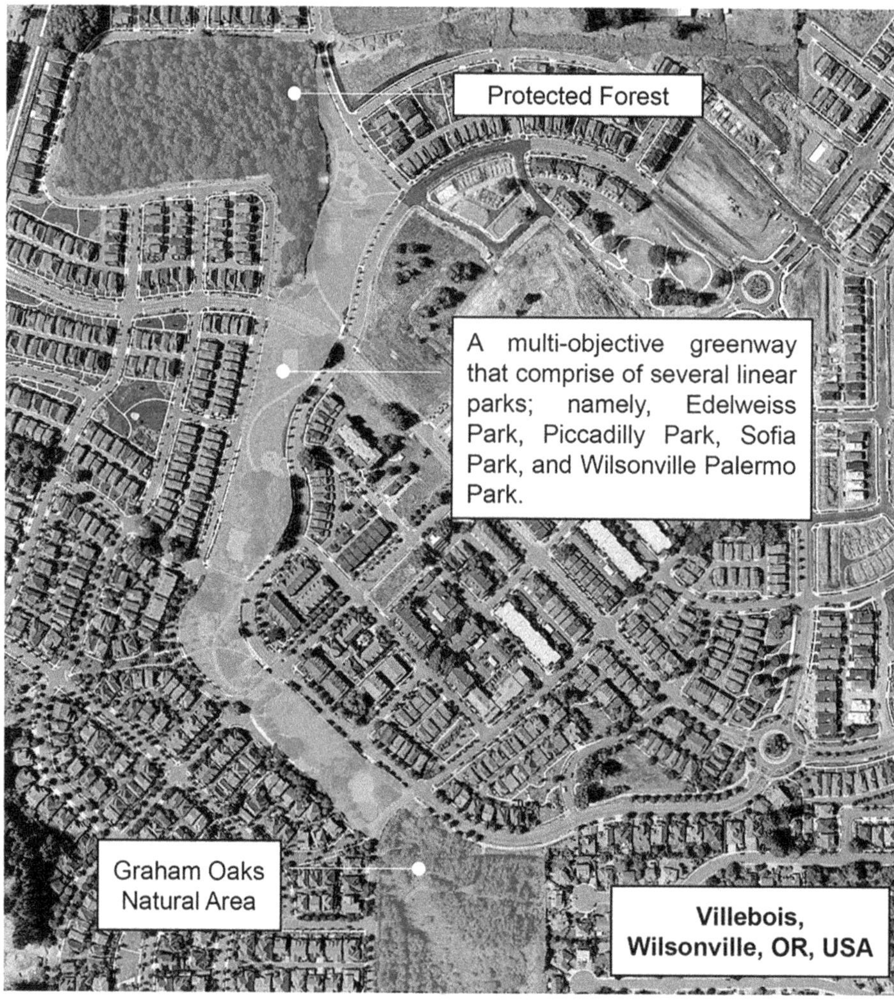

(Brown 2014). Furthermore, approaches to urbanism that characterize landscape as infrastructure use ecological and economic processes as formative design tools, and integrate natural and anthropogenic systems as well as programs to structure and sustain urban development (Nijhuis and Jauslin & Hoeven, 2015). In brief, landscape as infrastructure, a goal-oriented approach, is increasingly perceived as a medium that facilitates the formation and articulation of solutions that address many environmental, economic, and social problems found in many cities worldwide.

Among the urbanism discourses mentioned above, landscape urbanism, while aiming at (re)shaping cities in a way that place 'landscape' at the core of its planning, design, and management, when compared to New Urbanism, it is "a more fragmented matrix of discontinuous land uses" (Mossop, 2006, p.165). The emphasis here is to adapt the multifunctional landscape, rather than architectonic structures, to direct the development of urban areas for the following reasons:

- Landscape being "uniquely capable of responding to temporal change, transformation, adaptation, and succession" (Waldheim 2006, p.39).
- Via the lens of ecology, the scale and scope of factors affecting cities extend beyond the urban boundaries (Mostafavi, 2015).
- Instead of the revenue-driven development, congestion, pollution, and ill social effects produced by the transportation infrastructure and technology of high-density building, landscape exemplified by parks, greenways, street vegetation, and more reduce the urbanization ills (Corner, 2006).
- The landscape is everchanging and resilient compared to the fixed and definite urban infrastructure (Corner, 2009, p.30 as cited in Carmona, Tirsdell, Heath, & Oc, 2010, p.45).
- Ecological system's potential of providing a framework for urban intervention, while being autonomous and open-ended (Waldheim, 2016).
- The generation of landscapes provides the ability to use the ecological processes in design via the natural dynamic flows of energy, nutrients, water, and more (Mossop, 2006).

It can be inferred that many greenway projects of the 21st century are a network of linear landscape forms (re)linked to natural systems at various scales via ecological restorations of underperforming urban infrastructure to safeguard ecosystems' integrity, as well as overcome the current and future challenges to sustainability such as climate change, and automobile dependency.

Examples of this approach illustrate how 20th-century road-dominant cities could be transformed into resilient landscape-based systems. Many cities around the world concluded that the nuisance effects of freeways exceed their benefits. The movement from freeways to greenways was seen as a transition of focus from fast, uninterrupted, and automobile-oriented mobility to livability. The last quarter of the 20th century has seen multiple freeway demolition projects (Kang & Cervero, 2009). In Portland, Oregon in 1978 the Harbor Drive freeway was replaced by a 37-acre waterfront park; in San Francisco, the aftermath of the Loma Prieta earthquake in 1989 resulted in the removal (after great resistance) of the city's double-deck Embarcadero freeway along its waterfront in favor of scenic boulevards. Prominent examples of greenways in the 21st century are Cheonggye elevated highway in Seoul, South Korea; Madrid Rio project along the Manzanares river in Madrid, Rose Fitzgerald Kennedy Greenway, Boston, Massachusetts, US; Cultural Corridor Chapultepec, Mexico; Hamburg car-free green network in Germany; Donghaochong Greenway Project in Guangzhou, China; Vancouver's city and neighborhood greenways, Canada. Moreover, many more have extended these processes even further, shifting the prioritizations of streets, the most immediate public space for all, to favor the creation of livable and sustainable places for people via greenways, which has become an international movement.

In summary, greenways are a vital component of sustainable cities, whose form, functions, and values have evolved significantly in the 20th century and beyond. The table below summarizes this process:

SECTION 3: INTEGRATING GREENWAYS AND SUSTAINABLE URBAN MOBILITY PLANNING

"Show me your underperforming streets [...], and I will show you your future green-network." (Kullmann, 2013, p.57)

This section explains how efforts to adopt SUM planning (or SUMP) provide mutual benefits and opportunities to (re)develop greenway networks that support both in several fashions, which ultimately helps in overcoming automobile dependence. Interpretations and inferences mentioned in this section are based on the literature review of greenways' evolution (see Section 2) as a constituent of sustainability of SUM systems, in particular as non-motorized transport corridors, as well as recent literature in the transport sector. The chapter is concluded by providing an integration model of the planning process for greenways and SUM systems.

Programs initiated by the international and national coalitions and organizations such as the European Commission's Horizon 2020, the world's largest collaborative program for research and innovation, are facilitating the advancements towards overcoming societal challenges in the transport sector globally (in addition to many other sectors). Urban policies set by the *European Commission's Directorate-General for Mobility and Transport* (DG MOVE) and the *Directorate-General for Research and Innovation* (DG RTD) are transformed into projects with measurable outcomes by the *Innovation and Networks Executive Agency* (INEA). INEA is responsible for implementing parts of the EU funding programs for transport, energy, and telecommunications such as Smart green and integrated transport, CIVITAS initiative, Smart Cities and Communities, and many more (CIVITAS, 2019). Those

Table 1. A table that summarizes greenways' evolution globally (mostly in English-speaking countries) based on authors' interpretation of the major discussions made in Section 2 (By authors)

Variables	Before 1900	1900-1950	1950-1987	1987-2006	2006-Present
Generation Title	the origin of the greenway concept	Landscape Architects and Greenway Planning	Greenways in the wake of the Environmental Movement and Ecological Planning	Greenways beyond the attenuation of urbanization ills	Greenways as infrastructure
Selected Key Thinkers	Peter Lenne, Frederick Law Olmsted, George E. Kessler, Horace W.S. Cleveland, Charles Eliot, and John C. Loudon, George Perkins Marsh, and more.	Benton MacKaye, Ebenezer Howard, Patrick Geddes, Aldo Leopold, Lewis Mumford, John and Frederick Jr Olmsted, Peter J. Lenné, Henry Wright, Horace W.S. Cleveland, George E. Kessler, and more.	Philip H. Lewis Jr., Ian McHarg, Ervin Zube, William H. Whyte, Rachel Carson, Michael Hough, Anne Whiston Spirn, and more.	Robert M. Searns, Charles A. Flink, William L. Flournoy Jr., Paul Cawood Hellmund, Daniel Smith, Diana Balmori, Mark A Benedict, Edward T Mcmahon, Jack Ahern, Tom Turner, Charles E. Little, Frederick R. Steiner, Richard T. Forman, Anthony Walmsley, Julius Fabos, Byron Stigge, Hillary Brown, James Corner, Charles Waldheim, Pierre Belanger, Karl Kullmann, Gary Austin, Robert L. Ryan, István Valánszki, Sándor Jombach, Krisztina Filep-Kovács, Mark S. Lindhult, and more.	
Main Trends, concepts, and events that affected greenways' form, function, as well as planning and design	• Ancestral greenways to reference the axes and boulevards. • The US. Parkways: Stream Valley Park and Pleasure Drive for Carriages. • Utilitarian features that affected the economic and spatial development of cities. • Industrial Revolution and urbanization. • The axial walkways of Islamic gardens that brought unique identity and meaning, while providing several functional and aesthetic qualities.	• English Garden City Movement in 1898 • American Greenbelt Towns in 1920-30s • greenways for Non-motorized transport, separated from ones made for motor usage. • Suburbanization and the Automotive Industry. • The link between ecology and planning. • The rise of standardization and normative thinking. • Landscape architects play a vital role in its evolvement.	• Ian McHarg's Valley Floor, and Phil Lewis's Environmental Corridors. • Rails to Trails: results of the abandonment of the several city infrastructures in the 1960s and 1970s (and in some cases demolition by natural disasters), which are referred to as urban decay. • Public health and physical activity issues. • Landscape planning and suitability analysis. • The environmental Movement due to the prevalence of diseases, as well as air and water pollution post-WWII.	• US president Ronald Regan's Commission on American Outdoors Report in 1987. • Earth stewardship and sustainability. • Referred to as multi-objective greenways • The suburban regions of 78 large cities in developed countries had 85% faster growth patterns compared to their urban cores. • Integration between smart conservation and smart growth strategies. • Geographic Information System (GIS).	• landscape as a model and a building block for urbanism and infrastructure. • Greenways as part of the urban landscape that facilitate the establishment of solutions to integrate infrastructure with variable programming that can treat many problems found in cities worldwide. • Ecological restorations of operational post-industrial sites, streets, and structures. • Increased focus on greenway networks instead of signaler corridors within urban regions. • Climate Change.
Unified Characteristics between all generations	• A means to control urban and economic growth. • Preserve natural, cultural, and scenic resources. • Provide recreational grounds. • A medium that supports safe and healthy mobility alternatives. • Link between the city and the countryside, as well as between communities. • Linearity, continuity (to an extent), and connectivity. • Deliver community amenities that induce economic values such as the enhancement of the quality of life, increase of nearby property values, and tourism. • Offer ecological services including stormwater management. • Access to nature or to constructed natural-like landscapes. • Used in conjunction with planning and design initiatives.				
Project Examples	• Boulevard of Champs Elysees in Paris dating circa the 1700s. • Emerald Necklace, Boston, US.	• The 40-Mile Loop in Portland, US. • The open space plan for the Commonwealth of Massachusetts 1928.	• Wisconsin Heritage Trails Proposal of 1964. • Platte River Greenway, Denver, Colorado.	• Connecticut retirement village, US. • Villebois greenway in Wilsonville, Oregon, US.	• The High Line, New York, US. • Cheonggyecheon river in South Korea. • Madrid Rio along the Manzanares river.
Major Disciplines	Landscape Architects, City Planners, conservationist, historians, philosophers.	Landscape Architects, conservationist, urban planners, ecologists, historians, philosophers.	Landscape Architects, urban planners, urban designs, urban ecologists, biologists.	Landscape Architects, urban planners, urban designs, urban ecologists, botanist/plant ecologists, landscape ecologists, architects.	

programs are organized under ten topics or themes. Each topic houses a number of projects. The exchange of experiences and technologies via building (inter)national programs to consolidate and organize the efforts towards shared goals such as the independence of car-based planning practices and rehabilitation of its outcomes is a vital approach for the realization of these aspirations.

Even though all INEA's ten transport-related topics are linked at variable degrees to chapter's focus, perhaps the directly related research project is CIVITAS's *Sustainable Urban Mobility Planning: pathways and links to urban systems* also referred to as SUMP-PLUS (allocated funds of almost €4 million). One of the three-years project (9/2019-8/2022) goals is to:

"Demonstrate how cities can develop stronger ties with other policy areas necessary for holistic urban development, such as education, health, and land use planning while reflecting technological and contextual developments." (CIVITAS, 2019)

This section support this goal by capitalizing on greenways' potential to aid SUMP by being not only a guiding tool that preserves environmentally sensitives areas but also as a part of the non-motorized transport infrastructure. Many cities around the world have not yet adopted SUMP, especially since a solution for a city does not necessarily guarantee its effectiveness in other urban areas. That being said, the SUMP-PLUS is also an approach to develop context-specific solutions (via working in six co-created research labs located in different cities throughout Europe) that bridge the gap between many European cities of all sizes and their implementation of SUMP considering the differences found in culture, planning practices, public policies, and many more (CIVITAS, 2019). To conclude, the SUMP-PLUS research project should be replicated in other parts of the world, especially developing countries of the Middle East, Africa, and parts of Asia, to overcome the contextual development challenges and pave the pathway towards the adoption and implementation of SUMP.

How can greenways be integrated with SUM systems? For whom are transportation corridors designed? How can greenways enhance a city's transportation system? Will access to a greenway project be limited to non-motorized transport modes only? If not, what type of motorized vehicles will have access to the greenway? Charles A. Flink and Robert M. Searns (1993) suggest such questions can be resolved by:

1. Examining if the greenway project can be linked with existing or proposed non-motorized transport networks (i.e., bicycle routes) within the city or community of interest.
2. At each probable point of intersection with existing transport or utility systems, measure the safety level between existing and potential usages. Selecting a

suitable crossing type and design at each point will involve analyzing several factors such as road type and width, regulated vehicle design speed, and many more.

3. Identifying if future transportation plans are within the greenway project area and investigate if an underpass or overpass are possible options for new roads. Beyond that, determine if there are plans to modify the existing roads.
4. Investigating if the proposed roads include public transit services and explore its relationship with the proposed greenway project. In addition, determine if the transportation sector has allocated funds to enhance the connection between greenways and mass transit.

To maximize greenways' use and use-potential, their integration with SUMP must consider land-use planning as one of the strategies to overcome automobile dependency (Newman & Kenworthy, 1999). This is because "*transit shapes urban development by enhancing accessibility; attributes of land, such as residential and job densities and the degree of land-use mixing, affect travel demand.*" (Suzuki, Cervero & Iuchi, 2013, p.2). SUMP provide land reclamation opportunities at multiple scales within cities by reducing car ownership, parking requirements, and road capacities, which in return would enable the transformation of the mono-functional road infrastructure into public amenities via, for instance, (re)building a city's green infrastructure (includes greenways). At the same time, green infrastructure planning should serve as a foundational layer that guides the SUMP since it is influenced by ecological and economic processes as formative design tools. Furthermore, regional and community greenways' connectivity is not limited to green open spaces, but also town centers and major points of interests in a city, as Walmsley (1995) reference of additional links, which is something that influences SUMP as well (Crevero, Guerra & Al, 2017).

That being said, greenways can support SUMP by being a part of the non-motorized transport infrastructure. As presented in the fifth generation of greenways' evolution, more arterial corridors worldwide are being regenerated as a part of road contraction schemes. Unlike traffic calming, road contractions are about "shrinking the footprint of channel-ways given over to private cars and trucks and reassigning this space to other, less disruptive, more people-oriented uses, such as greenways, pedestrian zones, bike lanes, and public parks." (Crevero, Guerra & Al, 2017, p.143). In brief, unlike the car-centric planning of the 20 century that shaped urban environments based on the physical characteristics, movement, and capabilities of motor vehicles, greenway(s) and SUMP, when integrated at multiple scales, support the goal of overcoming automobile dependence (See Figure 5).

Stakeholders' involvement, sustainable funding streams, and public policies are the engine that operates greenways and SUMP, as well as facilitates its realization. It is important to note that governance plays a major role in the three aforementioned aspects since its role involves setting a shared vision and policies towards overcoming automobile dependency at multiple levels. Examples of international policies and initiatives include the CIVITAS initiative that promotes sustainable transport solutions in Europe, European Mobility Week, and the Partnership on Sustainable, Low Carbon Transport (SLoCaT). All these examples are just some of the global alliances of cities, and countries that aimed to advance the goals outlined in the Kyoto Climate Change Protocol, and the Paris COP-21 agreement in 2016 (Schiller & Kenworthy, 2017). Thus, effective governance of land-use and SUMP requires the establishment of a regulated hierarchical structure that balances the local, regional, national, and international interests, as well as reduces the segregation of responsibilities between the actors involved in the decision making process and implementation (Kennedy, Miller, Shalaby, MacLean, & Coleman, 2005).

In addition, aside from fund-raising campaigns, public and private sector funds, and grant programs, land value capture is an example of sustainable financial resources that aid the realization of land-use (i.e., greenways) and SUMP. The increase of land value as a result of planning permissions or urban redevelopment or specific public policy decisions is reinvested along with or the tax and fee-based revenues for the public good (Suzuki, Murakami, Hong, & Tamayose, 2015). In Bogota, due to the implementation of parks and non-motorized transport infrastructure, the city captures 30-50% of the increase in land value and then channels these funds into Metrovivienda projects and other social development purposes (Suzuki, Cervero, & Iuchi, 2013). Bogota's successful land value capture scheme is an example of an interdependent relationship and mutual benefits of greenways and SUMP integration. Despite the great potential of land value capture in securing funds for public benefits, it is usually accompanied by challenges on how, when, and where are those budgets should be spent. Thus, stakeholders' involvement in setting objectives, determining priorities, raising funds, and many more is critical in the effectiveness of land value capture strategies.

Stakeholders' involvement (i.e., via public-private partnerships, media communications, town meetings, public hearings, and many more) plays a central role in the integration process between greenways and SUMP. It increases the sense of ownership, promotes transparency, sets the right order of project priorities, increases the chances of sustainability, helps practitioners meet societal needs and expectations, provides educational and business opportunities, saves time and money in the long term, builds community trust, and more. For instance, one study in the 1990s and early 2000s showed that citizen's participation, in six European cities (Zurich, Basle, Amsterdam, Groningen, and Freiburg) had a key role in redirecting policies, investments, planning, and the public adoption of transit and non-motorized modes of transport (Schiller & Kenworthy, 2017). To recap on the topics discussed in previous paragraphs, the interdependence of greenways and SUMP must follow decisions that take into consideration the integration opportunities, as shown earlier via stakeholders' involvement, public policies, sustainable funding resources, and governance (See Figure 5).

A systematic review of the factors that influence the use of greenways (Paneerchelvam, Maruthaveeran, Maulan, & Shureen, 2020) reveals that these factors share many variables found in walkability and cyclability indexes Worldwide such as the Copenhagenize Index and the Asian Cyclability Index. Thus, when integrating greenways and non-motorized transport (NMT) planning, assessment of the walkability and cyclability, at the regional and local scale, of the area(s) of interest as well as greenway(s) to support as mobility corridors are vital to the integration process. At the same time, this process must coincide with efforts to define integration opportunities and challenges between greenways and NMT infrastructure, as well as existing and future public transit (as shown in Figure 5). Once the regional and local inventory and analysis are completed, the definition of development priorities and measurable targets ensues.

Unlike many planning models, the integration process shown in Figure 5 has the goal predefined. Thus, greenways and SUMP are means to achieve the goal, which is to overcome automobile dependency. That being said, the establishment of communication channels between stakeholders (as shown in Figure 5) must occur at multiple planning stages. However, it is stressed in this planning process after the preparation of development priorities because this is where the plans are communicated to stakeholders. Based on those communications, the plans might require redirections or readjustments to reflect on the provided feedback from stakeholders, not to mention the other benefits as a result of their involvement. Finally, the same holds with the adoption of evaluation and monitoring strategies, such as the impact assessment tools. The implementation outcomes or progress as a result of integrating greenways and NMT infrastructure must be benchmarked with objectives set in the walkability and cyclability assessments (ask for instance: How much did the integration outcomes improve the walkability and cyclability rates?). Furthermore, benchmarking the progress mentioned earlier should also include green infrastructure and SUMP planning objectives, as well as the goal of overcoming automobile dependence (ask for instance: To what degree did the integration between greenways and NMT infrastructure influence travel behavior?). In conclusion, Figure 5 is an attempt to illustrate the relationships, complexities, and the reiterative process involved in the integration between greenways and SUMP to overcome automobile dependency.

Figure 5. The integration process between greenway(s) planning as transportation corridors and sustainable urban mobility planning, specifically non-motorized transport planning (By authors)

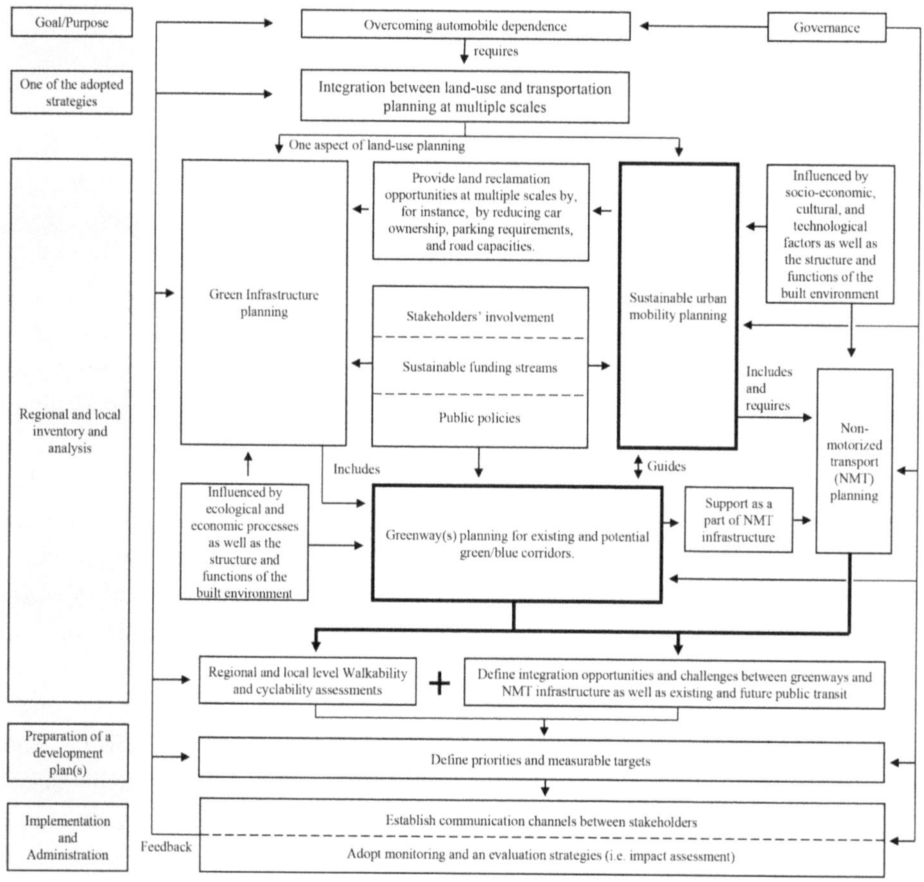

SECTION 4: CHALLENGES AND OPPORTUNITIES OF GREENWAYS PLANNING AS URBAN MOBILITY CORRIDORS

The previous sections have discussed how the design and function of greenways evolved in relation to various factors, including transportation. Today, cities continue to evolve, shaping the future of the planet and human society more rapidly than ever before (Urban Planet, 2018). If greenways hold the potential to solve urban mobility problems as well as achieving other ecological and environmental benefits, what is preventing their widespread implementation? It is important to anticipate potential challenges and opportunities for their ongoing evolution and implementation within the context of rapid global urbanization. Aside from issues related to financing greenway projects, land acquisitions, coordination among stakeholders, decision-makers' misplaced priorities, outdated policies and guidelines, insufficient scientific database, and communities' will and adaptability to change (Crawshaw, 2009; Mathur & Kacker, 2013), the following paragraphs will focus on some the planning challenges and opportunities of greenways as mobility corridors. In this section, the authors conclude by discussing several overarching considerations: the need to ensure physical connectivity of greenway networks, despite possible conflicts with other existing infrastructure and transport systems; address different climatic and cultural contexts (including user perceptions); connecting transport planning systems with greenspace/landscape planning, and capitalizing on the multifunctional benefits in addition to mobility.

CHALLENGES OF GREENWAYS PLANNING AS URBAN MOBILITY CORRIDORS

Even though the benefits of road contraction project types, in the form of greenways, are multifarious, when it comes to mobility, there are several challenges. To ensure a smooth flow of an expressway and/or arterial road traffic, their connectivity with other streets, private driveways, as well as buildings are limited (Carmona, Tirsdell, Heath, & Oc, 2010). The same holds with pedestrian crossings. They are placed in designated locations in the form of bridges or underpasses, guarded by barriers and railings to ensure safety. Furthermore, the concentration of activities/densities and maintaining a 5 minutes' walkshed contradicts the movement-centric road design where it does not need continuous urbanization. According to Loukaitou-Sideris 1993, "to accommodate both static and dynamic pedestrian behavior, a street should display both "place" and "link" qualities" (p.6). Therefore, considering the flow of other systems such as freight transport within the city is as important as providing humane and safe places where people can socialize, walk, and play.

Many successful greenways project examples managed to accommodate both through-movement and pedestrian access to places and services within a city. For example, the removal of Cheonggye elevated highway in Seoul, South Korea, to a greenway coincided with the implementation of a city-wide bus-rapid transit system that facilitated the through-movement (Kang & Cervero, 2009). Similarly, the restoration of the ecological, social, recreational, and economic values of the Manzanares river in Madrid Rio project, West 8 (consultation firm) and MRIO Architects developed a master plan that involved the reconstruction of the 43-kilometer M30 ring motorway and building of tunnels that facilitated the reclamation efforts

(Lomholt, 2011). Those tunnels provided a through movement for automobiles, while liberating the surface level from the deleterious effects of their existence to both people and the environment. In short, Cheonggyecheon and Manzanares rivers are both greenway project examples that proposed different strategies to include both 'place' and 'link' qualities in their regeneration efforts.

Another key challenge to reforming urban arterial roads and infrastructural ruins as greenways is the reduction of conflict and interruptions among the various modes of transit. Kullmann (2013) discussed the inevitable conflict between the green and grey networks and suggested utilizing the road shoulders for the formation of the green network. He adds, "ideally green-networks should be conceived of as a series of interlocking neighborhood loops that interdigitate with the local urban fabric, but also simultaneously 'tunnel' to bring the near and far closer together" (Kullmann, 2013). While the resolution of such conflicts is case-specific, generally, there are two major types of pedestrian crossings (1) at-grade, (2) above, or below grade. The first crossing type should always be considered first since it is simpler to implement, cheaper to build, faster, and easier to cross. However, a greenway intersection with important high-speed and high-volume highways, which is a likely scenario, often leaves no practical option but to build grade-separated structures in the form of over or underpasses (Ryan, Flink, Lagerwey, Balmori & Searns 1993). Even though conflict and interruption of movement between greenways and several road hierarchies are inevitable, especially in urban areas, the establishment of a holistic picture can facilitate setting out development priorities and stages that would minimize energy, time, and resource expenditures.

Climate conditions are also one of the major challenges of using greenways for transport purposes. The urban microclimate is affected by vegetation, air temperature, wind velocity, relative humidity, surface materials, shade, proximity of a water body, crowdedness, and spatial configuration (Nikolopoulou & Steemers 2003; Nicol, 2004; Nikolopoulou and Lykoudis 2006; Picot 2004; Cena and de Dear 2001 as cited in Kariminia & Ahmad, 2013). Outdoor thermal comfort is also linked to outdoor space quality, as well as human attitudes and behaviors (Lai et al., 2013 as cited in Eusuf[a] and Mohit and Eusuf[b] & Ibrahim, 2014). According to Gehl (1971) seminal book, Life between buildings: using public spaces, local sunny or shady conditions can dramatically influence visitors' decision to either stay or leave (Liang & Ng, 2012). Therefore, ensuring comfortable microclimate conditions is a major challenge (especially in unfavorable weather conditions) that can affect the use of greenways as soft mobility corridors.

Implementing greenway projects to serve as mobility corridors over an automobile-dependent city infrastructure, even in the form of a network, carry several challenges. While greenways enable non-motorized modes of mobility, they are incompatible with the car-centric city to function as transportation corridors effectively. As

Newman & Kenworthy (1999; 2015) repeatedly stressed, building a non-motorized transportation network must coincide with the implementation of a high-quality public transit system, as well as high density and mixed-use urban developments to overcome automobile dependency. There lies the challenge of incorporating greenways in a city's comprehensive plan to not only guide development but also to function as mobility corridors effectively.

Shifting public perception to be more amenable to adopt non-motorized modes of transport is another challenge that affects the use of greenways as mobility corridors. People who reside in automobile-dependent cities might view walking and cycling as a means to reach destinations, as impractical and time-consuming. Edwards & Tsouros (2006) argued that what influence being physically active and fostering active living include individual factors (i.e., age and gender), social environment (i.e., culture and income), built environment (i.e., urban design, open spaces, and transport), and natural environment (weather and topography). Thus, inducing a walking and a bicycling culture in an automobile-dependent city is a complex process that requires city-wide recalibrations in support of these transport modes. In short, seamless integration between various transportation modes, as well as walking and cycling advocacy, are all aspects that ensure that the success of greenways' function as soft mobility corridors.

Another dilemma with greenway planning is achieving a balance between corridor preservation efforts and human activities. Since many greenways offer public access, sometimes they become crowded by its several user types (e.g., joggers, bicyclists, dogs, equestrians, bird watchers, individuals using wheelchairs, hikers, and many more), creating conflict with ecologically sensitive areas. Furthermore, greenways not only support people's mobility needs, but also the movement of organisms throughout the landscape. As a result, there is a probable compatibility conflict between the functions of mobility/recreation and preservation in greenways. To address the challenge of encroachment of protected areas within greenways, governmental bodies in their multiple levels, as well as public and private agencies, need to act as partners and coordinate in forming integrated planning and management decisions. Examples of such collaboration can be in the form of providing regulations, education programs, and innovative design approaches (or simply signages and barriers). According to Ahern (1995), "The decision to adopt any particular greenway strategy should always be made on the basis of the particular biotic, abiotic and cultural factors in the local landscape as well as the values and perceptions of the landscape's inhabitants." (p.137). Furthermore, it is important to understand and monitor users' impact over time, manage, and understand users (Ahern, 2003; Flink & Searns, 1993; Smith & Hellmund, 1993). In short, use conflict within and around greenways necessitates the establishment of communication channels between all stakeholders.

OPPORTUNITIES GREENWAYS PLANNING AS URBAN MOBILITY CORRIDORS

Increasing fuel prices and vehicle ownership and operation costs, along with worsening traffic congestion in many cities, are providing an opportunity for a shift in how individuals and governments are approaching mobility. There is, therefore, the potential for greenways to become a far more prominent form of urban infrastructure at this time of transport transition.

Simultaneously, while the main purpose of transportation is to provide access to destinations and opportunities in an efficient manner, greenways offer an opportunity for health and well-being objectives to be simultaneously pursued (WHO Regional Office for Europe, 2017). Many scholars argue that the 'by-product' of movement (the series of spaces that an individual traverses to access a location), regardless whether initially or frequently, is more important than both the origin or destination of a travel journey (Carmona, Tirsdell, Heath, & Oc, 2010). Therefore, instead of nature-deprived sidewalks, greenways can be a scenic, healthy, vital, happiness-booster alternative where pedestrians and bicyclists can still reach their destinations while being exposed to nature or nature-like spaces and corridors.

Greenways, being multifunctional (achieve multiple policy objectives, and create efficiency gains), address several problems that are not limited to providing non-motorized transport infrastructure. These issues include, but not limited to, public health (i.e., physical activity and obesity), stormwater management, access to nature, urban pollution, biodiversity, natural habitats' protection, and many more (Ahern, 1995; Searns, 1995). Thus, there is potential for greenway projects to attract funding and support from a wide range of organizations, investors, and businesses.

At the same time, there is an opportunity for greenway projects to leverage the financial gains that come from increased property prices in areas in close proximity to them. An example of this is the Cheonggyecheon river in South Korea, where property values within a half-mile walkshed were shown an increase following the establishment of the greenway. Specifically, the commercial parcels within 100 meters from the Cheonggyecheon river witnessed 20% of land value post the greenway development (Kang & Cervero, 2009). In addition, as greenways gain traction over time, businesses abutting greenways benefit from its visitors and social programs via the consistent foot-traffic nearby their stores. In short, the multifunctional benefits of greenways can be direct, indirect, and incremental.

Technological advancements in electric bicycles and scooters and their increasing affordability present another important opportunity for greenway planning. These new forms of mobility overcome many of the challenges associated with the traditional bicycles (Fyhri and Heinen and Fearnley & Sundfør, 2017; Simsekoglu & Klöckner, 2019) such as distance, gradients, duration, physical ability, and weather.

By extension, innovations in these sectors are capable of changing users' perception of greenways and their potential as soft mobility corridors. According to Bird's (A major e-scooter US company) official website, 40% of car trips in the United States are under 3 miles, and 70% want new transportation options. The number of benefits and incentives to adopt this class of transport options is numerous, especially since they require less space to travel and park compared to conventional cars and trucks. Mainstreaming such technologies, as well as bike and scooter-share systems, can create opportunities to have more parks, social spaces, offices, playgrounds, and more. In short, e-bikes and e-scooters have numerous economic and environmental benefits (when paired with quality public transport systems) that can, directly and indirectly, influence the use and use-potential of greenways as soft mobility corridors.

CONCLUSION

This chapter aimed to describe how greenways can be a constituent of SUM systems, in particular, NMT infrastructure, to support the shared goal of overcoming automobile dependence. After explaining automobile dependence's detrimental impacts on natural and human systems, the authors of this chapter began to describe the potential of green infrastructure, particularly greenways, in addressing those problems. Next, Section 2 provided an updated literature review of greenways' evolution while stressing its role as transportation corridors in planning initiatives, policies, and case studies worldwide chronologically (mainly in English-speaking countries). Later in Section 3, the authors illustrated the integration process of greenways and SUMP as an approach to overcome automobile dependence while highlighting the relationships, complexities, and the reiterative process involved in their integration. Lastly, the chapter concluded by shedding light on some of the planning challenges and opportunities for developing existing and potential greenways to support as NMT corridors.

The chronological review of greenways' evolution provided an understanding of the sequence of events that influenced greenways' form and function, as well as planning and design, especially as mobility corridors. This effort can aid existing and future readers to have a holistic view, in retrospect, of how and why greenways planning evolved in the shape it is today while uncovering its potential, especially to support as NMT corridors. However, this chronological study can be further developed by the addition of greenways' evolution in other parts of the world, especially in the Middle East, Africa, and parts of Asia (see Figure 1).

In addition, as well as considering technological advances, perhaps the most significant difference between greenways of the past and those of the future will be concerning global location, with the associated differences in climate, culture, and economic context these different locations bring. The authors noted that most of the scholarly literature on greenways analyses American and European examples. However, pronounced rapid urbanization, accompanied by overwhelming levels of car dependence, is occurring not in these areas but other parts of the world such as the Middle East, Asia, and Africa (Dora and Hosking and Mudu & Fletcher, 2011). Many new and growing cities are in hot climates. How well do existing greenway designs translate to these climates? Several of the policy and practical considerations of creating greenways highlighted previously, such as the need for proper cross-departmental collaboration and resourcing, may be exacerbated in places where transport and green space are still treated as separate planning issues. Even if well connected and maintained greenways are constructed in such cities, will populations whose cultures and lifestyles are wholly car-dependent make use of them? More research is needed to understand how the successful greenways of the past can be adapted to meet the needs of different urban contexts. Fortunately, a history of built projects and scholarship illustrates the numerous potential benefits of attempting to do so.

Students and professionals in the landscape architecture, planning, and policy-related fields, as well as decision-makers, may use the integration process, provided in Section 3, as a model that minimizes the gap between the treatment of green infrastructure (one aspect of land-use planning) or more specifically greenways and transport planning as discrete subjects. At the same time, build a mind-map of the factors that can, directly and indirectly, influence the integration of greenways as a constituent of SUM systems. However, the integration can be further explored via the incorporation of different dimensions, and a more detailed level of explanation for each aspect. For instance, explore design or technological or financial or management related integration dimensions at multiple scales with more details.

Although the highlighted challenges and opportunities, in Section 4, of planning greenways to support as NMT corridors can be considered as common among many cities, future research can focus on some of those issues in greater detail, transferring knowledge via sharing context-specific experiences, and the tools to overcome those specific challenges. Despite those numerous challenges, because of greenways' multifunctional benefits, in addition to enabling non-motorized transport, it can aid in overcoming automobile dependency while adapting to current and future challenges posed by climate change and urbanization.

REFERENCES

Ahern, J. (1995). Greenways as a planning strategy. *Landscape and Urban Planning*, *33*(1–3), 131–155. doi:10.1016/0169-2046(95)02039-V

Ahern, J. (2003). Greenways in the USA: theory, trends and prospects. *Ecological Networks and Greenways: Concept, Design, Implementation*, 34–55.

Austin, G. (2014). *Green Infrastructure for Landscape Planning: Integrating human and natural systems*. Routledge. doi:10.4324/9781315856780

Bélanger, P. (2016). *Landscape as infrastructure : a base primer* (1st ed.). Routledge. doi:10.4324/9781315629155

Benedict, M. A., & McMahon, E. T. (2006). *Green Infrastructure: Linking Landscapes and Communities*. Island Press.

Benedict, M. A. E. T., & Mcmahon, J. D. (2002). *Green Infrastructure: Smart Conservation for the 21st Century*. The Conservation Fund. Retrieved from www.conservationfund.org

Bower, B. (1963). MacKaye, Benton, The New Exploration. *Natural Resources Journal*, *3*(2), 371. https://digitalrepository.unm.edu/nrj/vol3/iss2/14

Brown, H. (2014). *Next Generation Infrastructure, Principles for Post Industrial Public Works*. Island Press. doi:10.5822/978-1-61091-202-0

Brown, H., & Stigge, B. (2017). *Infrastructural ecologies : alternative development models for emerging economies*. The MIT Press. doi:10.7551/mitpress/10558.001.0001

Bryman, A. (2016). *Social research methods* (5th ed.). OUP Oxford.

Bueno, J. A., Tsihrintzisb, A., & Alvareza, L. (1995). South Florida greenways: a conceptual framework for the ecological reconnectivity of the region. *Landscape and Urban Planning, 33*, 249. Retrieved from https://ac-els-cdn-com.ezproxy.nottingham.ac.uk/0169204694020217/1-s2.0-0169204694020217-main.pdf?_tid=248f8fc4-c759-449a-90c1-3d142a6deb34&acdnat=1540031359_b766fac58a088ae00d5000a83a40ac41

Carmona, M., Tirsdell, S., Heath, T., & Oc, T. (2010). *Public Places Urban Spaces: The Dimensions of Urban Design* (2nd ed.). Elsevier Ltd.

Carson, R. (2000). *Silent spring*. Penguin.

Cervero, R., Guerra, E., & Al, S. (2017). *Beyond mobility : planning cities for people and places*. Island Press. doi:10.5822/978-1-61091-835-0

CIVITAS. (2019). Making European cities greener Towards clean and smart mobility. doi:10.2840/616004

Clewell, A., Rieger, J., & Munro, J. (2005). Guidelines for Developing and Managing Ecological. *Ecological Restoration*, *1*(December), 1–16. doi:10.1098/rspb.2013.2236

Corner, J. (2006). Terra Fluxus. In C. Waldheim (Ed.), *Landscape Urbanism Reader* (pp. 21–34). Princeton Architectural Press.

Crawshaw, P. (2009). *The Future of Greenways in Sydney*. Retrieved from https://www.be.unsw.edu.au/sites/default/files/upload/pdf/schools_and_engagement/resources/_notes/5A2_41.pdf

Davis, R., Duany, A., & Plater-Zyberk, E. (2002). *The Lexicon of New Urbanism*. Duany Plater-Zyberk & Company. Retrieved from https://www.dpz.com/uploads/Books/Lexicon-2014.pdf

Dora, C., Hosking, J., Mudu, P., & Fletcher, E. R. (2011). *Sustainable Transport: A Sourcebook for Policy-makers in Developing Cities*. Retrieved from https://www.who.int/hia/green_economy/giz_transport.pdf?ua=1

Edwards, P., & Tsouros, A. (2006). Active living, health and local leadership. In *Promoting physical activity and active living in urban environments: The role of local governments* (pp. 1–5). WHO Regional Office for Europe.

EGWA. (2000). *Greenways | Asociación Europea de Vías Verdes*. Retrieved January 20, 2019, from http://www.aevv-egwa.org/greenways/#

Eusuf, M. A., Mohit, M. A., Eusuf, M. M. R. S., & Ibrahim, M. (2014). Impact of Outdoor Environment to the Quality of Life. *Procedia: Social and Behavioral Sciences*, *153*, 639–654. doi:10.1016/j.sbspro.2014.10.096

Fabos, J. G. (1995). Introduction and overview: The greenway movement, uses and potentials of greenways. *Landscape and Urban Planning*, *33*(1–3), 1–13. doi:10.1016/0169-2046(95)02035-R

Fábos, J. G. (2004). Greenway planning in the United States: Its origins and recent case studies. J.G. Fábos. *Landscape and Urban Planning*, *68*(2-3), 321–342. doi:10.1016/j.landurbplan.2003.07.003

Farini Orleans, E., & Nijhuis, S. (2013). Flowscapes : exploring landscape infrastructures. Francisco de Vitoria University, Architecture Department.

Farr, D. (2012). *Sustainable Urbanism: Urban Design with Nature*. John Wiley & Sons. Retrieved from https://books.google.co.uk/books?id=meuuBgAAQBAJ&printsec=frontcover&source=gbs_ge_summary_r&cad=0#v=onepage&q&f=false

Flink, C. A., & Searns, R. M. (1993). *Greenways: A Guide to Planning, Design, and Development*. The Conservation Fund.

Fyhri, A., Heinen, E., Fearnley, N., & Sundfør, H. B. (2017). A push to cycling—Exploring the e-bike's role in overcoming barriers to bicycle use with a survey and an intervention study. *International Journal of Sustainable Transportation, 11*(9), 681–695. doi:10.1080/15568318.2017.1302526

Girling, C. L., & Kellett, R. (2005). *Skinny streets and green neighborhoods: design for environment and community*. Island Press.

Hellmund, P. C., & Smith, D. S. (2006). *Designing Greenways: Sustainable Landscape for Nature and People*. Washington, DC: Island Press.

Howard, E. (1902). *Garden Cities of Tomorrow*. Swan Sonnenschein & Co., Ltd. Retrieved from https://archive.org/details/gardencitiestom00howagoog/page/n31

Hung, Y.-Y., Aquino, G., Waldheim, C., Bélanger, P., Czerniak, J., & Geuze, A. … Robinson, A. (2013). Landscape Infrastructure: Case Studies by SWA (2nd ed.). Birkhäuser.

Imam, K. Z. E. A. (2006). Role of urban greenway systems in planning residential communities: A case study from Egypt. *Landscape and Urban Planning, 76*(1–4), 192–209. doi:10.1016/j.landurbplan.2004.09.032

Kang, C. D., & Cervero, R. (2009). From elevated freeway to Urban Greenway: Land value impacts of the CGC project in Seoul, Korea. pp.2772–2791. *Urban Studies (Edinburgh, Scotland)*. Advance online publication. doi:10.1177/0042098009345166

Kariminia, S., & Sh Ahmad, S. (2013). Dependence of Visitors' Thermal Sensations on Built Environments at an Urban Square. *Procedia: Social and Behavioral Sciences, 85*, 523–534. doi:10.1016/j.sbspro.2013.08.381

Kennedy, C., Miller, E., Shalaby, A., MacLean, H., & Coleman, J. (2005). The four pillars of sustainable urban transportation. *Transport Reviews, 25*(4), 393–414. doi:10.1080/0144164050011 5835

Kullmann, K. (2011). Thin parks/thick edges: Towards a linear park typology for (post)infrastructural sites. *Journal of Landscape Architecture, 6*(2), 70–81. doi:10.1080/18626033.2011.9723456

Kullmann, K. (2013). Green-Networks: Integrating Alternative Circulation Systems into Post-industrial Cities. *Journal of Urban Design, 18*(1), 36–58. doi:10.1080/13574809.2012.739545

La Rocca, R. A. (2010). Soft Mobility and Urban Transformation. *TeMaLab Journal of Land Use. Mobility and Environment*, *3*, 85–90. doi:10.6092/1970-9870/125

Lewis, P. H. (1996). *Tomorrow by design: a regional design process for sustainability*. Academic Press.

Liang, C., & Ng, E. (2012). Outdoor thermal comfort and outdoor activities: A review of research in the past decade. *Cities (London, England)*, *29*(2), 118–125. doi:10.1016/j.cities.2011.08.006

Linehan, J., Grossa, M., & Finnb, J. (1995). Greenway planning : Developing a landscape ecological network approach. *Landscape and Urban Planning*, *33*(1–3), 179–193. doi:10.1016/0169-2046(94)02017-A

Little, C. E. (1995). *Greenways for America*. Johns Hopkins University Press.

Lomholt, I. (2011). *Madrid Rio: Spanish Landscape Design*. Retrieved December 25, 2018, from e-architect website: https://www.e-architect.co.uk/madrid/madrid-rio-landscape

Loos, F., & van Vliet, M. (2014). Greenway Design: Context, Identity, and Sustainability. In *Landscape Record* (pp. 120–123). Design media Pub Ltd. Retrieved from https://www.ribabookshops.com/item/landscape-record-greenway-design-no.-6-2014/86565/

Loukaitou-Sideris, A. (1993). *Retrofit of Urban Corridors: Land Use Policies and Design Guidelines for Transit-Friendly Environment*. Academic Press.

Mathur, S., Menon, S. B., & Kacker, P. (2013). Greenways for a Greater Good-An Indian Perspective. *Proceedings of Fabos Conference on Landscape and Greenway Planning*. Retrieved from https://works.bepress.com/cgi/viewcontent.cgi?article=1001&context=mark_lindhult

Mossop, E. (2006). Landscapes of infrastructure. In C. Waldheim (Ed.), *Landscape Urbanism Reader* (Vol. 1, pp. 163–178). Princeton Architectural Press., doi:10.1017/CBO9781107415324.004

Mostafavi, M. (2015). Why Ecological Urbanism? Why Now? In *Ecological Urbanism* (pp. 12–45). Lars Muller Publishers.

Mostafavi, M., & Najle, C. (2003). *Landscape urbanism : A manual for the machinic landscape*. Architectural Association. Retrieved from https://copac.jisc.ac.uk/id/24979355?style=html

Mumford, L. (1963). *The highway and the city*. Harcourt.

National Association of City Transportation Officials. (2016). *Global Street Design Guide*. Island Press.

Ndubisi, F., DeMeo, T., & Ditto, N. D. (1995). Environmentally sensitive areas: A template for developing greenway corridors. *Landscape and Urban Planning*, *33*(1–3), 159–177. doi:10.1016/0169-2046(94)02016-9

Newman, P., & Kenworthy, J. (2015). *The end of automobile dependence: How cities are moving beyond car-based planning*. Island Press., doi:10.5822/978-1-61091-613-4

Newman, P., & Kenworthy, J. R. (1999). *Sustainability and cities : overcoming automobile dependence*. Island Press.

Nijhuis, S., & Jauslin, D. (2015). Urban landscape infrastructures Designing. In *Flowscapes: designing infrastructure as landscape* (pp. 13–34). TU Delft. Retrieved from https://books.bk.tudelft.nl/index.php/press/catalog/view/isbn.9789461864727/435/88-1

Paneerchelvam, P. T., Maruthaveeran, S., Maulan, S., & Shureen, S. F. (2020). The use and associated constraints of urban greenway from a socioecological perspective: A systematic review. *Urban Forestry & Urban Greening*, *47*. Advance online publication. doi:10.1016/j.ufug.2019.126508

Pourjafar, M., & Moradi, A. (2015). Explaining Design Dimensions of Ecological Greenways. *Open Journal of Ecology*, *5*(03), 66–79. doi:10.4236/oje.2015.53007

Ryan, K.-L., Flink, C. A., Lagerwey, P., Balmori, D., & Searns, R. M. (1993). *Trails for the Twenty-First Century: Planning, Design, and Management for Multi-Use Trails*. Rails-to-Trails Conservancy.

Salici, A. (2013). Greenways as a Sustainable Urban Planning Strategy. In M. Ozyavuz (Ed.), *Advances in Landscape Architecture* (p. 645). InTech. doi:10.5772/55757

Schiller, P. L., & Kenworthy, J. R. (2017). An introduction to sustainable transportation: Policy, planning and implementation. In *An Introduction to Sustainable Transportation* (2nd ed.). Policy, Planning and Implementation. doi:10.4324/9781315644486

Searns, R. M. (1995). The evolution of greenways as an adaptive urban landscape form. *Landscape and Urban Planning*, *33*(1–3), 65–80. doi:10.1016/0169-2046(94)02014-7

Sharma, A. (2015). Urban greenways: Operationalizing design syntax and integrating mathematics and science in design. *Frontiers of Architectural Research*, *4*(1), 24–34. doi:10.1016/j.foar.2014.11.002

Simsekoglu, Ö., & Klöckner, C. (2019). Factors related to the intention to buy an e-bike: A survey study from Norway. *Transportation Research Part F: Traffic Psychology and Behaviour, 60*, 573–581. doi:10.1016/j.trf.2018.11.008

Speck, J. (2012). *Walkable city : how downtown can save America, one step at a time*. North Point Press.

Suzuki, H., Cervero, R., & Iuchi, K. (2013). *Transforming cities with transit : transit and land-use integration for sustainable urban development*. World Bank. doi:10.1596/978-0-8213-9745-9

Suzuki, H., Cervero, R., & Iuchi, K. (2013). *Transforming cities with transit : transit and land-use integration for sustainable urban development*. World Bank. doi:10.1596/978-0-8213-9745-9

Suzuki, H., Murakami, J., Hong, Y.-H., & Tamayose, B. (2015). Financing Transit-Oriented Development with Land Values: Adapting Land Value Capture in Developing Countries. doi:10.1596/978-1-4648-0149-5

Taylor, S., Giang, C., Chau, P., & Aumann, P. (2017). *Cycling aspects of austroads guides* (3rd ed.). Austroads Ltd. Retrieved from www.austroads.com.au

Turner, T. (1995). Greenways, blueways, skyways and other ways to a better London. *Landscape and Urban Planning, 33*(1–3), 269–282. doi:10.1016/0169-2046(94)02022-8

Turner, T. (2006). Greenway planning in Britain: Recent work and future plans. *Landscape and Urban Planning, 76*(1–4), 240–251. doi:10.1016/j.landurbplan.2004.09.035

Turner, T. H. D. (1984). Landscape planning: The need to train specialists. *Landscape Planning, 11*(1), 73–79. doi:10.1016/0304-3924(84)90019-4

UN-Habitat. (2013). *Planning and Design for Sustainable Urban Mobility: Global Report on Human Settlements 2013. United Nations Human Settlements Programme (UN-Habitat)*. Routledge. Retrieved from www.unhabitat.org

United Nations Advisory Committee of Local Authorities, UNACLA Secretariat. (2013). *Transport and mobility snapshots*. United Nations Advisory Committee of Local Authorities (UNACLA) and United Nations Human Settlements Programme (UN-Habitat). Retrieved from www.unhabitat.org

Urban Planet. (2018). Urban Planet: Knowledge towards Sustainable Cities. Cambridge: Cambridge University Press.

Van Der Velde, R., & De Wit, S. (2015). Late twentieth century green infrastructures in Paris. In *Flowscapes: designing infrastructure as landscape* (p. 208). Research in Urbanism Series., doi:10.7480/rius.3.838

Waldheim, C. (2006). Landscape as Urbanism. In C. Waldheim (Ed.), *The Landscape Urbanism Reader - Review* (pp. 35–53). doi:10.1021/bi0013905

Waldheim, C. (2016). *Landscape as urbanism: a general theory*. Princeton University Press.

Walmsley, A. (1995). Greenways and the making of urban form. *Landscape and Urban Planning*, *33*(1–3), 81–127. doi:10.1016/0169-2046(95)02015-L

Walmsley, A. (2006). Greenways: Multiplying and diversifying in the 21st century. *Landscape and Urban Planning*, *76*(1–4), 252–290. doi:10.1016/j.landurbplan.2004.09.036

WHO Regional Office for Europe. (2017). *Urban green space interventions and health: A review of impacts and effectiveness*. Retrieved from http://www.euro.who.int/pubrequest

Zube, E. H. (1995). Greenways and the US National Park System. *Landscape and Urban Planning*, *33*, 17–25. Retrieved from https://ac-els-cdn-com.ezproxy.nottingham.ac.uk/016920469402011 4/1-s2.0-0169204694020114-main.pdf?_tid=c31cb0ae-9269-4e2a-a1fb-bb8c3179fc57&acdnat=1540157014_6f8cf8cc5000a1c3683b83f7750c397a

ADDITIONAL READING

Benedict, M. A., & McMahon, E. T. (2006). *Green Infrastructure: Linking Landscapes and Communities*. Island Press.

Cervero, R., Guerra, E., & Al, S. (2017). *Beyond mobility : planning cities for people and places*. Island Press. doi:10.5822/978-1-61091-835-0

Girling, C. L., & Kellett, R. (2005). *Skinny streets and green neighborhoods: design for environment and community*. Island Press.

Kang, C. D., & Cervero, R. (2009). From elevated freeway to Urban Greenway: Land value impacts of the CGC project in Seoul, Korea. *Urban Studies (Edinburgh, Scotland)*, *46*(13), 2771–2794. Advance online publication. doi:10.1177/0042098009345166

Kullmann, K. (2013). Green-Networks: Integrating Alternative Circulation Systems into Post-industrial Cities. *Journal of Urban Design, 18*(1), 36–58. doi:10.1080/13574809.2012.739545

Newman, P., & Kenworthy, J. (2015). *The end of automobile dependence: How cities are moving beyond car-based planning*. Island Press. doi:10.5822/978-1-61091-613-4

Searns, R. M. (1995). The evolution of greenways as an adaptive urban landscape form. *Landscape and Urban Planning, 33*(1–3), 65–80. doi:10.1016/0169-2046(94)02014-7

Walmsley, A. (2006). Greenways: Multiplying and diversifying in the 21st century. *Landscape and Urban Planning, 76*(1–4), 252–290. doi:10.1016/j.landurbplan.2004.09.036

KEY TERMS AND DEFINITIONS

Ecological Restoration: Human support for ecosystems and natural habitats' recovery process from degradation, damage, or destruction.

Green Infrastructure: A continuous and integrated network of natural and maintained open spaces and corridors that are planned, designed, and managed strategically at various scales to sustain the structure and functions of ecosystems while providing multifunctional benefits to people and places.

Landscape (Noun): All the visible natural and/or human-made elements in a geographical area that define its characteristics. Furthermore, the landscape is a composition of geophysical (i.e., landforms and water bodies), living (i.e., flora), human (i.e., vertical and horizontal structures), and transitory elements (i.e., weather conditions) in a given area or region.

New Urbanism: A planning and design approach that emphasizes the function and ethics of (re)building urban areas to be walkable, diverse, connected, mixed-used, aesthetically pleasing, and compact while protecting natural environments.

Non-Motorized Transport: Also known as active transport or human-powered means of transportation, which refers to modes such as walking, bicycling, skates, rickshaws, skateboards, scooters, and wheelchairs.

Sustainable Urban Mobility (SUM): Meeting travel needs of the current generation with minimal environmental impacts, and without affecting the ability of future generations to meet their own.

Urbanism: A study that aims to understand the interaction between urbanization, its processes and outcomes, and the people who live in such an urban area (i.e., a city).

Chapter 3
Smart Cars, Smart Cities, and Smart Sharing:
The Changing Nature of Urban Public Spaces

Celen Pasalar
https://orcid.org/0000-0002-3809-8299
North Carolina State University, USA

George D. Hallowell
North Carolina State University, USA

Yanhua Lu
North Carolina State University, USA

ABSTRACT

Streets, plazas, and parks are important components of a city that play a key role in affording socio-cultural, political, and economic activities for the benefit of society. The physical nature of these urban spaces facilitates sharing of resources, infrastructures, good, services, experiences, and capabilities. Recent socio-economic and technological changes have resulted in a new generation of city design and planning paradigms shifting the way that urban public and semi-public forms and spaces are designed, managed, and used. This chapter addresses the foundational changes brought by smart, or autonomous (AV), vehicles; smart city technologies; and the business models and associated technologies of sharing. The primary goal is to examine how these three socio-economic and technological changes may influence the use of current and future urban public space. It further informs designers on how urban spaces can provide opportunities to create new relationships of use and engaging public experiences through technology.

DOI: 10.4018/978-1-7998-3507-3.ch003

INTRODUCTION

Urban public spaces ensure the functional operation and resource sharing of a city. For centuries, streets, plazas, and parks have played a key role in affording cultural, social, political, and economic activities for the benefit of society. However, over the last two decades, several socio-economic and technological changes have resulted in a fundamental shift in the way we design, manage, and use urban public and semi-public forms and spaces. This chapter addresses three foundational changes: the introduction of smart, or autonomous (AV) vehicles; evolving smart city technologies; and the emerging business models and associated technologies of sharing. The authors explore how these three socio-economic and technological changes will affect the design and use of current and future urban public spaces, and offer new approaches for investigating and visualizing how they will alter cities in the future.

Autonomous Vehicles: During much of the 20th century, the expanding reliance on private cars formed and reshaped cities, often resulting in segregated, contained, and enclosed uses for streets and public spaces that provided minimal cohesion among people (Kunstler, 1994; Furman, 2017). However, the last ten years have seen a remarkable change occurring along city streets and public spaces in what is referred to as the public to private transect (Pasalar & Hallowell, 2019). As urban roads and parking areas are upgraded to AV or car-free environmentally, and human-centered designs, such as 'complete streets', the very nature of how people inhabit and move from public spaces through semi-public areas and finally into private zones is also being altered. Hence, a new strategy is proposed for exploring the concept of a private to public transect with the ability to examine the layers of transition from the public zone of the street right-of-way to the private interiors of buildings. This approach allows an urban transect to be measured and illustrated, so that as more and more autonomous vehicles, smart technologies, and sharing businesses alter the way streetscapes and other public spaces function, a record can be maintained over time.

Smart Cities: The increasing interest in smart cities, efficiency, safety, new sharing business models, faster services and new leisure activities, have all led to new and evolving devices such as self-driving cars, smartphones, and digital applications, virtual platforms, and so forth. A World Bank report (2012) suggests that 'smart cities' are about rethinking urban communities as inclusive, integrated, and livable. As part of ongoing smart city discussions, planners, designers, and economists are looking into these concepts and questioning the socio-cultural and economic effects of technology and sharing economies on the use and experience of public spaces. Urban life is inevitably and continuously undergoing dramatic change, leading to a reconsideration of how urban space is designed, transformed, and used. Hence, designers are considering how urban public areas in smart cities can provide affordances that help to create new relationships of use through technology and provide engaging public experiences within these spaces. Furthermore, designers should continue to address how smart, and shared technologies change the activities and 'formation of space' in city parks, plazas, and streets as places to gather and support communities. Creating human-centered smart cities that are able to bring people together by incorporating shared spaces and technologies that encourage socializing, collaboration, efficient mobility, and a healthy environment is essential. However, smart cities can run the risk of not being people-centered.

Sharing Economies: Because of new technologies and the evolving concept of sharing economies in the urban realm, the use of urban space continues to change rapidly. A sharing economy, often defined as collaborative consumption or a peer-to-peer economy, provides a process for underused resources to be easily shared or transferred to others to create more value or bring more benefits to society. As companies like Airbnb, Uber, and WeWork began sharing cars, homes, and office spaces, the idea of sharing became a new urban reality, enabled by smart apps and communication infrastructure in our everyday life. A new business model is also emerging that combines the use of autonomous vehicles with the principles of a sharing economy, reducing the total number of cars and parking spaces in the city. It is evident that technology associated with sharing economies has a significant impact on human interactions and business models. However, it is not yet clear how these sharing economies will alter the activities and use of space in city plazas, streetscapes, parks, and neighborhoods.

The following section provides a review of basic theory and terminology that has guided our understanding of urban spaces by city designers and planners for decades. It then examines how smart or autonomous vehicles, new directions in smart city development, and sharing economies and technologies may influence the way that urban public, semi-public and private spaces, are generated, used, and transformed.

BACKGROUND

The inevitable transformation in public, semi-public, and private areas due to the foundational changes of smart cars, smart technologies, and sharing economies is already underway in cities. If 'urban space' is a continuum from the most public, found in the Right of Way (ROW) of the street, to the privacy within offices and homes, this transect is being revised or blurred by the technologies and socio-cultural practices of the sharing economy. The use of space along city streets should be carefully considered as autonomous vehicles (AV) and car-sharing alter public transportation, parking, and drop-off patterns, as well as dominion over the road and nearby pedestrians. Smart and sharing technologies may also accelerate the way that

Privately-Owned Public Spaces (POPS) have blurred the line between public and private use and enjoyment of urban public spaces in plazas, parks, and even building interiors. At this point, planners, designers, and researchers are woefully unprepared for the changes of autonomous cars, smart technologies, and sharing economies will bring. It is incumbent on everyone in these fields to begin to research, design, and plan for the rapidly approaching changes arising from both semi-autonomous and driverless vehicles, as well as sharing and smart technologies. This should not be thought of as a cautionary tale, but rather a rare opportunity for inclusive and positive urban change.

As Norman (2013) states, technologies might change, but the fundamental principles of interaction are permanent, and should continue to apply in how our public spaces, such as streets, plazas, squares, and parks are able to support human comfort, enjoyment, and social engagement. Part of our goal as designers and researchers is to ensure that new smart technologies, sharing and automated vehicles succeed - not only in reducing the car-dominance in our cities, but also in helping to transform urban communities into greener, healthier, and more humanized places in which to live. Toward this end, the following sections will review long-standing theories on the nature of urban public space, and then discuss the anticipated changes brought about by smart cars, smart cities, and smart sharing.

The Changing Nature of Urban Public Space

It is within the public space of streets, squares, plazas, and parks that the complex mélange of urban life is created, observed, and reproduced. The identity and function of urban public and semi-public space is always a work in progress and never a finished product (Leary-Owhin, 2016). It is a physical reality of bricks and mortar, steel, and glass that is produced and reproduced continually through social interaction and human activation (Leary-Owhin, 2016). In that sense, Lefebvre (1991) problematized urban space, insisting it was not simply a neutral container, promoting its reconceptualization as both a social process and material product. Hillier (2007, p. 20) stated that "human behavior does not simply happen in space. It has its own spatial forms. Encountering, congregating, avoiding, interacting, dwelling, teaching, eating, conferring are not just activities that happen in space. In themselves, they constitute spatial patterns."

The success of cities has always been about shared space, interaction, experiences, and the exchange of goods and services (Agyeman et al., 2017). Traditionally public spaces, such as streets, plazas, and parks, provide people with opportunities to be among, see, and hear others, while being engaged in various activities (Gehl, 1971). According to Gehl (1971), there is a range of outdoor activities from low contact to high contact intensity. The lowest end of Gehl's scale entails necessary activities in a space slightly influenced by the setting, such as walking to destinations, waiting for transport, and so on. The middle range of influence is for optional activities, such as walking for leisure, sitting at a cafe or park bench, or watching other people in the same space. These optional activities and their occurrence are influenced by the affordances of the space. The highest end of the scale is social activities, which depend on the presence of people in space, such as conversations among people. These social activities are also supported by the affordances of the space, such as the existence of seating options, furniture for play or sport activities, and so on. According to Gehl (1971), spaces that support the highest level of social activities are typically considered as higher quality spaces that support a wider range of activities, as well as stronger social connectedness among people. However, recent studies are questioning how these historical paradigms of urban social interaction will change with smart technologies such as virtual viewing or smartphone routing systems (Pasalar & Hallowell, 2020).

In recent years, city dwellers have been reimagining and redefining their environments through various methods of urbanism, including guerilla, DIY, tactical, and pop-ups, maximizing mixed uses and activity types in public spaces (Agyeman et al., 2017). Rather than accepting places as they are, people are redefining what these places can become (Massey, 1995). By their nature, these social movements interpret the urban realm as a shared space.

One sharing concept that has had a dramatic effect on the way people use certain urban public spaces is the notion of Privately-Owned Public Spaces (POPS). Introduced as a part of zoning regulations in New York in 1961, POPS have often blurred the line between public and private use and enjoyment of urban spaces in plazas, parks, and even building interiors. In theory, they are dedicated to public use and enjoyment, but controlled and maintained by the private property's owners, often in exchange for bonus floor area or waivers. There are in excess of 550 POPS in NYC alone (Hilburg, 2018; Polyák, 2017), and they supply a range of affordances for relaxing, sitting, eating, and gathering with other people. POPS have been developed in a variety of scales and types, both outdoor and indoor, and present a diverse set of amenities. They are often the result of city zoning ordinances focused on the densest zones of a city so that those areas can still offer a measure of open public space and greenery. POPS can be hard-surfaced or green outdoor spaces, or they can be lobby areas in indoor atriums, where users can enjoy their lunch or gather for conversation. POPS will likely accelerate their evolution as smart cars, smart technologies, and sharing business models are introduced into these already somewhat nebulous public spaces.

Studies have examined the way that sharing economies have expanded, helped define, or in some cases, just confused POPS in cities such as New York, London, or Seoul (Garrett, 2015). The geographer Harvey stated that "the freedom to make and remake our cities and ourselves is … one of the most precious yet most neglected of our human rights" (Hatherley, 2012). A good number of urban theorists over the years, including Jacobs and Mumford (Jacobs, 1961; Mumford, 1961), have maintained that the place where cities are formed and remade is in public rather than the private realm. Garrett (2015) therefore contends that a crucial concern with POPS, such as plazas, parks, and building atriums, is that they may look public, but in reality, they do not perform that way. That is to say – the rights of the citizens within these spaces are severely restricted. Even though this may not concern someone sitting on a park bench on a sunny day, the outcome from ever-expanding POPS can affect everything from our personal psyche to our ability to gather and exercise free speech. Indeed, in some successful POPS, the shared urban space and function can become entirely controlled by private entities. The public to private transect can be revised or overridden by private functions. The enforcement agency for New York City POPS refers to this as "cafe creep". In the extreme, POPS can be unused spaces, vacant lobbies, or can even be illegally made inaccessible.

As Agyeman et al. (2017) state, the concept of sharing redefines smart cities as the future that "harness smart technology to an agenda of sharing and solidarity". As the cities cultivate a sharing culture and build capabilities in sharing enterprises and activities, they are becoming more accessible to all citizens by expanding physical and digital sharing infrastructure, incubating and supporting sharing startups, and sharing idle public resources (Agyeman et al., 2017).

Smart or Autonomous Vehicles and Changes in Land Use

The rise of car-sharing and autonomous vehicles is inevitable, and some believe that they will eliminate auto crashes and save the environment. Others are convinced they will force pedestrians away from the street, increase traffic, and create major gridlock. A fully autonomous vehicle (AV) is defined as "capable of completing journeys safely and efficiently, without a driver, in all normally encountered traffic, road and weather conditions" (Skinner & Bidwell, 2016, p. 6). Despite the divide over how this technology will, or should be implemented, it is incumbent on architects, landscape architects, and urban planners to begin to anticipate the effects of self-driving cars and ride-sharing systems and the potential benefits of less urban area devoted to vehicles and more space to people (Boyer, 2017; Gavanas, 2019). For example, will autonomous vehicles require drop-off zones that may support other uses such as retail or food kiosks, and how will they interact with pedestrians at street corners and crossings? As these questions are being raised, it is also important to focus on the redevelopment and reuse of urban forms and spaces that were previously held hostage to the private automobile, such as unused parking garages, curbside parking and high-occupancy vehicle lanes in roads that can be reformed into new parklets, retail or leisure areas, and even workspace.

Advances in transportation technology and sharing business models are already creating significant changes in the form and use of urban public space. Companies, such as Uber and Lyft, are providing an alternative to owning and using a private car. By connecting with nearby drivers, people are able to go anywhere they want easier, quicker, and more affordably. As an alternative to private taxi services, Uber and other ride-sharing businesses disrupted the transportation sector and the mobility behavior of urban users. Their sharing business models created a network of drivers using their personal cars and interconnecting them with nearby customers using smartphone applications. These sharing models also rely on navigation systems, such as Google Maps or Waze. Using smart routes suggested by such apps, Uber and Lyft are now designating pick-up points for multiple riders, thereby reducing the time and effort required to pick up various customers. As ride-share companies, such as Uber or Lyft, continue to develop new sharing business models, increasing numbers of cars are competing for curbside collection or drop off points. Ridesharing companies are now designating nodes as collection points, and access to those nodes is becoming important from both a financial and urban space perspective (McFarland, 2017). Among other things, this suggests a reconsideration of new transit-oriented development and the creation of new urban spaces designated as nodes that are also suitable for flexible transportation. These changes also provide a wide range of opportunities for rethinking transportation infrastructure, especially where curbside and pickup stations in the public right-of-way will become an important commodity, just as bus stop areas did in the last century. These advances require city stakeholders to reconsider how they manage the public property as well. As more travelers opt for ridesharing instead of using or owning their cars, the trip destinations are shifting from parking spaces and garages (i.e., as nodes to store cars) to streets and curbsides (i.e., as places for people to congregate), thus generating a congestion problem for roadways as well as deserted parking lots. This change to road edges and parking spaces can also be a positive outcome, if they are repurposed to meaningful public spaces with sufficient ecological wisdom.

With car sharing and the advent of autonomous vehicles (AV), land use designations, urban density, and human use of the urban landscape will all be substantially affected. Recent studies on smart city design suggest that AVs and sharing business models will help convert mobility from personal ownership of automobiles to a collective shared or service paradigm (Young & Lieberknecht, 2018). Along with the potential health and environmental benefits of AVs, designers must also prepare for potential negative effects, such as urban and suburban sprawl. Researchers have warned that a possible side effect of vehicle automation might be to encourage higher vehicle-miles traveled (Cox & Hart, 2016; Wadud et al.,

2016). There is a need for additional research to confirm how automated vehicles might impact the future of the built environment, and how soon this might occur given technology adoption rates and market penetration (Litman, 2017). Part of the concern is whether autonomous vehicles will encourage denser urban development or prompt more suburban sprawl, thereby possibly increasing negative health outcomes (Crayton & Meier, 2017, p. 250). Some studies contend that private ownership of AVs may actually lead to "an increase in the amount of land dedicated to new roadway infrastructure, thus encouraging dependency on private vehicles and community sprawl; increase congestion, trip length, energy consumption, air, and noise pollution; and counteract efforts to promote mixed-use facilities, uptake of transit, and active travel options" (Rojas-Rueda, 2017).

Discussions about car sharing and the advent of the autonomous vehicle (AV) often center on the rate of fatal accidents, air pollution, or reducing urban traffic. However, the incredible amount of space currently devoted to cars in the urban landscape should also be a critical factor for planners and designers to consider. The average private car is approximately 80 percent empty when it has only a single occupant as a driver. During a typical workday, private vehicles also tend to sit idle in garages, parking lots or on-street parking. Plumer (2016) contends that there are approximately one billion parking spots across the United States, with an average of four spaces for every car in operation. Considering the addition of the massive areas committed to roads and driveways throughout cities, "many downtowns devote 50 to 60 percent of their scarce real estate to vehicles" (Ibid, 2016). If a substantial percentage of urban land can be liberated from current parking use, the benefits could be game-changing to the way cities operate. The San Francisco Smart Cities Challenge (SFMTA, 2016, p. 4) began their effort to design for new sharing technologies and AVs with the following goals: (1) to increase technologies that reduce reliance on single-occupant vehicle trips by using new sharing models, (2) to phase in AV and connected technologies within that sharing model as soon as they become available and are shown to be cost-effective and safe. The ultimate SFMTA goal is to bring down travel costs, reduce or eliminate collisions, and reduce parking demand so that parking facilities can be repurposed into green space, affordable housing, parklets, and other city amenities.

A report from the Rocky Mountain Institute (Walker & Johnson, 2016) suggests that the era of private car ownership may start declining within the next ten years as new sharing and connected AV networks become more available and less expensive. Rather than owning a private vehicle, city dwellers could purchase or rent a ride-share vehicle. Certainly, that point of change will have the potential to revolutionize our streets and public urban spaces (Plumer, 2016). In cities where policymakers, planners, and designers have begun to liberate some of the street right-of-way (R.O.W) from parking and excessive lane numbers and widths, the potential for new public space is already being realized. Over the last decade, New York City has been converting many of its busiest streets into more human-centered public spaces (Luberoff, 2016), by recovering parking areas and reducing street R.O.W in order to create parklets, new urban plazas, bike lanes, and rental stands, and retail food areas. Parklets (see Figure 1) allow a phased approach, where street-side parking can be very quickly converted to public use areas.

Figure 1. Parklets and street R.O.W changes in New York for new human-centered urban activities.
Photo credit: Authors

Cities have also been rescuing R.O.W and parking areas for privately used public spaces, either permanently, such as protected seating areas and bike paths in Times Square, or temporarily for retail and dining plazas during evenings or weekends in Singapore (see Figure 2).

Figure 2. Permanent or temporary retail and foodservice spaces in New York and Singapore
Photo credit: Authors

Smart Cities and New Technologies

Cities are rapidly growing in population, thereby increasing the consumption of critical resources and the demand for unique services. It is predicted that 68% of the world's population will live in urban areas by the year 2050 (UNDESA, 2018). Hence, cities are seeking innovation in the management of their resources, such as water and energy, in order to accommodate this growth. Much of the past two decades of urban redevelopment efforts have centered on the reclamation of auto-centric spaces and forms through mega revitalization projects, such as the dramatic transformation of an elevated freeway and concrete paved waterway into a 3.6-mile long day-lit stream corridor, known as the Cheonggyecheon River linear park (Wang, 2014). More cities are also looking at the strategies to reduce traffic almost to zero and reallocate the space, formerly occupied by cars, to pedestrians and open/green spaces in order to fight the climate crisis. For example, Barcelona has been implementing urban planning and transport interventions like the superblocks that shifted away from the car-centered urban planning model and towards a people-centered approach by transforming residential districts into walkable, child-friendly neighborhoods with welcoming public spaces (Burgen, 2019).

Recently, open data, consumer-based technological applications, and the internet of things are now set to transform and shape cities well into the twenty-first century "smart" cities. The advocacy of smart cities highlights the role of technological innovations to make building operations more efficient, reduce energy use, streamline waste management system, and advance the transportation system. All of these measures revolve around the idea of greening cities while making mobility easier.

Twenty-five years ago, Mitchell (1995, p. 5) envisioned that the digital revolution and new civic and spatial arrangements would substantially change how we access economic and public services and engage in public conversations, activities, and experiences in our daily lives. Han and Hawken (2018) indicate that smart cities have mostly been considered in terms of the development of digital technologies and their business potential for cities' regulatory and economic benefits. Hence, the common definition of a smart city involves software-enabled infrastructures and networked digital devices and sensors that are used to produce knowledge-driven, competitive, and resilient urban systems (Kitchin et al. 2016). However, as cities invest in smarter and more efficient systems, sometimes technological solutions are disconnected from society's needs, aspirations, and spatial capacities (Acedo et al. 2017). A recent Smart City Challenge launched by the U.S. Department of Transportation (2015) asked cities across the US to share their bold ideas for how to create an integrated, revolutionary transportation system using data, applications, and technology to meet the needs of people of all ages and abilities. Thereby bridging the digital divide and allowing every resident to be connected to city resources.

The visionary applications received revealed a range of innovative approaches and technologies, including electric car-sharing, automated shuttles, on-demand delivery trucks with integrated mobility apps, as well as on-demand minibuses to provide affordable transportation options to underserved communities using a network of multimodal transportation hubs for mobility, economic development, and community activity. The proposals focused on how people and goods move, how they adapt to new technologies, connections between vehicles and infrastructures, connectivity between people and resources, as well as the new integrated data platforms that align decisions and dollars effectively. However, most of these proposals failed to effectively consider the spatial implications of these innovations and the potential influence on the character of streets and urban public spaces in the interim.

Focusing on the creation of people-centered livable communities is essential. Smart cities should provide social and physical infrastructure as part of the digital technologies driving urban innovation. They should enhance the character and sense of place in neighborhoods while they install technical and economic capacities. Advancing smart technologies are influencing the human experience in cities just as the evolution of the car, train and other transportation modes that changed spatial arrangements, mobility behaviors and demographics of cities over time (Han Hawken, 2018). Today's digitally mediated environments are influencing the way that people sense, use, and interact with the environment.

Sharing Technologies and Business Models

Recent sharing technologies represent the popular technology of this era and have been developed to control and provide profitable means to influence the use of both urban public and private spaces. They are also reshaping the nature of streetscapes as well as the home environment. For example, new sharing technology has transformed the concept of the bicycle into a bike-sharing model (see Figure 3) through which people check out a bike from a docking station, use it for a number of hours, then return it to another docking station in the city (Thompson, 2018). Although this concept was attempted in the 1960s, it failed back then due to a lack of technology to track down the bikes after they were rented for use. Today, the technology available through smartphones, such as GPS, Bluetooth, and mobile payment systems, bike-sharing has increased the efficiency of moving around in congested cities. As a result, the usage of shared bikes increased from 320,000 rides in 2010 to 28 million in 2016 (Thompson, 2018).

Another sharing technology that has changed socio-economic behavior is house sharing. Airbnb and similar online services, such as VRBO and Flipkey, have emerged as online services that connect people with hosts who offer short-term accommodation. This business model has enabled people to rent and experience

Figure 3. Docking station for bike-sharing in Raleigh and New York.
Photo credit: Authors

private homes and apartments in cities beyond the traditional hotel districts, typically considered in certain areas of downtown, attraction zones, business centers, and airports. However, recent studies have revealed that this widespread practice of house sharing has changed the stability and privacy of many quiet neighborhoods, impacting permanent residents' quality of life, the use of public spaces, and the overall sustainability of neighborhoods (Wachsmuth et al., 2018).

Coworking spaces have also emerged as community-oriented, shared spaces where freelancers, entrepreneurs, start-ups, and small teams work in the same environment to take advantage of an affordable and flexible space compared to a costly private office environment. As the demand for shared workspaces increases, the boundaries of these spaces are gradually stretching into the outdoors. Research has revealed that working outside improves mood, lowers stress levels, and makes people happy and healthier (Belanger, 2018). Hence, more pop-up coworking spaces are emerging in urban plazas or parks in cities like Boston, New York, and Raleigh by using mobile computers/tablets, smart phones, wi-fi connections, access to power/charging units, and so on. It is predicted that these businesses and sharing technology trends will continue to evolve and shape the future of parks and urban spaces that offer coworking components (Belanger, 2018).

Evolving sharing technologies and business models have the power to disrupt and reshape the neighborhoods, streetscapes, and urban public spaces for better and/or worse. It is worthwhile to consider the impacts of such technological disruptions and the resulting design expectations on urban spaces.

DISCUSSION, RECOMMENDATIONS AND NEW STRATEGIES FOR ANALYSIS

Visualizing the Urban Public to Private Transect

Madanipour (2003) promotes two important concepts related to both urban public and private spaces. He contends that the private and public spheres are concomitant, constantly reforming, and interacting with each other. However, as they push and pull against each other, the boundary between them is often blurred and reshaped. This constant struggle between public and private realms can create areas of a semi-public nature, or zones of vague ownership, use, and influence. This chapter proposes an innovative strategy to investigate and visualize the 'push and pull' between the private and public realms. This approach focuses on the idea that "public and private spaces are a continuum where many semi-public or semi-private spaces can be identified, as the two realms meet through shades of privacy and publicity rather than clear cut separation" (Gehl, 1971, p. 59; Madanipour, 2003, p. 210). The core concern here is less about whether urban space is becoming fragmented (Madanipour, 2003, p. 208), or how much privacy is desirable (Jacobs, 1961; Kilian, 1998; Sennett, 1977). The present discussion centers on how the edges and boundaries of this continuum, from the most public (such as the central corridor the street) to the privacy within homes, are being shifted or blurred by both smart city and car technologies, and by the socio-economic practices of the sharing economy. Adopting the concept of a continuum or transect to analyze and communicate the changes occurring in urban public and private realms is also a potentially valuable tool for this and future studies.

The most public end of the continuum is usually the center of the street in cities. The public to private *transect* is therefore a cross-section, extending from the street centerline out to the buildings and spaces on both sides of the road. The transect depicted in Figure 4 is specifically focused on streetscapes, but the same public to private continuum could also be utilized to portray other urban public spaces, such as plazas, parks, squares, and so forth. In the case of a public square, the transect would be a cross-section from the middle of the plaza to the inside of the buildings and private yards on the perimeter. The diagram in Figure 4 also illustrates a sense of what we might see in the past, present, and future, in terms of the public to private transect. Beginning at the bottom of the diagram, we see a typical street transect in a North American city in the 1990s. The street might be several lanes of traffic with designated parking on one or both sides. Forward-thinking suburban and urban communities may have even removed parking or narrowed driving lanes to allow for parklets or bike lanes, or to introduce wider sidewalks and planting areas.

This concept of the public to private transect can also be a useful strategy for examining the residential areas of cities and suburbs as well as the urban downtown core. In the case of a suburban street, a transect of the late twentieth century would have begun in the public middle of the two or four-lane road and extended out to private homes or small stores on either side. In that suburban setting, the transect would often slice through a row of parking, a zone of parked cars along the street, then through a strip of grass or trees and often a sidewalk. At that point in the transect, the public space of the street would transition to semi-public spaces such as front yards of homes or perhaps outdoor eating areas of street-side restaurants. In the case of front yards or house entrances, several techniques have helped identify and protect this public to private transition over the decades, including picket fences, hedgerows, or elevation changes in the form of stoops or porches (see Figure 5).

Within the 2020 transect in Figure 4, the diagram suggests that many cities and towns have begun to introduce Complete Street concepts, such as rain gardens, walking and biking lanes, wider sidewalks with fixed or flexible seating, and retail areas for pop-up food stands or vending machines. The described changes are depicted in the diagram as wider areas of the original street R.O.W being converted to those new uses. Figure 2 illustrates the changes taking place in Singapore and New York as street R.O.W is recovered from daily traffic and parking. These semi-public zones of commercial food and retail vendors are slowly moving out into what was one of the busiest streets in New York City - Broadway. The once distinct edge of the sidewalk in Bussorah Street, Singapore (see Figure 2) has slowly melted into a mix of tightly controlled private restaurant tables and narrow channels of pedestrians, only allowing vehicles and trucks for delivery at certain hours of the day. In the year 2020, changes caused by smart city technologies, smart cars, and technologies or business models of sharing have also been leaving their mark on the public to private transect. Figure 1 indicates some of the transformations that are taking place along city streets in 2020, but as noted earlier, sharing in bicycles, scooters, and designated drop off zones for Uber and Lyft vehicles are also changing the transect (see Figure 3). Office sharing technologies and business models are also blurring the edge between semi-private workplaces within buildings and the more public space of lobbies, streets, and urban plazas outside (Waber, Magnolfi & Lindsay, 2014). WeWork and Industrious office sharing businesses are experimenting with the conversion of traditional outdoor public plaza areas for shared office tables and desks. In 2018, the co-working company Industrious partnered with a retailer to introduce their "Be An Outsider at Work" campaign (Berlanger, 2018). The temporary experiment featured a pop-up co-working space in Madison Square Park, NY, and included a makeshift conference room in a custom trailer with transparent walls.

By 2040, as AVs and vehicle sharing businesses become fully embedded in the way urban transportation systems and public spaces may evolve, the section perspective at the top of Figure 4 presents a possible result as proposed by the authors. One recent study predicted that by 2040 AVs will account for up to half of all road

Figure 4. The Public to Private Transect
Source: Authors

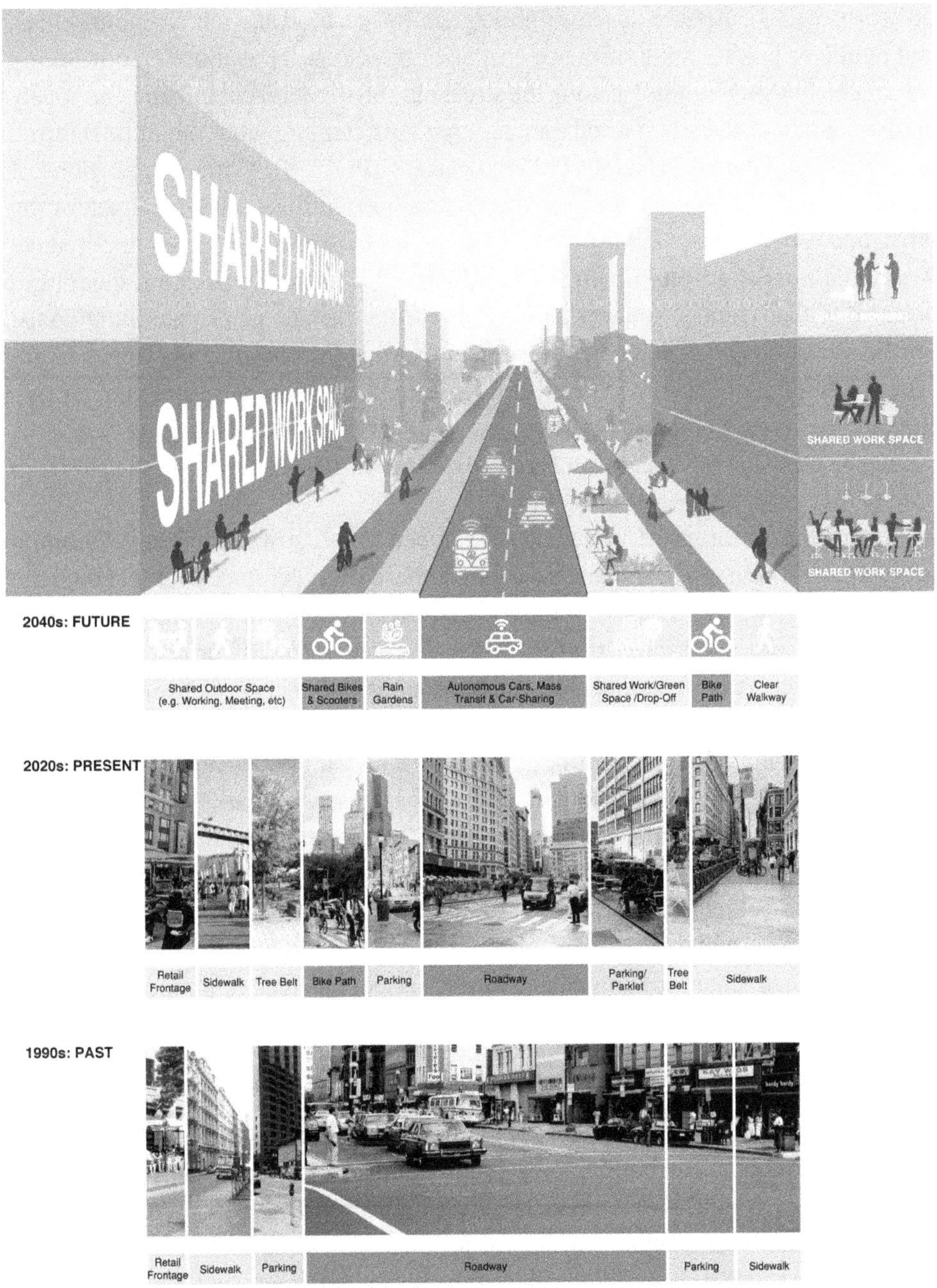

travel (Litman, 2017). This means that if an equivalent amount of curbside parking is recovered, then what started as small areas of recovered parking in 2020, could fully evolve into a mixture of semi-public green space, food and retail vending areas, and public or private outdoor office spaces or pavilions. The adjoining coworking and shared housing buildings along the street may also extend out into the previously public domain of the street or city square; perhaps creating what might be referred to as Publicly Owned Privately Utilized Spaces (POPUS). Certainly, portions of publicly owned and operated urban spaces have been utilized and managed for the profit and use of private individuals or businesses for centuries. Cities like Venice, Amsterdam, or Bruges (see Figure 5) have always allowed the demarcation and control of parts of their central squares and plazas for the benefit of private individuals. Further research is needed to show whether sharing and smart technologies might usurp more permanent control over those spaces through actions like "cafe creep". But only by researching and developing design parameters and planning standards can we be prepared to successfully take advantage of these changes in the public to private transect if and when they occur.

As designated drop-off nodes are developed and distributed along the urban streetscape for AVs and carshare businesses, these drop-off nodes can become the modern coffee shop, and bus stop combined, providing Wi-Fi, coworking or meeting spaces, interconnected mass transit, and reinvented forms of urban semi-public gathering space. Architects and planners have suggested many possible reinventions and reuse of car-centric forms and spaces in cities of the future (Baumgardner, 2015; Lubell, 2016), including converting unused parking garage structures into housing and office space, and parking lots into new urban plazas or parks. Part of the task of forward-thinking designers and planners will be to leave behind the car-centric mindset of the last century, and incorporate new technologies of smart city, sharing and automated vehicles. We are only limited by our ability to think outside the box.

There will be challenges or unintended consequences as these smart cars, smart cities, and smart sharing advancements affect changes to public spaces and streetscapes of cities and suburbs. Typical mechanisms to protect areas of privacy, such as raised porches, steps, fences, bollards, and so forth, will have to be reinvented as the public to private transect shifts. As depicted in Figure 5, urban mechanisms, such as raised stoops, fences, and lowered yard areas, can provide a porous but effective barrier between public and private. For centuries, public streets, squares, and courtyards around the world have protected semi-private space for exclusive commercial interests, such as restaurant seating or commercial vending. As the 2040 transect slides back and forth, designers may need to reinvent how transitions between semi-public, public, and private are secured.

At the extreme private end of the transect, new sharing technologies and business models that were discussed earlier, such as Airbnb or WeWork, may also change the character of neighborhoods, and even the very nature of building interiors, private homes and backyards. For example, in the historic residential Treme neighborhood of New Orleans, residents have complained that the neighbors they once knew are disappearing, replaced by the rapid gentrification of house-sharing businesses like Airbnb (Peck & Maldonado, 2017). Once quiet and desirable neighborhoods in the city have essentially become tourist districts inhabited by visitors from all over the world, driving up rents and home prices. The influx of short-term rental units within an existing apartment or condo buildings can also directly challenge the privacy of adjoining apartments within the same building.

Several cities have responded by prohibiting house-sharing businesses from buying properties within certain zones. With enough forethought and use of a research strategy, such as using the transect in Figure 4, it may be possible for city managers and planners to prepare for this eventuality by determining how short-term rental housing units are distributed and designed throughout the city.

FUTURE RESEARCH DIRECTIONS

This chapter highlights the foundational changes brought about by the introduction of smart or autonomous vehicles; smart city technologies; and the business models and associated technologies of sharing. As advancing technologies and sharing economies permeate through cities, and influence functional urban operations, more changes in the use of public spaces and related social behaviors are expected. Ongoing smart

Figure 5. Protecting the edge between public and private, Brooklyn and Bruges.
Photo source: Authors

city global discussions are exploring the idea of urban spaces that can make use of technology to reduce energy consumption and pollution, make transportation more efficient and smart and transform urban areas into attractive spaces accommodating the latest innovations for safety, energy efficiency, and the comfort of its users. As new technological approaches are implemented, planners, designers, and economists are looking into strategies that aim to redefine the functional aspects of urban spaces in order to promote healthier city lifestyles and community well-being.

There are also spatial implications from these foundational changes, and the pressing need to envision the design of future urban public spaces with a human-centered focus. Further examination of case studies on how technological changes may influence the use of current and future urban public spaces, associated human behaviors, and the evolving nature of public-to-private transect is needed. This will be particularly useful to inform designers on how urban spaces can be formulated to create new relationships of use as we continue to experience the disruptions from smart technologies, sharing, and their associated business models.

Further applied research is also needed to understand how automated vehicles and associated smart technologies will influence people's mobility behavior, as well as the consequences of urban landscapes. A century ago, the automobile transformed the cities extensively, resulting in unexpected problems, such as congestion, pollution, sprawl, and roads designed to move vehicles rather than people. Associated with these challenges experienced over time, routing and wayfinding apps were developed. However, recent research indicates that routing and wayfinding apps have also created problems by directing commuters to normally quiet neighborhood streets overflowing with rush-hour jams and late-night speeders (Surico, 2018; Pasalar & Hallowell, 2020). With the evolving interest in driverless vehicles, it is important to research and design the future design of urban spaces, streets, and neighborhoods in order to minimize the impact on the quality of life and pedestrian safety. As the demand for driverless cars increases, more road infrastructure and parking lots might be freed up for pocket parks, parklets, and greenways, contributing to people's mental and physical health (Jolma Architects 2018). It is worth studying different design scenarios and the potential impact on our urban experience and stored perceptions.

CONCLUSION

Urban streets, plazas, and parks have always played a key role in affording cultural, social, political, and economic activities for the benefit of society. However, this chapter has examined three fundamental socio-economic and technological advances that are creating a foundational shift in the way we design, manage, and use the public built forms and spaces in cities. These three foundational changes are smart or autonomous (AV), vehicles; smart city technologies; and the business models and associated technologies of a sharing economy. The primary goal of this chapter is to examine how these three socio-economic and technological changes will likely affect the design, operation, and use of current and future urban public spaces. A discussion about these three fundamental processes is no longer just about imagining a possible urban future. The future has already begun.

This inquiry contends that the adoption of car-sharing and autonomous vehicles (AVs) is already well underway. Therefore, it is essential for architects, landscape architects, and urban planners to understand, integrate, and design for the coming effects of self-driving cars and ride-sharing systems. Far from a cautionary tale, this is a tremendous opportunity for inclusive, human-centered, and positive urban change in the way we design public and semi-public spaces in the future. Both car-sharing businesses and AVs will allow cities to devote less area to vehicles and more space to people. For example, this study examined how autonomous vehicles are already developing drop-off zones, and how these areas might be designed for new combined forms of public space, such as coworking, integrated public transit, retail, food service, and so forth. The redevelopment and reuse of urban space recovered from the private automobile, such as unused parking garages, curbside parking, and high occupancy vehicle lanes, also presents marvelous opportunities for designers and planners to create new human-centric and environmentally sensitive urban public spaces. At present, though, research, design, and planning tools have not yet fully developed for studying the increasing changes brought on by autonomous vehicles or sharing and other smart technologies.

This chapter also emphasized the importance of creating human-centered smart cities that are able to bring people together by incorporating shared spaces and technologies to encourage socializing, collaboration, efficient mobility, and a healthy environment. Emerging research from around the world is promoting the value of enabling a diverse range of city stakeholders through evidence-based research and planning. The idea is to keep people at the center with the support of smart city technologies and infrastructure. This concept is often referred to as the responsive city, essentially making the city more livable for all of its people. The availability of smart technology to improve our environment and daily living may usher in an era where open data, consumer-based technological applications, and the internet of things are set to transform and shape city building well into the 21st century. While we continue to develop tools that enable public participation in the improvement of our environment, it is also imperative for us to understand how consumer-based technological applications are changing our behaviors and the level of interaction with our environment. Planners, policymakers, and researchers are already making use of the latest advances in technology to solve urban problems through the deployment of urban data systems, information and communication, and the internet of things.

It was noted that new technologies, business models, and the evolving concept of a sharing economy are also dramatically affecting the formation and use of urban space. This chapter stated that sharing businesses and technologies can provide a process for underused resources to be easily shared or transferred to create more value and benefits for society. From Airbnb and Uber to shared bicycles and coworking office space, the idea of sharing is becoming a new economic paradigm enabled by smart apps in our everyday life. Urban planners, economists, and designers are beginning to study these concepts and questioning the socio-cultural and economic effects of a shared economy on the use of urban public space. City life is inexorably and constantly changing, and this requires constant research and reconsideration of how urban space is designed, transformed, and used. The unintended results of the sharing economy and technology may create urban public spaces that are unsafe or unsuccessful, and therefore will require close attention to the ways we reconfigure our current streets and public spaces, and even our own homes (Furman, 2017). This chapter has promoted the idea that there are clear spatial implications from sharing technologies and services, and therefore a need to envision the design of future urban public spaces with a human-centered focus. Designers, planners, and policymakers should study how urban public spaces can provide affordances through sharing technology and business models that are able to engage the public within these spaces. Sharing economies can help provide activities and affordances in city parks, plazas, and streets as places to gather and support communities.

Finally, this chapter introduced the idea that the public and private realms are a complex and constantly changing continuum that meets through shades of privacy and publicity rather than at clearly cut edges or boundaries (Gehl, 1971; Madanipour, 2003). But as the public and private spheres push and pull against each other, the boundary between them is often blurred and reshaped, especially over longer periods of time. Given the complex nature of this continuum, we have suggested a novel strategy for researching and communicating the various components of this concept. Figure 4 proposed a transect to analyze and document the changes occurring within the urban public and private continuum, both in detail and over long spans of time. We noted that since the public end of the continuum is usually the center of the street, the proposed public to private transect is proposed as a cross-section, from the street or plaza centerline out to the private buildings and spaces on all sides. This transect thus extends from the most public areas along a street, through various boundaries of semi-privacy to the most intimate spaces in homes. Utilizing this transect as a research and communication strategy allows an ongoing study of the ambiguous changes brought about by AVs and car-sharing, and smart city technologies, and sharing economies. This transect would also allow the research process to set a benchmark and then examine the change in areas, such as parking and drop-off zones, vague POPS developments, and even how streets, sidewalks, and porches are altered by the technology and economics of sharing.

The fundamental changes in the formation and use of urban public spaces brought about by socio-economic and technological advances in AVs, smart city technologies, and sharing economies are by no means at an end. As autonomous vehicles and future advancements in smart city and sharing technologies continue to evolve, more changes will be seen in human behaviors and the use of the urban environment. They will contribute to a transformation of the ways that streetscapes and urban open spaces are designed and used. This rapid advancement in AVs, smart city, and sharing technologies is an inflection point for designers and planners. They must be agile enough to study and accommodate the resulting changes as they occur. This chapter suggests a strategy for researching and illustrating current and future changes occurring in cities as a consequence of AVs, smart city technology, and sharing economies by examining the public to private transect in urban spaces, streets, and buildings. This strategy also provides the ability to set baselines and study changes to public spaces that will help guard against challenges and benefit from new opportunities.

ACKNOWLEDGMENT

This research received no specific grant from any funding agency in the public, commercial, or not-for-profit sectors.

REFERENCES

Acedo, A., Painho, M., & Casteleyn, S. (2017). Place and City: Operationalizing Sense of Place and Social Capital in the Urban Context. *Transactions in GIS*, *21*(3), 503–520. doi:10.1111/tgis.12282

Agyeman, J., & McLaren, D. (2017). Sharing cities. *Environment*, *59*(3), 22–27. doi:10.1080/00139157.2017.1301168

Baumgardner, K. (2015). *Beyond Google's cute car*. Retrieved September 10, 2019, from http://offcite.org/wp-content/uploads/sites/3/2015/10/Baumgardner_Autonomous_Cars.pdf

Belanger, L. (2018). *'Put them near a park' and other things this coworking founder finds companies miss when picking an office space*. Retrieved September 1, 2019, from https://www.entrepreneur.com/article/316938

Boyer, B. (2017). Are designers letting autonomous vehicles shape future cities? *Medium*. Retrieved September 9, 2019, from https://medium.com/dashmarshall/are-designers-letting-autonomous-vehicles-shape-future-cities-153c4b42a999

Burgen, S. (2019). *Barcelona's car-free 'superblocks' could save hundreds of lives*. Retrieved September 15, 2019, from https://www.theguardian.com/cities/2019/sep/10/barcelonas-car-free-superblocks-could-save-hundreds-of lives?fbclid= IwAR0rb9bGTNUMhUZKkqIOE9ndAIZgT9TCfgeg VREP1qkum6GMsManIANpnY4

Cox, C., & Hart, A. (2016). *How autonomous vehicles could relieve or worsen traffic congestion*. Retrieved October 1, 2019, from https://www.here.com/sites/g/files/odxslz166/files/2018-12/HERE_How_autonomous_vehicles_could_relieve_or_worsen_traffic_congestion_white_paper.pdf

Crayton, T. J., & Meier, B. M. (2017). Autonomous vehicles: Developing a public health research agenda to frame the future of transportation policy. *Journal of Transport & Health*, *6*, 245–252. doi:10.1016/j.jth.2017.04.004

Furman, A. (2017). Exploring affordances of the street. *International Journal of Sustainable Development and Planning, 12*(3), 606-615.

Garrett, B. (2015, Aug. 4). The privatisation of cities' public spaces is escalating. It is time to take a stand. *The Guardian*. Retrieved September 14, 2019, from https://eprints.soton.ac.uk/380084/1/pops-privately-owned-public-space-cities-direct-action

Gavanas, N. (2019). Autonomous road vehicles: Challenges for urban planning in European cities. *Urban Science*, *3*(2), 61. doi:10.3390/urbansci3020061

Gehl, J. (1971). *Life between buildings: using public space* (Jo. Koch, Ed.). The Danish Architectural Press.

Gehl, J. (2011). *Life between buildings: using public space*. Island Press.

Gibson, J. J. (1977). The theory of affordances. Hilldale.

Han, H., & Hawken, S. (2018). Introduction: Innovation and identity in next-generation smart cities. City, Culture. *Society*, *12*, 1–4.

Hatherley, O. (2012). Rebel cities: from right to the city to the urban revolution by David Harvey–review. *The Guardian*, 12.

Hilburg, J. (2018). New York City launches interactive map of its privately-owned public spaces. *The Architect's Newspaper*. Retrieved October 20, 2019, from https://archpaper.com/2018/09/new-york-city- privately-owned-public-spaces-pops-map/

Hillier, B. (1996). *Space is the machine: A configurational theory of architecture*. Cambridge University Press.

Ison, S., & Mulley, C. (2014). Parking: issues and policies. In M. C. J. Bliemer, C. Mulley, & C. J. Moutou (Eds.), *Handbook on transport and urban planning in the developed world*. Emerald.

Jacobs, J. (1961). *Death and life of great American cities*. Random House.

Jolma Architects. (2018). *How autonomous vehicles are influencing urban design*. Retrieved October 20, 2019, from https://land8.com/how-autonomous-vehicles-are-influencing-urban-design/

Kilian, T. (1998). Public and private, power and space. In A. Light & J. Smith (Eds.), The production of public space. Lanham, MD: Rowman and Littlefield.

Kitchin, R., Lauriault, T. P., & McArdle, G. (2016). Smart cities and the politics of urban data. In S. Marvin, A. Luque-Ayala, & C. Mcfarlane (Eds.), *Smart urbanism:Utopian vision or false dawn* (pp. 16–34). Routledge.

Kunstler, J. H. (1994). *The geography of nowhere: The rise and decline of America's man-made landscape*. Simon and Schuster.

Leary-Owhin, M. E. (2016). *Exploring the production of urban space: Differential space in three post-industrial cities*. Policy Press.

Lefebvre, H., & Nicholson-Smith, D. (1991). *The production of space* (Vol. 142). Blackwell.

Litman, T. (2017). *Autonomous vehicle implementation predictions*. Victoria Transport Policy Institute.

Lubell, S. (2016). Here's how self-driving cars will transform your city. *Wired Magazine*. Retrieved October 25, 2019, from https://www.wired.com/2016/10/heres-self-driving-cars-will-transform-city/

Luberoff, D. (2016). *Reimagining and reconfiguring New York City's streets*. Transforming Urban Transportation.

Madanipour, A. (2003). *Public and private spaces of the city*. Routledge. doi:10.4324/9780203402856

Massey, D. (1995). Places and their pasts. *History Workshop Journal*, *39*(1), 182–192. doi:10.1093/hwj/39.1.182

McFarland, M. (2017). *Cities warm up to designated Uber, Lyft pick-up spots*. Retrieved October 20, 2019, from https://money.cnn.com/2017/11/16/technology/uber-lyft-designated- pickup-spots/index.html

Mitchell, W. (1995). *City of bits: space, place, and the infobahn*. The MIT Press.

Mumford, L. (1961). *The city in history: Its origins, its transformations, and its prospects* (Vol. 67). Houghton Mifflin Harcourt.

Norman, D. (2013). *The design of everyday things*. New York: Basic Books.

Pasalar, C., & Hallowell, G. (2018). *Toward human-centered smart cities: understanding emerging technologies and their effect on the urban experience*. Paper presented at AMPS, ARCHITECTURE_MPS, University of Arizona.

Pasalar, C., & Hallowell, G. (2020). Toward human-centered smart cities: Understanding emerging technologies and their effect on the urban experience. In J. Bean, S. Dickinson, & A. Ida (Eds.), *Critical Practices in Architecture: The Unexamined* (pp. 317–339). Cambridge Scholars Publishing.

Pasalar, C., & Hallowell, G. D. (2019). Sharing economies, technologies, and the changing nature of urban public space in medium-sized cities. *Terr@ Plural*, *13*(3), 418-433.

Peck, E., & Maldonado, C. (2017). How Airbnb is pushing locals out of New Orleans' coolest neighborhoods. *Huffpost*. Retrieved December 30, 2019, from https://www.huffpost.com/entry/airbnb-new-orleans-housing _n_59f33054e4b03cd20b811699

Plumer, B. (2016). *Cars take up way too much space in cities. New technology could change that*. Retrieved October 20, 2019, from https://www.vox.com/a/new-economy-future/cars-cities-technologies

Polyák, L. (2017). *A bundle of rights and obligations: Privately owned public spaces*. Retrieved October 20, 2019, from A https://cooperativecity.org/2017/11/01/privately-owned-public-spaces/

Rojas-Rueda, D., Nieuwenhuijsen, M., & Khreis, H. (2017). Autonomous vehicles and public health: Literature Review. *Journal of Transport & Health*, *5*, S13. doi:10.1016/j.jth.2017.05.292

Sennett, R. (1977). *The fall of public man*. Knopf.

SFMTA - San Francisco Municipal Transportation Agency. (2016). *City of San Francisco: Meeting the smart city challenge.* Retrieved October 20, 2019, from https://www.sfmta.com/sites/default/files/projects/2016/SF%20Smart%20City%20 Challenge_Final.pdf

Skinner, R., & Bidwell, N. (2016). *Making better places: Autonomous vehicles and future opportunities.* WSP-Parsons Brinckerhoff Engineering Services.

Surico, J. (2018). *What Happens When a City Bans Non-Resident Drivers? Citylab.com.* Retrieved October 20, 2019, from https://www.citylab.com/transportation/2018/04/ the-small-town-that-took-on-waze/558215/

The Conversation. (2017). *Here's how driverless vehicles will utterly transform how our cities look.* Retrieved October 20, 2019, from https://futurism.com/heres-how-driverless-vehicles-will-utterly-transform-how-our-cities-look

Thompson, C. (2018). *The vehicle of the future has two wheels, handlebars, and is a bike.* Retrieved October 20, 2019, from https://www.wired.com/story/vehicle-future-bike/

UNDESA. (2018). *The 2018 revision of the world urbanization prospects.* Retrieved October 10, 2019, from: https://www.un.org/development/desa/en

U.S. Department of Transportation. (2015). *Smart city challenge: lessons for building cities of the future.* Retrieved September 20, 2019, from https://www.transportation. gov/policy-initiatives/smartcity/smart-city-challenge-lessons-building-cities-future.

Waber, B., Magnolfi, J., & Lindsay, G. (2014). Workspaces that move people. *Harvard Business Review, 92*(10), 68–77. PMID:25509577

Wachsmuth, D., Chaney, D., Kerrigan, D., Shillolo, A., & Basalaev-Binder, R. (2018). *The high cost of short-term rentals in New York City. A Report from the Urban Politics and Governance Research Group School of Urban Planning.* McGill University.

Wadud, Z., Mackenzie, D., & Leiby, P. (2016). Help or hindrance? the travel, energy, and carbon impacts of highly automated vehicles. *Transportation Research Part A, Policy and Practice, 86*, 1–18. doi:10.1016/j.tra.2015.12.001

Walker, J., & Johnson, C. (2016). *Peak car ownership: the market opportunity of electric automated mobility services.* Rocky Mountain Institute.

Wang, L. (2014). *How the Cheonggyecheon river urban design restored the green heart of Seoul.* Retrieved November 1, 2019, from https://inhabitat.com/how-the-cheonggyecheon-river-urban-design-restored-the-green-heart-of-seoul/

World Bank. (2012). *Who needs smart cities for sustainable development?* Retrieved October 20, 2019, from https://www.worldbank.org/en/news/feature/2012/03/20/who-needs-smart-cities-for-sustainable-development

Young, R., & Lieberknecht, K. (2016). From smart cities to wise cities: Ecological wisdom as a basis for sustainable urban development. *Journal of Environmental Planning and Management*, 1–18.

Young, R. F., & Lieberknecht, K. (2019). From smart cities to wise cities: Ecological wisdom as a basis for sustainable urban development. *Journal of Environmental Planning and Management*, *62*(10), 1675–1692. doi:10.1080/09640568.2018.1484343

KEY TERMS AND DEFINITIONS

Affordances: An object's properties that show the possible actions users can take with it, thereby suggesting how they may interact with that object. It refers to all action possibilities depending on users' physical capabilities. A chair not only "affords" being "sat on," but also "thrown," "stood on," etc.

Automated Vehicle: A vehicle capable of sensing its environment and operating without human involvement completing journeys safely and efficiently.

Complete Streets: Streets designed and operated to enable safe access for all users, including pedestrians, bicyclists, motorists, and transit riders of all ages and abilities. Complete Streets make it easy to cross the street, walk to shops, and bicycle to work. They allow buses to run on time and make it safe for people to walk to and from train stations.

Privately-Owned Public Spaces (POPS): It is a term used to describe a type of public space that, although privately owned, is legally required to be open to the public under a city's zoning ordinance or other land-use law.

Privately-Owned Publicly-Utilized Spaces (POPUS): A potential expansion of how portions of publicly owned and operated urban spaces have been utilized and managed for the profit and use of private individuals or businesses for centuries.

Public-to-Private Transect: A cross-section, extending from the most public urban areas such as a street centerline or plaza centroid out to the private spaces and buildings on both sides of the road or public square. The edges and boundaries of public and private spaces are, therefore, a continuum between the two realms meeting through constantly changing shades of privacy and publicity rather than at any clear boundaries of separation.

Sharing Economy: An economic model based on sharing underused assets from spaces to skills to stuff, for either monetary or other types of benefits.

Smart City: A smart city is an urban area that incorporates information and communication technologies to enhance the quality and performance of urban services such as energy, transportation, and utilities in order to reduce resource consumption, wastage, and overall costs. The overarching aim of a smart city is to enhance the quality of living for its citizens through smart technology.

Chapter 4
The Dawn of Walkability

Mohamed Abedo
German University at Cairo, Egypt

Mohamed Salheen
https://orcid.org/0000-0003-1885-4373
Ain Shams University, Egypt

Abeer Elshater
https://orcid.org/0000-0002-5061-6861
Ain Shams University, Egypt

ABSTRACT

This explanatory research investigates the impact of configurations in the urban fabric on walkability regarding connectivity and global integration. This study aims to examine this effect by comparing street networks in two residential neighbourhoods in Heliopolis and New Cairo cities. The research methods use DepthMapX and Walk Score to provide relative indicators about distinctive features that enhance the walkability in the case studies. The results show significant differences between the walkability patterns and pedestrians gate counts in two case studies. In study areas, the pedestrian gate count, connectivity, and global integration give signs about people interactions during the day time. The findings figured out that the spatial configurations of street networks—that create urban fabric—play a vital role in enhancing the walkability. The possibility of simulation, to a great extent, can investigate other cases in future research with real situations on the ground. The results confirm a worthy choice for the space syntax technique as a predictor in analysing walkability.

DOI: 10.4018/978-1-7998-3507-3.ch004

INTRODUCTION

Walking is considered the oldest way of transportation. In the nineteenth century, transportation technology had a significant transformation. Before that, most cities were built in ways that supported walkability (Jeffrey & Newman, 1999). According to the Behavioural Risk Factor Surveillance System (BRFSS) (2003), walking is the most common form of physical activity among 41% of adults in the United States compared to 4% who are more engaged with other activities.

Globally, private-car ownership became a norm during the twentieth century. Public transportation and urban walkability lost their first places of transportation means. This issue leads to an increase in energy costs, less fuel availability, more congestion and pollution, and other environmental impacts. Globally, initiatives follow the fact that public transportation, private car dependency, and walkable cities concepts attract more attention by city planners and designers (Ewing & Handy, 2009). On a local level, the new urban development strategy in the Greater Cairo Region (GCR) creates sprawled new urban communities along four decades with two million inhabitants (Ahram Online, 2010). This development led to more dependency on motorized transportation using more power, costing more resources, and causing more pollution.

Several pieces of research addressed issues related to pedestrian environment from different aspects: socially, safety-wise (Mohamed, 2019), visually (Elshater, Abusaada, & Afifi, 2019; Lynch, 1981; Lynch, 1960), imageability-base (Lynch, 1960; Lynch, 1981) and functionally (Elshater, 2012; Jabbari, Fonseca, & Ramos, 2018; Jeffrey & Newman, 1999). However, the need for investigating these aspects comparatively through case studies addressing the impact of the urban fabric of Egyptian cities on the pedestrian moves is more needed.

As mentioned in Ahram Online newspaper based on the Egyptian report released in 2010 by the Central Agency for Public Mobilisation and Statistics (CAPMAS), car ownership rates in Egypt have reached 7.4 privately owned cars per 100 people after being 6.7 in 2009. Cairo had the most significant share of this percentage, with 20.4 per 100 persons in 2010 compared to 16.9 in 2009 (Ahram Online, 2010). Moreover, the transportation expenses to income ratio have increased by more than double per family in five years. The statistics moved from 3.65% in 2008 to 6.27% in 2013 (CAPMAS, 2014). Other pieces of research in this domain show that increase of car ownership affects negatively on creating a walkable environment (Elshater, 2012; Elshater, 2016; Ewing & Cervero, 2010; Koohsari, Takemi Sugiyama, & Owen, 2014). However, a limited number of studies provide digital analysis for measuring the walkability regarding the Egyptian fabric.

This paper aims to investigate aspects affecting walkability comparatively in two Egyptian cases: Heliopolis and New Cairo City. The study addresses questions on the relation between the configuration of urban settings and walkability. To what extent can urban fabric affect walkability? What convenient methods are used to measure walkability? What are comparative spatial analyses used to explore the enhancement of urban structure to walkability? The contribution of the current study is examining comparatively two cases from Cairo city by utilising a mix of spatial analysis tools using space syntax and walk score.

BACKGROUND

This section discusses the term walkability and factors affecting the walkability. To interpret a walkable neighbourhood, the authors here firstly define the term precisely and identify the factors and affecting criteria. In this domain, Southworth (2005) refers to walkability as the extent to which the built environment upkeeps and supports walking by providing for pedestrian comfort and safety, connecting people with varied destinations within a reasonable amount of time and effort, and offering visual interest in journeys throughout the network.

Forsyth (2015) defined nine themes that affect walkability (Figure 1) in a study that focuses on dense fabric and supportive amities. Some definitions of walkability focus on the means or conditions by which walkability is achieved. These definitions include areas that are traversable, physically-enticing, compact, or safe. In the same study, he defines walkability by its outcomes or performance, like making places lively and friendly, improving transportation means, or encouraging people to exercise. The last set of themes defines walkability as a proxy for urban areas. In other studies, walkability scholars define walkability as multi-dimensional, while others consider it a holistic solution to various urban problems (Forsyth, 2015).

The first set of themes includes dimensions related to the walkable environment, which are the minimum means to create walkability (Elshater, 2012; Forsyth, 2015). Traversable factor describes situations that have the minimum physical conditions that allow people to go from one place to another without facing significant obstructions (Bently, Alcock, Murrain, McGlynn, & Smith, 1985). Compact describes the shortest distances to destinations for people walking for utility. Safe has a multi-dimensional definition such as crime rates and traffic safety, both mainly related to potential harm to the person in a walking action (Adel, Salheen, & Mahmoud, 2016). The second cluster of themes includes dimensions about the outcome of walking. Shopping areas and mixed-use neighbourhoods are usually perceived as lively and sociable because they offer an attractive and pleasant experience to walking people (Elshater, 2016). For those who do not have access to cars because of income, age, or disability, walking is seen as a sustainable transportation option. Aside from that, using these means

Figure 1. The compact fabric and amenities within the nine themes affecting the walkability and walkable places. Source: the authors based on Forsyth (2015)

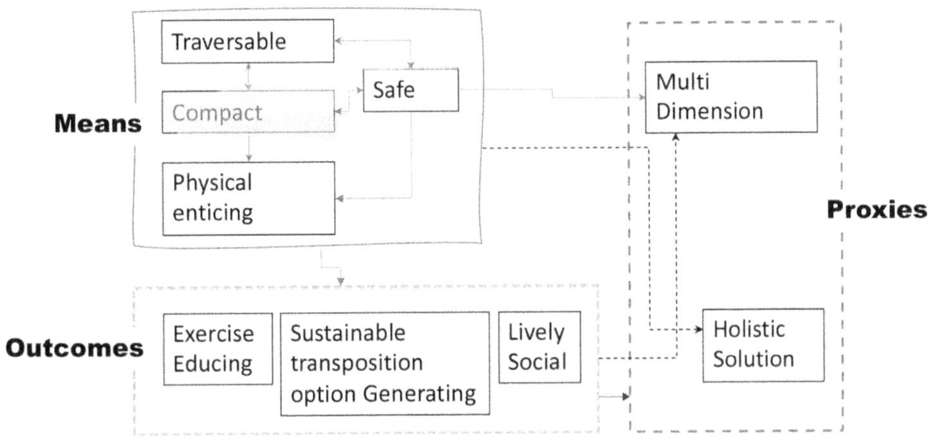

is the way to achieve social equity components and environmental preservation of sustainable urban form (Elshater, 2012; Mohamed, 2019).

The final set of themes is perceived as a proxy for the better walkable environment. In this regard, walkability is seen as a multi-dimensional mean that has measurable conditions. Many of the beforementioned definitions are seen as one integrated package representing a holistic solution. This approach is less about walkability and more about creating an urban environment that is more human scale, healthier and happier, generally speaking, creating an excellent place to live (Forsyth, 2015).

During the past few years, many tools for measuring the quality of walking environment have emerged (Abusaada, Vellguth, & Elshater, 2019; Haq & Berhie, 2018; Kim & Penn, 2004; Refaat, 2019). Scholars called this as walking audit instruments that are now widely used amongst US researchers, local governments, and community groups (Adel, Salheen, & Mahmoud, 2016; Jabbari, Fonseca, & Ramos, 2018). Some of these audit tools require the measurement of physical features such as block length, building height and street, and sidewalk width (Mohamed, 2019). However, these auditing tools cannot explain the personal experience of the walking individual (Arisha & El-Moneim, 2019). According to Reid Ewing & Susan Handy (2009)—in researching walkability measuring tools, 51 perceptual qualities define the perception of the walking environment besides its physical being. They chose eight qualities as most important: imageability, human scale, enclosure, transparency, complexity, legibility, linkage, and coherence. Out of these eight qualities, five of them were successfully operationalized based on a rating made the expert panel of video clips of 48 commercial streets across the United States (Ewing & Cervero,

2010). Other researchers give each of these eight qualities a consensus qualitative definition and an operational definition (Arisha & El-Moneim, 2019; Bereitschaft, 2018; Elshater, Abusaada, & Afifi, 2019; Ewing & Handy, 2009).

According to Ewing and Handy (2009), imageability is the quality of a site that makes it distinct, memorable, and recognizable. A place with high imageability as one with specific physical elements that capture attention, create a long-lasting impression, and evoke feelings. Enclosure refers to the degree to which streets and other public spaces are visually defined by buildings, walls, trees, and other vertical elements. Areas are proportionally related to their height, where the width of the space between vertical elements, while they define human scale as physical elements with a proportional size, texture, and articulation to proportions of humans. Scholars also set transparency as the limit to which people perceive or see beyond the street edge. This aspect also tackles the threshold in which people can view human activity beyond the street edge. Finally, they see a sophisticated space as the one with the visual richness that has an environment with a variety of physical elements such as numbers and typologies of buildings, diverse architecture and ornaments, different landscape elements, street furniture, and human activity.

There are many factors affecting walkability (Ewing & Handy, 2009; Abusaada and Elshater, 2020). Figure 2 shows some of these factors that we can recognize in the physical features, under design quality and individual reaction. This paper focuses on the effect of the compact urban fabric. Although there are countless scholarships in urban planning research focus on walkability, its definitions and elements still need more investigation regarding the Egyptian context. Further study is more need to the correlation with the urban structure, and neighbourhood layout is somehow mistreated, especially in traffic simulation models (Adel, Salheen and Mahmoud 2016, Arisha and El-Moneim 2019, Kim and Penn 2004). Pedestrian traffic is often overlooked in some of these models as well (Xiongbing & Grammenos, 2013). A study that looked at seven typical neighbourhood layouts from the last 100 years, these were two traditional grid layouts, two post-war suburban expansion layouts, two traditional neighbourhood layouts, and the last one was the fused grid.

To make sure of excluding the influence of variables instead of urban fabric configuration, several researchers choose all layouts within an area of 800 x 800 meters. The exits and local facilities are located at four corners. Also, they set population density to 16.2 households or 33 persons per hectares, which is a medium-level density for suburban neighbourhoods. The results of these studies show a great significance in the walk share of trips within layouts of higher intersection densities (Xiongbing & Grammenos, 2013; Bahrainy & Khosravi, 2013; Jabbari, Fonseca, & Ramos, 2018), (Figure 3).

Figure 2. The factors affecting walkability, source: (Ewing & Handy, 2009)

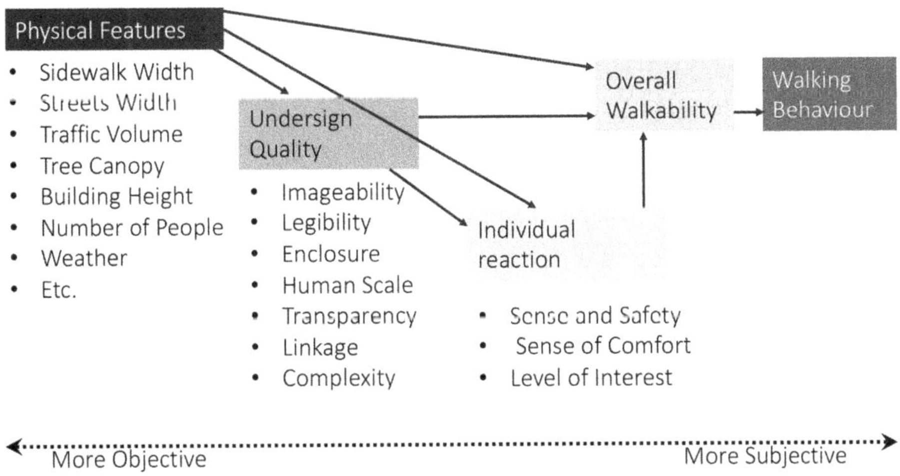

MATERIAL AND METHODS

The current case studies are two neighbourhoods in Heliopolis and New Cairo City, which we chose to compare the walkability in two districts. The reason behind selecting these two cases is verifying the argument of falling or rising the walkability in two different fabrics in Cairo. In the two cases, the structure and composition of urban fabric are dependent variables (Carlson, 2006; Forsyth, 2015), and the

Figure 3. Intersection density and walk share of trips, source: (Xiongbing & Grammenos, 2013)

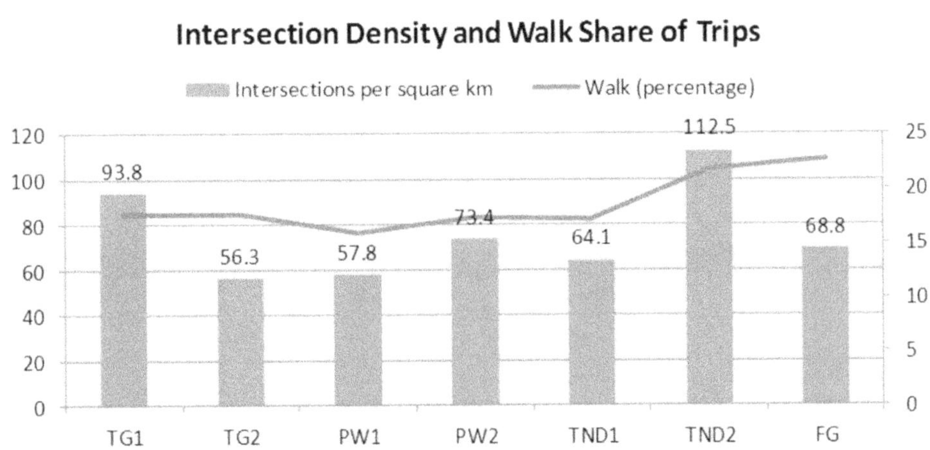

kinetic perceptions of pedestrians are the independent variables (Elshater, Abusaada and Afifi, 2019; Hillier and Hanson, 1989). The urban fabric here as a dependent variable is being tested and measured using independent variables of users' kinetic perception (Hillier, 2015; Hillier & Hanson, 1989). The change in the users' attitude through walking here is our independent variables.

Meanwhile, the effect on the urban structure is a dependent variable. We investigated the sets of variables in a selection of two neighbourhoods that are branded as an expansion in the desert. These two cases have some similarities in the morphological structure and differences in the architecture styles. Figure 4 shows these two neighbourhoods located around are Street 17, Cairo Governorate, and Al-Ahram Street in Heliopolis.

Figure 4. The study areas of Heliopolis and New Cairo, source: The authors of the present research

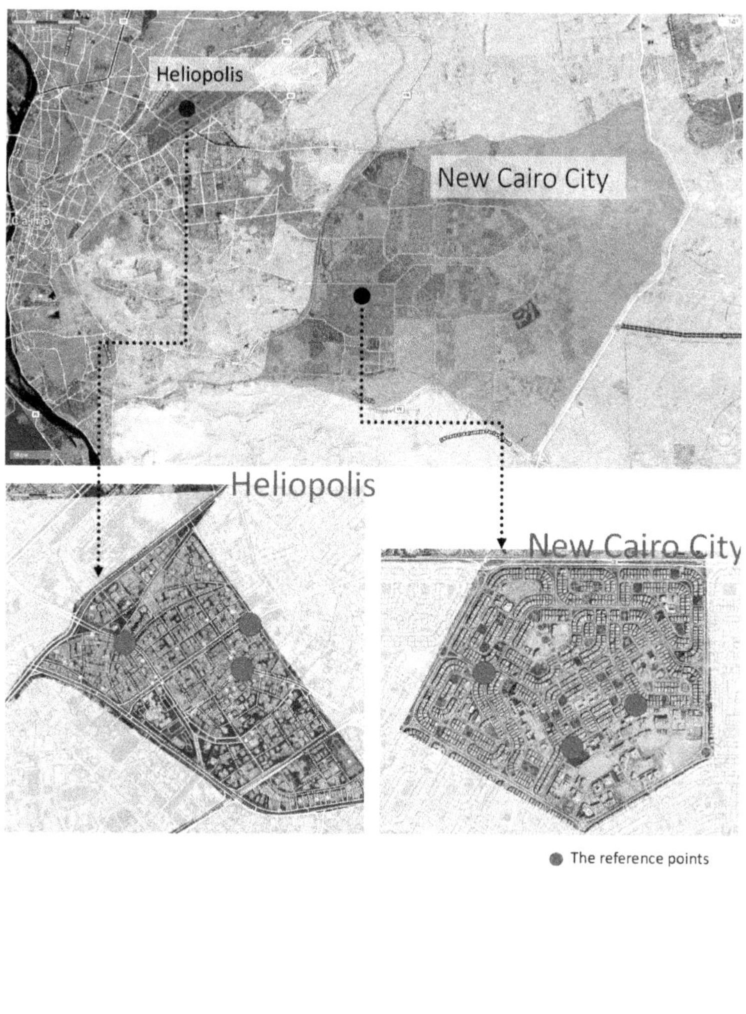

We set-up here the evidence to investigate the urban structure in the two instances comparatively. We explored these differences statistically in two neighbourhoods in Heliopolis and New Cairo City by using space syntax software and Walk Score interference. A potentially unrealistic assumption arises from the fact that the dense fabric in late development enhances the walkability while the new developments diminish the ability of persons to walk. For the current work, it is sufficient to point out that the comparative analysis of these cases can highlight the direction of new development to compact development.

The earliest records of Heliopolis go back to the first decade of the 20th century when Belgian banker and business tycoon—Baron Edward Empain—bustled to Cairo in anticipation of the economic boom. In 1905, Empain built the Cairo electric railway and established Heliopolis oasis company (Dobrowolska & Dobrowolski, 2006). Empain efforts climaxed in 1907 with the building of the new town of Heliopolis, in the desert ten kilometres from the centre of Cairo, located between the airport and the city centre. Baron Empain acknowledges the design of Heliopolis as a paradise or city of luxury and leisure. City observers can instantly feel the influence of Haussmann's Paris in the planning of Heliopolis. It was designed to have a garden type city. The master plan contained public gardens, parks, and playgrounds; restrict building rules were set (not more than half of the private lots could be built up), and buildings heights were also set (Dobrowolska & Dobrowolski, 2006).

New Cairo City has been built around Cairo to alleviate the congestion in the Downtown. It was established in the year 2000 (Elshater, Abusaada, & Afifi, 2019). The new urban community covers an area of 86,533 Hechtars, 15km away from Cairo Downtown, and five km away from Nasr City. It is planned to accommodate five million residents. This city has a variety of land-uses (e.g., residential, educational services, touristic amenities, industrial and commercial activities). The city has 1.5 million inhabits and could eventually host a population of five million living on 46,346 residential land lots containing 302,304 residential units. For in-depth spatial analysis, the researchers here choose small scale neighbourhoods in both Heliopolis and New Cairo City.

Through using Space Syntax theory, DepthMapX software (Hillier, 2015; Hillier & Hanson, 1989) and on-site pedestrian gate count, a comparative spatial configuration analysis, the researchers investigate the street networks on the two residential neighbourhoods. The research uses space syntax software to measure the kinetic perception of pedestrians. Space syntax was founded and developed in the 1970s at the Bartlett Unit for Architectural Studies at University Collage London (UCL) (Hillier & Hanson, 1989). Thes studies are produced a reliable technique that can be used to define and analyse patterns of architectural and urban spaces (Elshater, Abusaada, & Afifi, 2019; Jabbari, Fonseca, & Ramos, 2018). It tries to interpret human behaviours and social activities from a spatial configuration point of view (Hillier, 2015). The connection between any two spaces taking into account their relation to at least a third space, and at most all other areas in a system was termed as spatial configuration (Hillier, 2015). Space syntax estimates the distance between spaces topologically, expressing the space that can be accessible to all other spaces with the fewest number of changes in turns as the most integrated.

Integration calculates how deep or shallow each line is from all lines up to *n* steps away. The type of integration differs according to the radius *n*. That is, if (n) includes all levels, then the integration obtained is called global integration; if *n* is up to three steps, then the integration obtained is radius-3 integration. Connectivity is integration when *n* equals one.

The analysis technique contains several steps. Axial maps of study areas were drawn, which represents the most extended sets of lines till all entire spaces are covered these maps are then processed using DepthMapX software to get values of integration (with different radii), connectivity, control, mean depth. The higher the radius, the more global the extent of the measure (Kim & Penn, 2004). The axial lines are represented from red to blue——red means the highest value of the parameter, while blue means the lowest. For instance, for the integration value, the red lines mean the most integrated (i.e., the fewest changes of direction from others on average); the blue ones mean the most segregated (i.e., the least mixed). A pedestrian gate count was conducted here to confirm the outputs resulted from space syntax mapping analysis. Moreover, the researchers of the present work chose three points at the intersection randomly in different places in the study area to make sure it's well-covered (see red dots in Figure 4). The selection was to make sure the sample size is representative, and the count is conducted on three days (beginning, intermediate days, and the weekend) three times a day (morning, noon, and night).

Using the Walk Score is an online interface that gives a score from 1 to 100 to measure the walkability of a particular area (Bereitschaft, 2018; Elshater, 2019). This interface ranks the walkability of any site based on the distance to amenities, with points are given based on walking distance. Places that have facilities within a five-minute walk are assigned maximum points, while those requiring more than a 30-minute walk are ranked with no points. A location is described as walker's paradise if it has a Walk Score of 90-100 points, while any address with a Walk Score lower than 50 points is designated "car-dependent." This interference also calculates the walking distance from any given point which we used to determine the exact borders of the study area which is a 15-minute walk distance from the centre of our selected study locations (Bahrainy & Khosravi, 2013; Duncan, Aldstadt, Whalen, Melly, & Gortmaker, 2011; Handy, Cao, & Mokhtarian, 2005). As Bereitschaft (2018), Koschinsky, Talen, Alfonzo and Lee (2017) argue the contribution of Walk Score as an easy interface both inside and outside of the academia of urban planning and design, the present work focuses on the final results of data that report the walkability level in the case study.

RESULTS AND DISCUSSION

We describe here the results regarding the pedestrian gate count and Walk Score. Our findings from space syntax analysis showed a significant difference between the two study areas (Table 1). In Heliopolis, the first time slot of the day always had people who walk while going to work in the study area. The second slot of time on the same site had the highest pedestrian count, had a mix of people coming from work, doing shopping, or just passers-by. The last time slot had mainly people coming for shopping or just taking a walk.

Meanwhile, in New Cairo, it was somehow different. The dependency on cars, on the one hand, was very high during the day, especially in the morning slots the three selected samples of the day. On the other hand, the night slot in Heliopolis had the highest pedestrian count. It had the fewest count in New Cairo City. People tend to walk less at night in New Cairo City for safety reasons.

Our results of Walk Score pedestrian gate count are in line in terms of the differences in walking settings in the two cases (Walk Score, 2015). In Heliopolis, the selected nodes are significantly and daily errands and can be accomplished on foot. The walk score was 93 out of 100. In New Cairo, the result of a walk score indicated that some errands could be accomplished on foot with a score of 54 out of 100.

The present study confirmed the findings of the impact of urban fabric on enhancing walkability. In this, Heliopolis shows global functional integration in the study area, as shown in Figure 5, the most integrated segments are the primary, longer streets. In contrast, the least integrated are the shorter streets. These results match the pedestrian gate count that was done on the same streets. Notably, these streets have most of the core activities in these neighbourhoods. The colour-grading of the streets is also significantly related to their dominance and importance in the districts. For New Cairo, the primary and main streets show high integration values while the twisted streets—especially the ones between the residential blocks—show low integration values, which are also reflected in the pedestrian gate count (Figure 6).

The results gained from Walk Score (Figure 7) and space syntax (see Figure 5 and Figure 6) about the two cases tie well with the results obtained by others (Elshater, Abusaada, & Afifi, 2019; Adel, Salheen, & Mahmoud, 2016). These results go beyond previous research, showing that the walkability has a direct connection with the nine factors addressed by Forsyth (2015). The two cases of Heliopolis and New Cairo proved significant differences in the tendency of people to walk. Others

Table 1. The pedestrian gate count in the case studies

Date	Time of Day	Heliopolis			New Cairo		
		Gate 1	Gate 2	Gate 3	Gate 1	Gate 2	Gate 3
Sunday, June 2, 2019	10:00 am	21	12	23	5	6	2
	5:00 pm	49	22	57	15	13	6
	9:00 pm	66	34	42	19	7	1
Tuesday, June 4, 2019	10:00 am	26	16	27	12	8	4
	5:00 pm	44	31	42	33	21	9
	9:00 pm	72	58	69	21	9	3
Friday, June 7, 2019	10:00 am	15	11	9	7	13	3
	5:00 pm	66	37	25	37	27	5
	9:00 pm	75	59	42	17	11	3

have shown that supporting the amenities alongside the walking sidewalks beside the dense fabric improves the walkability (Jabbari, Fonseca, & Ramos, 2018; Haq & Berhie, 2018).

In Table 2, the spatial configurational values indicate that Heliopolis has the highest global integration (Rn) values at 1.9584. The followed result comes for the district in New Cairo at 1.2687. Also, Heliopolis has the highest connectivity value at 4.2954, while the New Cairo study area has a connectivity value of 3.4235. This is translated into that Heliopolis is the most integrated. The maps of Heliopolis have many axial lines with high integration values, which means that those lines or streets are movement generators and expected to have more visitors. Hence, they have a lot of commercial uses.

One concern about the findings of using Space Syntax and DepthMapX in showing comparatively the walkability was that how the urban system functioned. The configuration of amenities, for instance, one method also lets the results here inconvenient (Creswell & Clark, 2011). The limitation also extends to test the effect of urban fabric on walkability in other contexts and settings.

It is by now generally accepted that the urban fabric is not the only factor that affects the walkability in the two case studies. An essential answer to the research questions associated with walkability and urban fabric is the direct connection between them. Indeed, space syntax has a full contribution to investigating this issue. However, recent studies proved the importance of using the missed method to investigate that issue (Elshater, Abusaada, & Afifi, 2019; Jabbari, Fonseca, & Ramos, 2018; Refaat, 2019).

Figure 5. Heliopolis connectivity analysis on the left, and global integration on the right, source: the authors of the present work

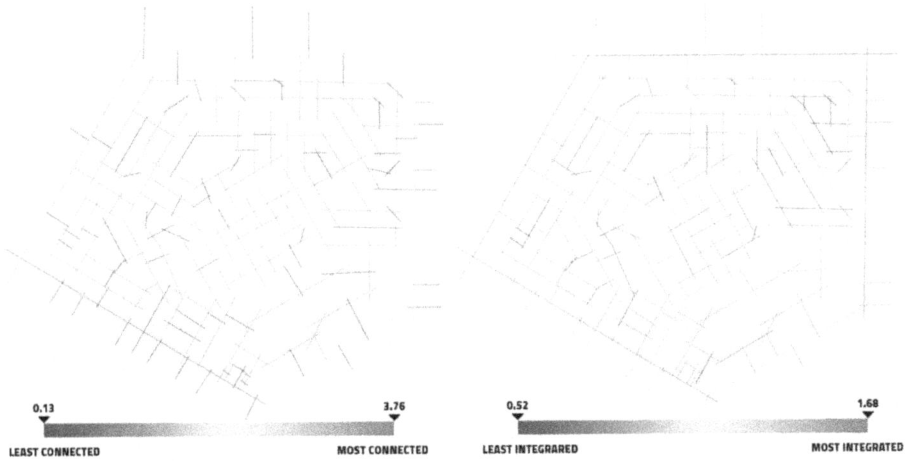

Figure 6. The connectivity of New Cairo City analysis on the left, and global integration on the right, source: the authors of the present work

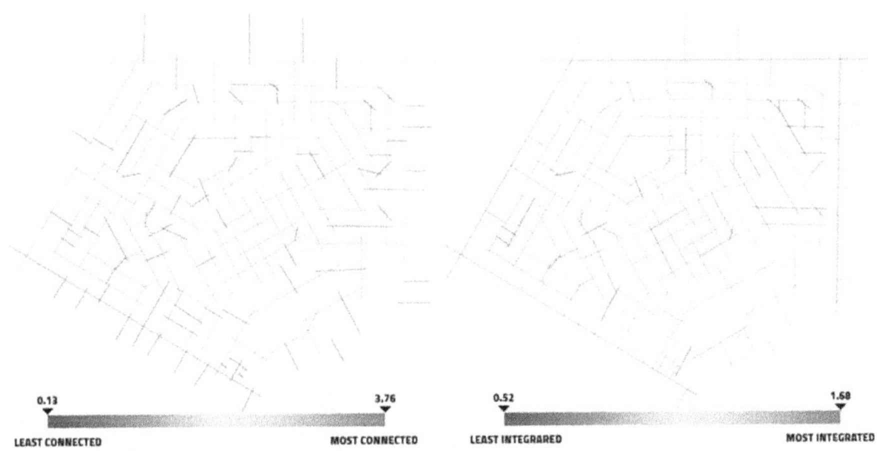

CONCLUSION

This paper discussed the effect of urban fabric on walkability. We select here two cases to compare the differences in the urban structure that ease the pedestrian flow. It analyses the walkability using space syntax theory and DepthMapX software. Space syntax theory with DepthMapX software was used in the present

Figure 7. The amenities and Walk Sore provided in New Cairo (left and Heliopolis (right), source the authors based on (Walk Score, 2015)

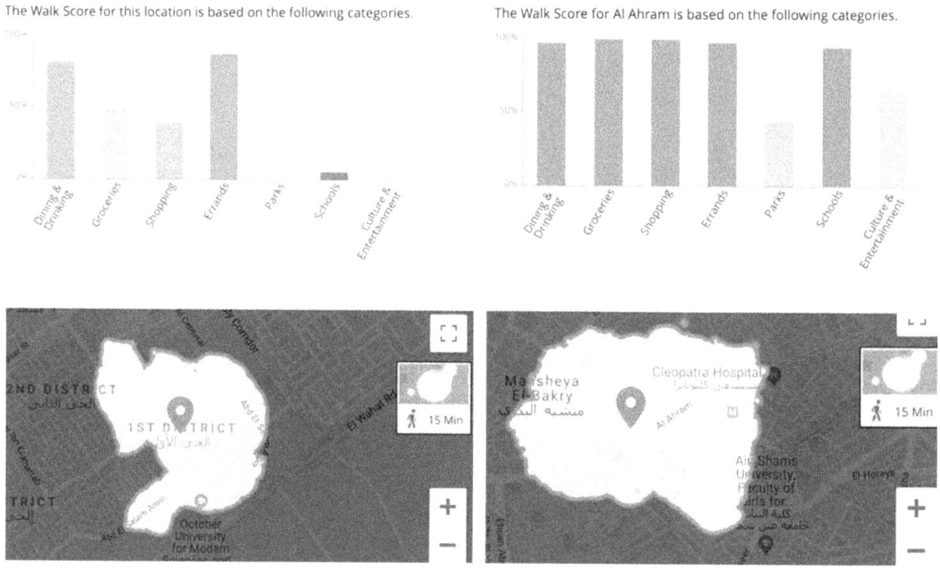

work to perform spatial configurational analyses in two study areas. Maps showed significant differences in integration values and connectivity values. Results match the situation on the ground to a large extent. Main roads were the busiest, while inner and secondary roads tend to have fewer pedestrians. The results also show that Heliopolis has a more significant ratio of mixed-use zones than New Cairo City, which encourages people to walk. The results also confirmed that the pedestrian gate count could show the great significance in pedestrian counts and walking patterns of people. This study concludes by arguing that pedestrians in Heliopolis tend to walk more at night than pedestrians in New Cairo City due to safety reasons and connectivity of street networks.

The future direction of research can extend to apply the methods used in other settings with a different social background of walking behaviour. For more precise results about the relationship between the urban fabrics and walkability, the possibility

Table 2. The spatial configurational values for the study areas

Attribute	Heliopolis study area	New Cairo Study area
Global Integration (Rn)	1.9584	1.2687
Connectivity	4.2954	3.4235

of broader investigation can compare the present researchers' results about the two cases in different settings. As our research focuses on the compact fabric as a theme affect walkability, further studies should investigate other themes related to the means that easing the walking, such as traversable ability, safety, physical enticing. In the same vein, additional research should investigate the expected outcomes from walking, such as physical exercise and sustainable transportation. The selection of gate location is not clear as well and not readable from the provided maps.

The results here show a limitation on using space syntax that depends on axial analysis and global integration for pedestrian kinetic perception. As literature nowadays describes the local integration in the space syntax literature as a good predictor; however, the results received from using it need to strengthen Segment analysis. The limitation here is this method depends on topological only, while several pieces of the research reported to the effect of geometric and metric distance as essential factors. Further development of this work can use Segment analysis with metric radius 400m and 800m to have a liable and valid to predict of the pedestrian movement. Furthermore, the segment analysis can provide a more concrete base for the selection of the gate-count locations.

REFERENCES

Abusaada, H., & Elshater, A. (2020). Effects of Urban Atmospheres on Changing Attitudes of Crowded Public Places: An Action Plan. *Int. Journal of Community Well-being*, 1-51.

Abusaada, H., Vellguth, C., & Elshater, A. (Eds.). (2019). *Handbook of Research on Digital research Methods and Architectural Tools in Urban Planning and Design*. IGI Global. doi:10.4018/978-1-5225-9238-9

Adel, H., Salheen, M., & Mahmoud, R. A. (2016). Crime in relation to urban design. Case study: The Greater Cairo Region. *Ain Shams Engineering Journal*, *7*(3), 925–938. doi:10.1016/j.asej.2015.08.009

Ahram Online. (2010). *Egypt vehicles up 13.5 percent in 2010*. Ahram Online.

Arisha, A. R., & El-Moneim, N. M. (2019). Space Syntax Beyond Cairo Street Markets: An Application of a Theory on Socio-Spatial Contemporary Practices. In H. Abusaada, C. Vellguth, & A. Elshater (Eds.), *Handbook of Research on Digital Research Methods and Architectural Tools in Urban Planning and Design* (pp. 250–281). IGI Global. doi:10.4018/978-1-5225-9238-9.ch012

Bahrainy, H., & Khosravi, H. (2013). The Impact of Urban Design Features and Qualities on Walkability and Health in Under-Construction Environments: The Case of Hashtgerd New Town in Iran. *Cities (London, England)*, *31*, 17–28. doi:10.1016/j.cities.2013.01.001

Bently, I., Alcock, A., Murrain, P., McGlynn, S., & Smith, G. (1985). *Responsive environments: A manual for designers*. Elsevier.

Bereitschaft, B. (2018). Walk Score® versus Residents' Perceptions of Walkability in Omaha, NE. *Journal of Urbanism*, *11*(4), 412–435. doi:10.1080/17549175.2018.1484795

CAPMAS. (2014). *Income, Expenditure & Consumption survey*. CAPMAS.

Carlson, R. (2006). *A concrete introduction to real analysis*. CRC Press. doi:10.1201/b15894

Creswell, J. W., & Clark, V. L. (2011). *Designing and conducting mixed methods research*. Sage.

Dobrowolska, A., & Dobrowolski, J. (2006). *Heliopolis: Rebirth of the City of the Sun*. American Univesity in Cairo Press.

Duncan, D. T., Aldstadt, J., Whalen, J., Melly, S. J., & Gortmaker, S. L. (2011). Validation of Walk Score® for Estimating Neighborhood Walkability: An Analysis of Four US Metropolitan Areas. *International Journal of Environmental Research and Public Health*, *8*(11), 4160–4179. doi:10.3390/ijerph8114160 PMID:22163200

Elshater, A. (2012). The New Urbanism Principles vs. Urban Design Dimensions towards Behavioral Performance in Egyptian Neighborhood. In Future Communities: Socio-Cultural & Environmental Challenges (pp. 826-843). Giza: AicE-Bs-ASIA Pacific International Conference on Environment-Behavior Studies.

Elshater, A. (2016). The Ten-Minute Neighbourhood Is[not] a Basic Planning Unit for Happiness in Egypt. *Archnet-IJAR*, *10*(1), 344–357. doi:10.26687/archnet-ijar.v10i1.878

Elshater, A. (2019). Food consumption in the everyday life of liveable cities: Design implications for conviviality. *Journal of Urbanism*, 1–29.

Elshater, A., Abusaada, H., & Afifi, S. (2019). What makes livable cities of today alike? Revisiting the criterion of singularity through two case studies. *Cities (London, England)*, *92*, 273–291. doi:10.1016/j.cities.2019.04.008

Ewing, R., & Cervero, R. (2010). Travel and the Built Environment: A Meta-Analysis. *Journal of the American Planning Association, 76*(3), 265–294. doi:10.1080/01944361003766766

Ewing, R., & Handy, S. (2009). Measuring the Unmeasurable: Urban design qualities related to walkability. *Journal of Urban Design, 14*(1), 65–84. doi:10.1080/13574800802451155

Forsyth, A. (2015). What is a Walkable Place? The Walkability Debate in Urban Design. *URBAN DESIGN International, 20*(4), 274–292. doi:10.1057/udi.2015.22

Handy, S., Cao, X., & Mokhtarian, P. (2005). Correlation or Causality between the Built Environment and Travel Behavior? Evidence from Northern California. *Transportation Research Part D, Transport and Environment, 10*(6), 427–444. doi:10.1016/j.trd.2005.05.002

Haq, S., & Berhie, G. (2018). Space syntax investigation of Lubbock, a grid-like American city and some insights into isotropic layouts. *Journal of Urban Design, 23*(1), 5–22. doi:10.1080/13574809.2017.1369872

Hillier, B. (2015). *Space Is the Machine: A Configurational Theory of Architecture*. Space Syntax.

Hillier, B., & Hanson, J. (1989). *The Social Logic of Space*. Cambridge University Press.

Jabbari, M., Fonseca, F., & Ramos, R. (2018). Combining multi-criteria and space syntax analysis to assess a pedestrian network: The case of Oporto. *Journal of Urban Design, 23*(1), 23–41. doi:10.1080/13574809.2017.1343087

Jeffrey, K., & Newman, P. (1999). *Sustainability and cities: Overcoming automobile dependence*. Island Press.

Jin, X. (2010). *Modelling the influence of neighbourhood design on daily trip patterns in urban neighbourhoods*. Retrieved from Memorial University Research Repository: https://research.library.mun.ca/9553/

Kim, Y., & Penn, A. (2004). Linking the Spatial Syntax of Cognitive Maps to the Spatial Syntax of the Environment. *Environment and Behavior, 36*(4), 483–504. doi:10.1177/0013916503261384

Koohsari, M. J., Takemi Sugiyama, A. T., & Owen, N. (2014). Associations of Leisure-Time Sitting in Cars With Neighborhood Walkability. *Journal of Physical Activity & Health, 11*(6), 1129–1132. doi:10.1123/jpah.2012-0385 PMID:25368941

Koschinsky, J., Talen, E., Alfonzo, M., & Lee, S. (2017). How Walkable Is Walker's Paradise? *Environment and Planning. B, Planning & Design*, *44*(2), 343–363.

Lynch, K. (1960). *The Image of the City*. MIT Press.

Lynch, K. (1981). *Good City Form*. The MIT Press.

Mohamed, A. A. (2019). Space Syntax Approach for Articulating Space and Social Life. In H. Abusaada, C. Vellguth, & A. Elshater (Eds.), *Handbook of Research on Digital Research Methods and Architectural Tools in Urban Planning and Design* (pp. 223–249). IGI Global. doi:10.4018/978-1-5225-9238-9.ch011

Refaat, A. M. (2019). Space Syntax Theory and Its Contribution to Urban Design. In H. Abusaada, C. Vellguth, & A. Elshater (Eds.), *Handbook of Research on Digital Research Methods and Architectural Tools in Urban Planning and Design* (pp. 203–222). IGI Global. doi:10.4018/978-1-5225-9238-9.ch010

Southworth, M. (2005). Designing the walkable city. *Journal of Urban Planning and Development*, *131*(4), 246–257. doi:10.1061/(ASCE)0733-9488(2005)131:4(246)

The Behavioral Risk Factor Surveillance System. (2003). Walking trends among U.S. adults. *American Journal of Preventive Medicine*, 95–100. PMID:12880875

Walk Score. (2015). *Walk Score®*. Retrieved 2018, from Walk Score Methodology: https://www.walkscore.com/methodology.shtml

Weinberger, R., & Sweet, M. N. (2012). Integrating walkability into planning practice. *Transportation Research Record: Journal of the Transportation Research Board*, *2322*(1), 20–30. doi:10.3141/2322-03

Xiongbing, J., & Grammenos, F. (2013, May 21). *Taking the Guesswork out of Designing for Walkability*. Retrieved from Planetizen: https://www.planetizen.com/node/63248

Chapter 5
The Car-Free (Day) Movement:
Transformation of Space and Place in Bandung, Indonesia

Frans Ari Prasetyo
Independent Researcher, Indonesia

ABSTRACT

As various cities around the world are implementing car-free policies, the need to understand it from a dynamic point of view becomes more pronounced. In effect, by invoking the organic nature of urbanism, a complexity involving the growth of cities in relation to their environment and human society emerges. Seeking to contribute to an understanding of the production processes of space in the contemporary street, the discussion of the planning future of cities, and perspectives on urban transformation, this chapter aims to build an understanding of the production of spaces for public life in Indonesian cities from the perspective of planning, production, and culture in car-free (day) movement in Bandung. This chapter contributes to the process of spatial production in car-free (day) and implies a reflective paradigm of practice and its potential to illustrate in planning the street transformation-productions of public spaces within the current process of globalization in car-free scheme.

INTRODUCTION

Forms of urban development, which are located in the global circuit, encompass the flow of capital accumulation but are also characterized by local history and landscapes as well as certain values and relationships (Bishop, 2003; Douglass, 2008; Harvey, 2012). Global citizens are experiencing a space of relationship in a

DOI: 10.4018/978-1-7998-3507-3.ch005

globalized world through networks that continuously move in the realm of experience, practice, imagination, and memories. I just think it related to spatializing and visualizing Indonesian urban politics by uniting the imagination of urban, public, and political spaces and places.

Indonesian cities are diverse but always begin with Jakarta. The country's capital, it is also the center of Indonesian politics and capitalist modernization sites that provide various streams of transformation, including the city space that image pursues and sometimes imposes on various other cities with global cities term, regulation or connection, including Bandung. One of them is through the Car-Free movement as a global practice. Bandung's Car-Free (day) (CFD) began in 2009, following Jakarta's lead; the first CFD in the capital took place in 2007. As an urban movement formed by formal-informal power, which civilians can test to enjoy the boundaries enforced by this spatial intervention, including landscape as a host, and risking the image of the city.

Car-Free as transformation investment in the urban space with a frame of public space and a package of sustainable environmental issues that simultaneously govern political and economic negotiations in the framework of developing shared knowledge in a city through infrastructure. Infrastructure, as a significant component of urbanism, reveals how people experience space. The pedestrian movement in Copenhagen follows at the same time -the 60s- as Jane Jacobs wrote *The Death and Life of the Great American Cities* (1961) and developing was experiencing unimaginable Car-Free movement in global cities. It began when, in response to the oil crisis of the early 1970s, Car-Free for four weeks was held in Switzerland. It took twenty years in New York in 1991 that created the first International Conference on Auto-Free City by transportation alternative and started experiencing unimaginable development in global cities.

This situation shows how the perception of the past, anxiety about the rate of urban change in the present, and hopes for the future of car-free development have been manifested in the infrastructure landscape in cities in different histories and practices. Seeing everyday infrastructure means investigating the intimacy of power in people's lives by emphasizing the individual body's role in the diffusion of power in society as well as witnessing and examining different exercises and the effects of power in their daily lives citizens and in their streets as daily infrastructure.

The street is constantly throwing up new objects and new social formations that pique our curiosity and nudge us toward interpretive acts (Barker, 2009a). We can see how Walter Benjamin ((2002)416–55) became a flâneur, a person who wanders the streets or pedestrian zones in order to expose himself to urban life directly. Pedestrian zones usually involve a special legal procedure and demand that urban planners employ higher standards for design. Consequently, the physical installation of such a zone takes longer and is more expensive than a time-limited ban on car traffic.

This chapter examines the car-free movement through a case study of the changing dynamics of street development in Bandung. Taking Car-Free as the analyzed infrastructure allows this author to use a different approach compared to works of globalization, which have criticized as being overly broad perspectives that fail to consider daily practices in a local context. Car-Free is the most global sophisticated regulation and strategy in the urban life agenda. At its most sophisticated, this approach can offer us a historical understanding of how the role of space changes as technologies, shifting production conditions, and allied representations give rise to new patterns of capital accumulation and income distribution (Harvey, 1989).

Place and space by means of architecture, infrastructure, and spatial training become a medium for preparing diverse knowledge and narratives and recording their history. Space is not static, so if it can be conceptualized in terms of binding social interactions, then Car-Free is itself not a matter of moving in space but moving through space with inertia and social interaction as its energy of movement. Car-Free provides the infrastructure for the above processes, including creation, agency, and cross-border accumulation in a space and place with crowds. Their emergence at certain times and spaces causes a disruption that triggers various parties to use the space as an arena of contestation with economic, political, social, and cultural motives.

Friedmann (2007) thoroughly discusses how placemaking appears at the state level, as the city is smaller both geographically and demographically and with a more straightforward government structure and system than the state, as a mini version of a country. In the Indonesian context, however, cities possess an autonomous authority that is centered on decentralization. Bandung enjoys this status as a mini-country thanks to decentralization. This status allows the city to create images of place by constraining everyday life in the way it determines which activities are allowed at certain times of the day or night, who may or may not be seen on the street, which forms of public behavior are permitted and which are not, what kind of traffic may circulate, and what sort of structures may be built, regulated and for what purpose.

Car-Free, therefore, is placemaking that constructs and is based on certain settings. The Movement appears as a third space where time-space compression is experienced, and where the human perspective in that space today is closely related to material progress that touches almost every aspect of our lives. Car-Free has become an urban environment project that is growing in popularity globally, creating new spatial and aspatial formations. More stakeholders envision Car-Free as a narrative of contemporary evaluative social urbanism and an effort to develop the politics of spatial access and mobility therein related to the use of physical and social infrastructure and controls that reinforce the power of the community and the state.

Dago Street is at the center of my case study. Since 2010, I have been researching how the Car-Free (day) movement is practiced in this colonial-era neighborhood and famous thoroughfare in the capital city of Bandung, West Java. In the first part of this chapter, I provide a brief description of Dago Street, followed by a historical overview of the dynamic flow of the area's development. Nassauer and Opdam (2008) point out that landscape ecology should also focus on design, not only patterns and processes. Car-Free is essentially packaging space for reasons of landscape ecology and building livable cities, as a method of space transition and temporary transformation in conjunction with the urban studies' methods of patterns and processes. After the public consumes it, time and space revert to its quotidian existence as a street with the infrastructure design.

Scholars often associate the place with the world of past and location and space with our present and future. From one perspective, the place is therefore nostalgic, regressive, or even reactionary, and space is progressive and radical. Car-Free ideas are about how the world works than to an accounting of its concrete geographical everyday life realities as a progressive overcoming of local places by national and global spaces. Dago Street as local place whereas Car-Free (Day) is a national – global space.

Car-Free (Day) is an urban issue rooted in-vehicle use adapted and transformed into a global issue, but in practice, it depends on local identity, whether it is Bandung or Berlin. The global-local attraction gives birth to a new space production identity in forming a component of urban composite culture. The transformation of infrastructure and street culture manifested in Car-Free (Day) concentrates and rationalizes time and activities so that the opportunity to create public open spaces becomes the center of social public activity and acculturation in the formation of images, existence, sources of knowledge, information and moral values.

This shows that there has been a progression of street development structures in urban life and that the street has been a privileged site at times for the deployment of new governmental rationalities and populism with urban (street) activation. However, history shows that municipal policy has often sought to shore up various types of supposedly pre-existing forms of scenario and strategy urbanism as a means to achieve capital and public control. In more global cities, Car-Free develops into an infrastructure scenario that provides unique cultural logistics that exceed its structural capacity through demonstration strategies of production and consumption. In urbanism, however, various degrees of stakeholders' arrogance in each urban movement evolves into a gamble of claims to imaginary territories related to the space that consumption produces and strongly related to the history of city or street expansion.

These divergent strategies of the rule have given rise to a peculiar situation whereby, on the one hand, the street is a formal infrastructure of a quasi-bureaucratic state (city) and, on the other, is relatively autonomous informal. In the remainder of the chapter, I examine the changing dynamics of the street in this formal-informal domain. In practice, public space is a terrain of conflict and struggle between different functions and interests. Rapid and contested processes of urbanization, industrialization and de-industrialization, and the heightened class inequalities associated with these developments have shaped urban spaces in Indonesia, including Bandung (Douglass, 2008); (Kusno, 2000:146-148); (Mc Gee, 2002:637-652); (Nas, 2003).

When one is in other areas of public Bandung, locals refer to Dago Street as an elite street because of its history, its colonial architecture, and its status as a high-end commercial and social area. People refer to the public as a "*Jalan Gaul*" or "*Daerah Gaul*", terms that can be loosely translated as "popular-famous-lifestyle street" or simply as "hype area". Grand residential houses have been transformed into elite offices and commercial or fashion shops. As a city landmark, the local government uses Dago Street to attract the public and bring tourists to the city. These circumstances make this area a top priority in various Bandung development scenarios in the past, present, and future.

This does not portray the city street as a place where one goes to take in the view, but rather as a battleground where the boundary between public and private is constantly being challenged, fought over and negotiated. Taking the street as the fundamental unit of analysis within the Car-Free movement—a transformation site where urban space and society converge, collide, and contradict-contestation makes abstract global phenomena visible in the reality of daily interactions. The contradiction of heterogeneous space reveals the overlap of dominant and opposition practices through the emergence of sites (Foucault, 1967). Space of representation (Lefebvre, 1991), practical tactics (de Certeau, 1984), and power (Foucault, 1967) as a discourse to understand space as a place of struggle, quarrels, and negotiations. Street with Car-Free can seing this situation and condition.

A city street or center where an area-wide ban limits motor traffic to that which is considered to be functionally necessary refers to administrative regulatory measures based on road traffic laws. The problem that immediately follows if that of setting criteria for exemptions and enforcing them. Inevitably, there is a need for political sensitivity and administrative competence if the scheme is to be effective and fair, and both are preconditions of civil society acceptance of government regulation.

Car-Free (or car-limited) areas are governed by consist of measures that can be applied over areas larger than pedestrian-only streets, and this suggests several points of distinction. Car-limited centers require a shift away from the car to "city-friendly" modes. The Car-Free movement is seen as grand space, or spatial crossroads in exchange, trade, diffusion, and connectivity in the city. In Bandung, the Car-Free (day) movement has helped to shape the trajectory of the city's democratization. Place meanings are highly beneficial for our understanding of the ongoing changes in our environment, should be integrated into conservation strategies, and should act as a beneficial manipulation strategy.

Hall (2012) works that short circuit our understanding of how diverse individuals and groups establish meeting points and assertions of divisions in the context of accelerated urban change. Utilizing spatial analysis and human-environment relations involves competing conceptions of space and place strategy. The claims that the nature of space and its effects depend on the economic conditions under which space is "produced" (Harvey, 1973), however, conceptions of space and to associate them with temporal changes in the content and/or disposition of space (Jones, 2009). A production-consumption street like Dago Street can reveal how car-free movement is working in a street-road corridor.

DAGO: A BANDUNG STREET

Bandung is a city situated on a plateau in the highlands of West Java, located at 768 meters above sea level. The colonial government of the Dutch East Indies originally developed Bandung as a resort city in the 1800s, due in part to its rich natural setting and comfortable climate (Voskuil, 2007). The city remained a small administrative outpost and market town until the 1880s before its growth increased after it was received its new status as a municipality *(Gemeente)* in 1906. Dutch East Indies Governor-General J. B. van Heutsz separated the city from Bandung Regency, and there were further plans to replace Batavia (today known as Jakarta) with Bandung as the new capital city (Kunto, 1984). The latter continued to grow up until and after the Second World War and continued to hold an essential position in the country after Indonesia gained its independence in 1945. Ten years later, the city was the host of the first Asian African Conference, an important step within the Non-Aligned Movement, a significant force in world politics during the Cold War period.

During colonial times before independent 1945, the area along the Dagostraat (Dago Street) became an elite Dutch residential settlement, and the road still boasts rows of heritage buildings in the classic Dutch colonial style associated with the Dutch Indies Architectural Community (NIAK—Nederland Indie Arsitecture Krink). In the eighteenth century, the Bandung plateau was an agricultural region with two main seats of government, one in the southwest and one in the northeast. Together with the plantation landlords, many European expatriates—who found Bandung to be a pleasant resort—came and settled in the northern suburbs, which offered a hilly vista with beautiful landscapes. At the same time, people from the surrounding areas *(pribumi)* also poured into Bandung and most settled in the southern area, which was relatively flat (A.K.M. Tarigan, 2015). Chinese residents congregated in a camp in the center of town (Barker, 2009b).

Bandung still has two seats of government in the form of two old alun-aluns, one in Dayeuh Kolot (southern Bandung) and Ujung Berung (eastern Bandung).

However, since the colonial era, the city is more practically separated into two parts: northern Bandung and southern Bandung. The northern former is known for its colonial infrastructure, and the southern is known for its indigenous people *(pribumi)*. This division naturally created social segregation among urban dwellers based on political power, wealth, and ethnicity (Siregar, 1990). While the northern area is well planned and developed, the southern area is characterized by urban sprawl and uneven development. Despite these, separation of northern and southern still works.

Dago Street in northern Bandung has a long history as an important location for commerce, socializing, transport, and recently as a residential area, reflecting the wider history of Bandung's urban development. The neighborhood is named for the major thoroughfare that passes through it, still commonly known as 'Jalan Dago' despite being officially renamed *'Jalan Ir. H. Juanda'* in 1970. The name *'Jalan Dago'* (Dago Street) is variously attributed to an epithet of Prince Suryadiningrat, the younger brother of *Sultan Agung* of Mataram, whose descendent Prince *Wiradadaha* became the First Regent of Sukapura after suppressing the Dipatiukur rebellion in the Sunda region (Prasetyo, 2013).

Dago Street has a colonial legacy and has been used as a heritage road for commercial and tourism purposes and is characterized by the remnants of infrastructure and colonial planning. This paved road is lined with shops selling clothes, mobile phones, and electronics as well as containing a high school, university, hotel, restaurants, pharmacy, hospital, the local branch of an international bank, and, of course, (elite) residents. In the 1970s, between the 1990s, the Dago area was converted into a commercial center. The street has a wide sidewalk lined with many big trees, lights, and flowers. The latter have cultural claims—Bandung is a flower city.

On the one hand, Dago Street reveals a romantic city through a performance of colonial city architecture and landscape, modernity, and production-consumption transformation. On the other, it is a buffer against urban slum villages that provide the labor needed to support the work of colonialism. We understand Production-Consumption to have multiple roles as a political, economic, and cultural touchstone, and to be an active ingredient in the construction of place and space as identity and urban regeneration. This situation is ongoing; it is just that the kampong around Dago Street, such as Kampong Manteos, Kampong Cisitu, Kampong Babakan Siliwangi, Kampong SadangSerang, Kampong Tamansari, and Kampong Sekeloa, continue to provide labor resources for Dago Street, including low-cost housing and rents for labor.

In the planning and design of this small colonial city in the pre-independence period, Bandung was intended to be home to a maximum of 600,000 people. Today in 2019, it is Indonesia's third most populous city after Jakarta and Surabaya, with 2.5 million registered residents (Kota Bandung, B. P. S., 2018). The political authorities still maintain a pattern of colonial development in their planning, especially in the northern part of the city, evidenced by Dago Street. Since the 90s, Bandung has been developed into an important national center for art, popular culture, and the creative industries. A smart city as a think has played a significant role in the development and transformation of the global city. The city as an entity has replaced the country as an arena for the representation of popular values.

Dago Street is a congeries of micro-architecture ecologies or places separated by distinctive colonial histories, urbanisms, and social practices in which connectivity and mobility within the region are more of a response to the management of environmental, transportation and social risks than a simple outcome of street initiatives production-consumption. The Car-Free (day) movement in Dago Street transforms street function with the car-free scenario, livable cities, and brings climate change politics into the street, engaging individual and public emotions with place and crowds, social networks, and public cohesion. This is how an example of how globalization effects transform place and locality.

CITY AND STREET TRANSFORMATION

The place is socially constructed and operating, including interactions between people and groups, institutionalization of land use, political and economic decisions, and the language of representation (Saar, 2009). This view of place presents it as a series of locales or settings where quotidian life activities take place. The location is not just a mere address, but where social life and the environment are transformed. Vorkinn and Riese (2011) stress the importance that the sense of place plays in maintaining the quality of the environment.

Would be such social settings from everyday life like workplaces, homes, shopping malls, cafes, factory outlet, mini markets, hospitals, universities, schools, churches, mosques, vehicles, whose structuring of social interaction helps forge values, and attitudes and behaviors? Streetlife can be terrain for creativity and innovation. De Certeau (1984:96) argues the pedestrian affirms the fragmented, plural, and creative quality of urban life simply by how she or he moves through space.

Placemaking deals primarily with wider scales, and authors investigating global influences on place rarely engage with the topics of place meaning or identity. The Car-Free movement is a global phenomenon designed to activate the street and produce meaning and a new identity but also views global processes as only having a concept, material, and ideological influences on the place and not related to ever-changing place meanings, production, and consumption in everyday life. Place meanings and created identity related to place attachment are functional, emotional, and social meanings as well as containing aspects of class dimension. On the one hand, people inflict landscape changes; on the other hand, when space changes, lifestyles, cultures, and attitudes are transformed (Antrop, 2003). Changing places causes changes in people's values, but people's values can also change the (street) landscape.

Car-Free frames are constantly in transition and are remade by creating common ground for collective action and shaping people's ideas about place. New meanings are created by bringing out common experiences, interests, and values. The meanings that a person or collective attaches to a place are similarly influenced by personal experiences, education and global networks, politics, and information to construct subjectivity and activity in the place. The meaning of the Car-Free movement may be influenced by an individual's activity in the street, sidewalk, using public space, and car use. The individual's participation in some organizations will influence the meaning of the Car-Free movement for him or her, for example, members of an environmental NGO or climate change collective may be concerned with a state or local policy towards global agenda (ex. Sustainable Development Goals-SDG's), and its position in society. These street-level movements are transformative and helping to constitute an urban fabric that is the unpredictable sum of singular, creative acts rather than a function of the disciplinary logic of the kind of panoptical view privileged by cartographers, architects, and planners (Barker, 2009a).

The place is also used to invoke certain feelings like comfort, security, belonging, being anchored, self-expression, and freedom to be oneself (Smaldone, 2005). All of these meanings can sometimes also be seen as unifying certain groups, like car-free (day) bike, runner, or skate collectives gathering in the street because it provides them with the feeling of comradery in coming together in a public space. Some places may acquire meaning through certain activities that take place within their boundaries (Yi - Fu Tuan, 2001). Feelings invoked by some places can also play a role in forming and maintaining place connections and place identity.

In this framework, the place may be thought of as an open articulation of connections and identities of subjects and of places constructed through interrelations that not only challenge notions of past authenticities but also hold open the possibility of change in the future (Massey, 1999). Lefebvre (1991) introduces the third term, and Soja (1996) later argues that space is understood as a physical and social landscape that is imbued with meaning in everyday place-bound social practices and emerges through processes that operate over varying spatial and temporal scales.

For Habermas (1990), the public sphere is egalitarian, inclusive, pluralistic and open, enabling inter-cultural dialogue and the identification of shared interests, while remaining free from state coercion or domination by political and economic elites. In this sense, public space can be identified as part of an autonomous community sphere of social solidarity, as distinct from both the capitalist market economy and the bureaucratic state apparatus (Prasetyo, 2014). Harvey (2003) explains the metaphysical ability of capitalism as a reproduction of space where capital accumulation cannot be agile accelerated in spatial fix sovereignty so that capitalism needs additional space—a Spatio-temporal fix. This capital and territorial logic in their journey are intertwined for cooperation, but there is also a conflict because territorial logic (spatial) is static, while capital logic contained in financial logic (political economy) is dynamic in the state and city policy. Space and time appear on Car-Free as an indication of human existence, while the mastery of capital, space, and time form a substantial link in the form of relations, capital, and social chaos. We can see changes in the use of space, including the chaos of activities on Car-Free (Day) in Bandung, which include a variety of work motives and the interests of actors who seem to do the work of urbanism.

State or city policy also influences national and other scales of placemaking. For instance, if the state and/or the city has been implementing neo-liberal policies, this assumes that communities take care of local facilities (Martin, 2003). This kind of policy also creates more segregation in a city's spatial character. As a result, different districts obtain various meanings. The Bandung city government considers the existence of the Car-Free (day) movement policy as a tool to increase the city's happiness index.

In 2017, for example, the Mayor of Bandung, Ridwan Kamil (2013–18), claimed that Bandung Happiness Index was 73.42, higher than Gross National Happiness (GNH) Indonesia (70.69) and other global cities (Kota Bandung, B. P. S., 2017). There are no official city policies in the form of regulations attached to Car-free (day) in Bandung, not even an informal policy to close the street for public space that is claimed as Car-Free. This particular street transformation only changes the street's function, replacing the vehicle and opening the space to everything but vehicles, and there is no happiness index available from this space also. One of the claims from the happiness index is only that many residents come to this CFD and look happy.

The street makes a social space with its particular cast of characters, its forms of social organization, and its vernacular process and form of socio-spatial politic. It is also the site of particular kinds of politics and forms of knowledge as well as for distinctive genres of music, literature, performance, and visual art (Barker, 2009). Knowledge creation and circulation are invariably situated somewhere (Schatzki, 1991). Cracked, hybrid, and liquid spaces replace integrated space, coherent, and easily identified as part of the work of modernity.

This inventory shows that although the street may often be seen as a space of freedom, it may, in fact, be a space that is overdetermined and overregulated by a municipal regulation. To approach the street as a democratic space of diverse encounters, the most crucial quality of place is that it has the power to order and to focus on human intentions, experiences, and actions spatially. Historical events can create various meanings in places. Alexander (2002) discusses the history of space event that is used for far broader contexts and spaces.

As we have seen in this history of Dago Street, the street has undergone a transformation that is part of the historical landscape of the Car-Free movement that has become a contemporary agenda of global cities and positions Bandung as one of these cities. Based on Write spectrum (2005), my argument as Car-Free in Bandung is organized as a temporary, small-scale initiative—only part of Dago Street is closed.

Figure 1. Car-Free (day) Movement maps in Dago street-Bandung, Indonesia.
Source: (Prasetyo, 2014:2017)

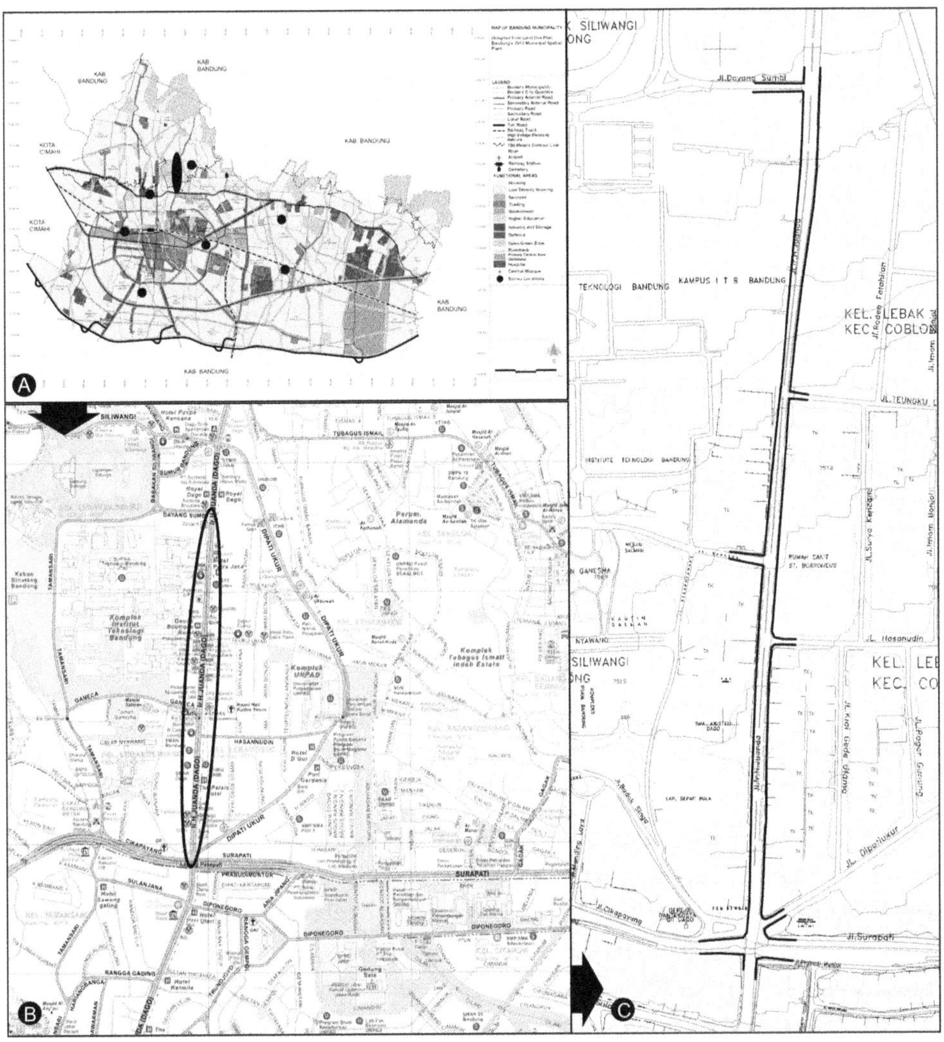

The Car-Free character that influences place, and people's identity is a common culture. Culture involves a typical socialization process that takes place through sharing similar production-consumption experiences, signs, symbols, and policies. An individual's identity is formed through a contribution from many places and their meanings (Manzo, 2005), and linear with the collective identity and public identity in the Caf-Free area.

Car-Free (Day) emphasizes and plays a new role in the destruction of space in the time of urbanism, where this site politically compresses space and time solely for the sake of specific images. Further, this movement provides an escalation of political, economic, and cultural space with accumulation and commodification. This process argues—or more often confirms—that it has gained new momentum in reaching new stages related to the existence of space and time, which are represented in urban street infrastructure elements. The phenomenon of Car-Free (Day) is experiencing uncertainty. It is in a paradox of its own space and time related to the identity and politics of space, structural agencies in relation to society, and the state as well as socio-cultural relations in the class hierarchy, including identity formation.

Investors who support new construction projects and particular renovations find rational homogeneity in the emergence of an ecological class of commercial use of public space as a hegemonic environment of economic units. Eventually, space loses its original identity, and a new identity with a thick package of capital penetration replaces it. This condition is evident in Dago Street—the location of the Car-Free (Day) and the city's main axis. Incidentally, it is also an economic axis that owners of capital built, which makes it is a real daily consumption tourism location during the weekly Car-Free Day. The production of this public space provides other efforts so that the essence that occurs is obviously blurred and only becomes a shield or a wrapper of reality. Reengineering a city through a "show" of certain spaces has resulted in urban transformations that absorb surplus capital and social capital through a contest of spaces and their publics.

CAR-FREE(DAY) MOVEMENT

As the street often stands as a synecdoche for the city, a focus on the street serves much the same function as does a broader focus on city life: it provides a conceptual or empirical anchor for interdisciplinary conversations about the dynamics of modern public life (Barker, 2009a). The Car-Free movement provides post-modern city space with a deep relationship between the power of geographical topography and that of power held around the thesis of urban autonomy, public space, and globalization temptations embedded in the dominance and exclusivity of urban infrastructure. In the post-modern paradigm, concepts such as solidarity, identity, society, and differences are no longer based on contact or spatial factors but rather relational in substantial configuration frames.

The changing role of locality in relation to place-making still matters. Molotch et al. (Molotch, 2000) reason that locality still matters since similar outside forces can have very different consequences depending on the local context. The Car-Free movement in Dago Street provides a unique role in the context of the movement's locality of the development or practice that exists in all cities in the world. Placemaking has complicated the relationship with place meanings. On the one hand, sharing similar experiences and having common purposes unites citizens, and on the other hand, this activity transforms previous meanings.

Car-Free (Day) in Dago Street provides a boost to aspects of the vision process of the urban ecosystem by encouraging a sense of place, which then drives change into space of flow, which in turn transforms into space of culture. These changes are clearly seen in the Dago Street corridor where Car-Free (Day) has been a weekly event on Sundays since 2010, despite changes in the actual timing; initially, the street was Car-Free at 06:00 am to 12:00 am from Cikapayang Park to Simpang Dago along a 1.51-kilometer route, which was changed a year later to 6:00 am to 10:00 am along a 1.2-kilometer route.

Figure 2. Spectrum Car Free Development in various cities, including Bandung.
Source: The Spectrum car-free movement, the author adopted from (Write,2015)

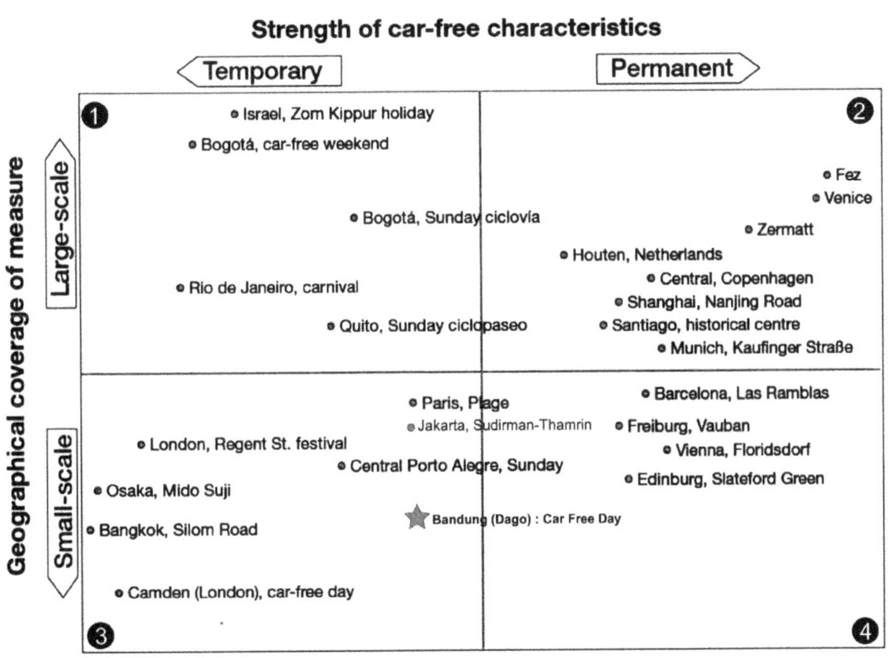

The CFD area uses a street corridor, which is a public space, but along the length of this road there are private spaces, such as the home yard, shops, and offices. When the CFD, a private space here becomes a temporary public space. This private area is leased informally to other stakeholders, who need it to carry out activities, while the CFD is working. This area is transformed into a variety of activities, such as street vendors that 80% of them sell food and beverages, and others sell sports equipment, such as shoes, t-shirt running, and bicycle merchandise. Nevertheless, some others use this private space as an arena for aerobics, or social campaigns to sell cars or motorcycle. Rent space prices also vary depending on the owner, but on average, rent it out during the CFD takes place within four hours at a price of around USD 50-150.

Space's placement of the position and the role of Car-Free in Bandung, which seems modern and imaginative in its efforts as a policy and virtue, in reality, does not work that way, primarily because residents are not accustomed to such spatial interventions as a part of this kind of contemporary urbanism. This flow then provides the spatial polarization and contestation (space and public) that happens therein. An arena has lost its territorial roots due to the erosion of its culture regarding the effectiveness of the fermentation of place and space through the intervention of the 'bacteria'—the globalization of space.

When we consider the history of the Car-Free (day) movement in cities around the world, including in Bandung, we may conclude that it missed its target of activating space and the public. In Bandung, it does not encompass one full day or the full street corridor. One might also question its status as 'Car-Free' since people still use vehicles such as cars and motorbikes in the street area, which causes congestion on other roads. It also cannot be considered environmentally friendly since it does not reduce pollution in the area. Instead, Car-Free (day) in Bandung is simply an escalation of new public spaces and for the public's urban leisure and flaneur.

One can see art practices through various forms of performance, dance, music, performance arts, installations, and street art by various communities, agencies, collectives, and businesses. Car-Free Days take place on Sunday morning, a time that people may reserve for sporting activities, especially Car-Free packaged in nuances of sustainable urban environmental practices for its citizens. In Car-Free (day), the public rides bicycles, skateboards, use rollerblades and play freestyle football, cheerleaders' performances attract attention—all accompanied by *dangdut* music, played loudly to attract and more often than not, please the public. Music in this space is linked to specific agendas—each individual, community group, or agency chooses a genre of music to meet their needs. For example, music for sports is mixed with other loud music or sounds for a campaign or commercial events or product-launching promotions. Hundreds of people sell food and clothing, use cars or work stalls. When viewing this landscape as a whole, we can see that there is a power that makes this situation persist in Car-Free (day) movement—formal or informal power. There are 30 types of stakeholders or actors with several more sub-stakeholder categories related to the relationships, activities, or roles that are performed within a period of four hours on Car-Free (Day) in Bandung.

Figure 3. Car Free (day) movement in Dago Street –Bandung.
Photo by Frans Ari Prasetyo (20/04/2013)

Today, society consists of a market economy (capitalism), a bureaucratic system (the state), and social solidarity (society). The locus of public political space lies in social solidarity, which must be imagined as an autonomous space that distinguishes itself from both the market and the country. State and politician use Car-Free (Day), as a space for politicization, as an arena of mastery over various motives of politics to create new values for space and politics. For example, it becomes a public political space that is positively associated with democracy through the major and government election process, making the public space a political battleground for contestants.

The commercialization of Car-Free (Day) is a shift from games and communal celebrations to commercial entertainment provided by capital and state capital owners as well as political interests. This shift is an essential stage in the development of an urban middle-class culture that leaves us with the impression of the emergence of 'new' civil and democratic values, but they are actually popular values. These are often translated into populist programs. Ironically and hierarchically, the Car-Free day is an example of a populist program.

The emergence of capital interests in the economic frame and formal and informal power arising from various stakeholders as well as formal-informal, tangible-intangible mechanisms that overlap in Dago Street during Car-Free (Day) form a temporary public space as the third space that is appreciated and anticipated in zoning changes (multi-use zoning). The car-Free day has tangible and intangible power control over the public and public spaces. This is related to the area of power, which involves formal and informal power over an area. The projection of humans' intense desire for control over space as a political praxis that rationalizes space and is abstracted spatially or utilitarianly as an asset of capital.

The struggle with formal and informal power for public space creates new segregation in public space. In this context, Car-Free (day) becomes an arena for heterogeneous, multicultural, and political public contestation to label it as a collaboration space then. Following Lefebvre (1991), who understands space as a practice of producing reality and interacting with time and events, Soja (1996) pays special attention to space's formation through the third space. In Bandung, Car-free (Day) on Dago Street is an urban transformation that changes the street landscape and the road's meaning. This transformation appears in three forms. First, is perceived space, which includes the emotional and behavioral bubbles that invisibly surround people's bodies as well as the complex spatial organization of practices that shape action spaces in architecture, residents, kampongs, streets, cities, nations, and even in global geopolitics.

Secondly, conceived space, which refers to our knowledge of spaces that are primarily produced by discourses of power and ideology that professionals and academics such as planners, engineers, and researchers construct. Third space refers to space where all of the spaces are and where the subjectivity and objectivity, the abstract and the concrete, and public and private authority. Third space is a space as directly lived, with all of the related intractability. Spaces of representation contain all other real and imagined spaces (Soja, 1996).

This author argues that Car-Free (day) in Bandung is part of this Third Space, where the exploration of the socio-cultural dimension as the accumulation of urban society values with creativity patterns so that the emergence of a movement called Car-Free (day) as a unique urban cultural context of Bandung's public spaces (Prasetyo, 2014). The production of space is very dynamic, but it opposes the linear view of time and space itself. The mode of production builds space relations and then produces new spaces in accordance with the interests of production, as seen in the form of Car-Free (day) as part of the interests of urban transformation in building images for a certain awareness.

CAR FREE AS SPACE, PLACE, AND FLOW

Now, the world is increasingly placeless as space-spanning spatial connections and flows of activation and information result in people creating the rootedness of a wide range of processes anywhere in a particular activity. Today, it seems that space is progressively conquering place (Friedman, 2005), but space and place are not clearly distinguished from one another analytically, or their meaning is reversed (de Certeau, 1984). Implicitly, however, the language of space is privileged. Though there is not necessarily a connection of this type, the usage of space is very common with place standing in for the local (and traditional) proportion and location/space representing the global (and the modern). Amin (2004), for example, argues that all geographical concepts—place, territory, and scale—that stress internal coherency and boundedness reify rather than reveal the logic of socio-spatial relations in a globalizing world. "In this emerging new order, spatial configuration and spatial boundaries are no longer purposively territorial or scalar, since the social, economic, and political inside and outside are constituted through the topologies of actor-networks, which are becoming increasingly dynamic and varied in the spatial constitution" (Amin, 2004).

The Street as a ground provides free landscaping along with surrounding infrastructure and supports the process of capital accumulation in various modes of thinking and associated with spatial overlays. The privileging of place as simply location has continued. The term place carries with it is not only the meaning of spatial location but also those of social position and moral order (Tuan, 1974). Other conceptualizations, particularly those that privileged space over the place, tend to leave little likelihood of political change coming from bottom-up political organizations (e.g., Hardt, 2000).

Car-Free (day) movement, sees the street itself as simply incidental to more profound non-spatial processes such as class gap, perceptual capacity and orientation, capital accumulation, or commodification. For Jacobs (1961), the street is important, not so much because it adds a poetic and creative dimension to the structuring of city life, but because it provides the social basis for civic engagement and political mobilization (Barker, 2009a). The street can be space, but it can also be a place. On behalf of the people, Car-Free (day) on the street challenges those attempting to recapture place-space with modern capitalism. Through Car-free (day), this sort of configuration of contextual changes can give an opening to a new political space or a redefined old one as interests, identities, and political heuristics (yardsticks) shift and upset established space affiliations.

The emergence of public space in Bandung as an initial discourse of the need for public space legitimizes the emergence of the Car-Free (day) movement, which was initially a response to global trends in urban environmental issues and then packaged as a public space. Car-Free (day) is a valuable public contest as an administrative location, socio-cultural location, including lifestyle location, economic location, and political location for city residents. Here we can borrow from de Certeau (1984), who explains the relationship of domination and contestation through other avenues for spatial regulation. Car Free change public relations and regulate spatial planning through spatial control, which gives rise to contestation between them within a certain time.

CONCLUSION

The Car-Free hypothesis presents an image of a future city, livable, and environmentally friendly that repossesses street space from vehicles for humans in the global city. This hypothesis, tinged with fear and revolutionary hope, raises the question of what will happen to city development in the coming decades. The Car-Free (day) movement emerges from cartographers, planners, architects, regulators, scientists, and a host of others in their particular office, homes, workplaces, and commercial in this street in the city, all joining this performance of street with its spatial rhythm and production-consumption practice. Car-free (day) as a transportation system engineering, especially cars, is carried out in global urban centers with planners who have dreams about livable cities but work in specific areas with certain social classes. Because of spatial politic, class sentiment ensures and segregation is maintained.

In the Bandung context, Car-Free (Day) takes place on an elite road, in an ex-colonial residential area that is transformed into a commercial area. It has shown certain processes in spatial engineering that conducted trials to update according to the needs of its global city image, but they want to maintain social-cultural conditions in certain class formations. The area surrounding Dago Street in the city's north remains as an elite area with a colonial atmosphere and is a representation of the city. Here the city government create image city as a city sweetener infrastructure and create new social classes in the production-consumption of the movement with gentrified, and leisure-time physical activity. The change from space of place to space of flow and space of culture in Car-Free (Day) on Dago Street through public contestation results in a struggle for space in various contexts of interest.

An examination of Dago Street's history raises serious questions about the underlying assumptions of the Car-Free hypothesis. While there can be no doubt that Dago Street is part of the city's heritage, a site of work, living, and shopping, the city has been very active in shaping public space and public activities in its streets. The space produced by economic transactions and state policies has colonized everyday life by means of both spatial practices in commodification and bureaucratization and representations of space through discourses of planning and surveillance.

Broadly speaking, these forms can be characterized first as a progressive city development albeit gradual, deepening of a street's image and power, and including a city's formal-informal circuit. The bureaucratization of the street began in the colonial period and continues to this day, with different formations, but the goal of taming and controlling the streets remains unchanged. Those seeking a movement against this colonization of concrete space (or place) to reclaim such spaces of everyday life could look to insurgent counter-discourses based on spaces of representation that build on memories and residues of an older authentic existence and new practices in concrete space.

Car-free (day) in Bandung reclaimed local streets and public spaces for dance, sports, biking, the informal sector, street food, street festivals, art-music performances and installations, political and religious campaigns, social programs, product promotions, and shared use of space for anything activity. You can imagine a situation that is very crowded, complex, but formally controlled by the state through the city government and police but also by *ormas*[1] and *preman*[2]. During Car-Free (days) in Bandung, there was a deep union between the state and society so that the public sphere was suppressed. The state began to play a more fundamental role in daily life by turning the public sphere into a site for contestation of the state apparatus as a public show for the state's existence in public spaces. The public sphere (should) act as the only place/space where the state power cannot intervene.

The fact, Car-Free (Day) became the legitimacy that was raised by the Bandung city authorities to cover the limitations of open public space by juggling the Dago Street corridor into an open public space wrapped by sustainable environmental issues as an effort to manipulate the city's image through certain activities. In reality, Car-Free (Day)s legitimate form is a transformation of urban infrastructure by playing the roles with the public from street-based on issues for constructing lifestyles and the projection of a shiny city image. Urban transformation reengineered a city through specific movements that absorb surplus capital and social capital through the third space contestation.

There is a manipulation of the image of the city through certain activities so that the city can be seen as a sustainable and environmentally friendly city. The Car-Free movement supports this. It is a real example of the legitimacy of the city government's power to cover the limitations of open public space by juggling Dago Street into an open public space wrapped by sustainable environmental issues that actually only follow the trends of other global cities in bringing up the issue of car-free (day) in Bandung, and failed to reduce vehicle emissions for a sustainable environment and global warming issue.

In examining the Car-Free movement, we must consider its structure, its politics, and its creative tendencies, in studying the interplay between these tendencies of the street conditions under which public culture may be transformed and intervention.

At the same time, the Car-Free movement has served the interests of capitalists by providing them with place and space along with a means to protect landscape (street corridor) behind heritage and tourism in which capital accumulation is taking place. The Car-Free movement as a new generation is a more extended part of a city's development agenda defined by capital interests with the public package. Car-Free (Day) as a public space and global urban environmental trend has only progressed to the euphoria stage to date.

ACKNOWLEDGMENT

The author is very grateful to Alicia Filipowich and all editor-reviewer for discussion and some substantial editing of this article.

REFERENCES

Alexander, J. C. (2002). On the Social Construction of Moral Universals: The 'Holocaust' from War Crime to Trauma Drama. *European Journal of Social Theory*, *5*(1), 5–85. doi:10.1177/1368431002005001001

Amin, A. (2004). Regions unbound: towards a new politics of place. *Geografiska Annaler, B, 36*, 33-44.

Bakker, L. (2018). Security Blurs and Citizenship: Consequenses in Indonesia. In E. G. Tessa Diphoorn (Ed.), *Security Blurs: The Politics of Plural Security Provision*. Routledge. doi:10.4324/9781351127387-4

Barker, J. (2009). Introduction. *City & Society*, *21*(2), 155–162. doi:10.1111/j.1548-744X.2009.01019.x

Barker, J. (2009b). Negara Beling: Street level authority in an Indonesian slum. In G. v. K. a. J. Barker (Ed.), State of authority: The state in society in Indonesia (pp. 47–72). Ithaca: Cornell University Press.

Benjamin, W. (2002). *The Arcades Project*. Harvard University Press.

Bishop, R., Philips, J., & Yeo, W. W. (2003). Perpetuating cities: Excepting globalization and the Southeast Asia supplement. In J. P. W. Y. R. Bishop (Ed.), *Postcolonial Urbanism: Southeast Asian Cities and Global Processes* (pp. 1–36). Routledge.

de Certeau, M. (1984). *The Practice of Everyday Life* (S. Rendall, Trans.). University of California Press.

Douglass, M. (2008). Civil society for itself and in the public sphere: Comparative research on globalization, cities and civic space in Pacific Asia. In K. C. H. G. L. O. M. Douglass (Ed.), *Globalization, The City and Civil Society in Pacific Asia* (pp. 27–49). Routledge.

Foucault, M. (1967). *Of Other Spaces*. Academic Press.

Friedman, T. (2005). *The World is Flat: A Brief History of the Twenty-First Century*. Farrar Straus and Giroux.

Friedmann, J. (2007). Reflections on Place and Place-making in the Cities of China. *International Journal of Urban and Regional Research, 31*(2), 257–279. doi:10.1111/j.1468-2427.2007.00726.x

Habermas, J. (1990). *The Structural Transformation of the Public Sphere: an Inquiry into a Category of Bourgeois Society*. MIT Press.

Hall, S. (2012). *City, Street and Citizen: The Measure of the Ordinary*. Routledge. doi:10.4324/9780203118597

Harvey, D. (1973). *Social Justice and the City*. Johns Hopkins University Press.

Harvey, D. (1989). *The Condition of Postmodernity: An Enquiry into the Origins of Cultural Change*. Blackwell Publisher.

Harvey, D. (2003). *The New Imperialism*. Oxford University Press.

Harvey, D. (2012). *Rebel Cities: From The Right To The City To The Urban Revolution*. Verso.

Jacobs, J. (1961). *The Death and Life of Great American Cities*. Random House.

Jessop, B., Brenner, N., & Jones, M. (2008). Theorizing sociospatial relations. *Environment and Planning. D, Society & Space, 26*(3), 389–401. doi:10.1068/d9107

Jones, M. (2009). Phase space: Geography, relational thinking, and beyond. *Human Geographies, 33*, 802–822.

Kota Bandung, B. P. S. (2017). *Indeks Kebahagiaan 2017* [Bandung Happyness Indeks]. Retrieved from https://www.bps.go.id/publication/2017/12/05/1f99cefd596c449b93405fcd/indeks-kebahagiaan-2017.html

Kota Bandung, B. P. S. (2018). *Kota Bandung dalam angka* [Bandung in numbers]. Retrieved from https://bandungkota.bps.go.id/

Kunto, H. (1984). *Wajah Bandoeng Tempo Doeloe*. Granesia.

Kusno, A. (2000). *Behind the Postcolonial: Architecture, Urban Space and Political Cultures in Indonesia*. Routledge.

Lefebvre, H. (1991). *The Production of Space* (D. Nicholshon-Smith, Trans.). Blackwell.

Manzo, L. C. (2005). For better or worse: Exploring multiple dimensions of place meaning. *Journal of Environmental Psychology*, *25*(1), 67–86. doi:10.1016/j.jenvp.2005.01.002

Marston, S. A., Jones, J. P. III, & Woodward, K. (2005). Human geography without scale. *Transactions of the Institute of British Geographers*, *30*(4), 416–432. doi:10.1111/j.1475-5661.2005.00180.x

Martin, D. G. (2003). Place-Framing' as Place-Making: Constituting a Neighborhood for Organizing and Activism. *Annals of the Association of American Geographers*, *93*(3), 730–750. doi:10.1111/1467-8306.9303011

Massey, D. (1999) *Power-Geometries and the Politics of Space-Time*. Hettner Lecture 1998. Heidelberg, University of Heidelberg-Institute of Geography.

McGee, T. (2002). Jalan, Jalan: Invading, destroying and reconstructing the Southeast Asian city. *Bijdragen tot de Taal-, Land- en Volkenkunde*, *158*(4), 637–652.

Molotch, H., Freudenburg, W., & Paulsen, K. E. (2000). History Repeats Itself, But How? City Character, Urban Tradition, and the Accomplishment of Place. *American Sociological Review*, *65*(6), 791–823. doi:10.2307/2657514

Nas, P. P. (2003). The streets of Jakarta: Fear, trust and amnesia in urban development. In P. N. R. Jaffe (Ed.), *Framing Indonesian Realities* (pp. 275–294). KITLV Press.

Nassauer, J. I. O. P., & Opdam, P. (2008). Design in science: Extending the landscape ecology paradigm. *Landscape Ecology*, *23*(6), 633–644. doi:10.100710980-008-9226-7

Prasetyo, F., & Iverson, S. M. (2013). *Art, activism and the 'Creative Kampong': A case study from Dago Pojok, Bandung, Indonesia*. Paper presented at the PlanInk International Conference, Malang, Indonesia.

Prasetyo, F. (2014). *Car Free Day: Contestation of 'Third Space' as a Phenomenon in the Production of Urban Public Space in Bandung* (Unpublished Thesis). ITB-School of Architecture, Planning and Policy Development-Department of Regional and City Planning, Bandung, Indonesia.

Prasetyo, F. (2017). Car Free Day: Transformasi Ruang dan Globalisasi Urbanisme Kontemporer di Bandung. *Jurnal Pemikiran Sosiologi -UGM, 4*(1), 1-24.

Prasetyo, F., & Sean, M. Iverson (2014). *Playing Under The Fly Over: A Collaborative Creative Community in Bandung*. Paper presented at the 5th Arte Polis International Conference and Workshop, Bandung, Indonesia.

Saar, M. a. H. P. (2009). *The Dimensions of Place Meanings Living* Rev (Vol. 3). Academic Press.

Siregar, A. (1990). *Bandung, the architecture of a city in development: urban analysis of a regional capital as a contribution to the present debate on Indonesian urbanity and architectural identity* (Unpublished doctoral dissertation). Katholieke Universiteit Leuven, Leuven, Belgium.

Smaldone, D., Harris, C., & Sanyal, N. (2005). An exploration of place as a process: The case of Jackson Hole, WY. *Journal of Environmental Psychology, 25*(4), 397–414. doi:10.1016/j.jenvp.2005.12.003

Smith, N. (1984). *Uneven Development*. Blackwell.

Soja, E. (1989). *Postmodern Geographies: The Reassertion of Space in Critical Social Theory*. Verso Press.

Soja, E. (1996). *Thirdspace: Journey to Los Angeles and Orher Real-and-Imagined Places*. Blackwell Publishers Inc.

Tarigan, A. K. M., Sagala, S., Samsura, D. A. A., Fiisabiilillah, D. F., Simarmata, H. A., & Nababan, M. (2015). Bandung City, Indonesia. *Cities (London, England), 50*, 100–110. doi:10.1016/j.cities.2015.09.005

Tuan, Y.-F. (1974). Space and Place: Humanistic Perspective. *Progress in Geography, 6*, 211–252.

Tuan, Y.-F. (2001). *Space and Place: The Perspective Experience*. University of Minnesotta Press.

Vorkinn, M. R., & Riese, H. (2011). Environmental Concern in a Local Context: The Significance of Place Attachment. *Environment and Behavior, 33*(2), 249–263. doi:10.1177/00139160121972972

Voskuil, R. P. G. A. (2007). *Bandoeng, Beeld van een stad*. Planologi-ITB and Jagaddhita.

Write, L. (2005). *Car Free Development*. GTZ.

ENDNOTES

1. *Ormas* are literally 'community organizations' (*organisasi kemasyarakatan*). *Ormas* a local level provision of security that appears beyond the control of official authorities and differentiates between citizens along religious or ethnic line can cause inequality and tensions as much as the abuse of government power against which *ormas* agitate. See Bakker, L. (2018). Security Blurs and Citizenship: Consequenses in Indonesia. In E. G. Tessa Diphoorn (Ed.), *Security Blurs: The Politics of Plural Security Provision*: Routledge.
2. *Preman* is a member of an Indonesian organized gang, encompassing street level criminals up through crime bosses. Premans are often perceived negatively throughout Indonesian society due to associations with violence and criminality.

Chapter 6
Confronting the Conundrum of Shared Space Street Design:
The Design, Development, and Delivery of the University of Central Lancashire's Masterplan

Robert Michael Turner
University of Central Lancashire, UK

Ehab Kamel
https://orcid.org/0000-0002-6737-9356
University of Central Lancashire, UK

Amal Ramadan
https://orcid.org/0000-0001-9096-1259
University of Nottingham, UK

ABSTRACT

This chapter discusses the challenges and debates related to the concept of shared space street design via demonstrating the case of the ongoing Masterplan development scheme of the University of Central Lancashire (UCLan), in Preston, North West England, United Kingdom. Based on hands-on experience, being involved with the project on multiple layers, in project management, working, and living in Preston City, the authors employ observational analysis methods to explore and reflect on the challenges UCLan Masterplan has faced, how it learned from the city's most recent Shared Space development (the Fishergate Project), and they further expand on their prediction on how the project may influence the transformation of Preston's public realm. This chapter aims to start a debate on how Shared Space tactics can lead to near-to car-free urban zones and contribute to the enhanced people-focused urban experience.

DOI: 10.4018/978-1-7998-3507-3.ch006

INTRODUCTION

The Institution for the Diffusion of Knowledge was founded back in 1828, and continued to contribute to the development of Preston and the North West England under this state until 1932, when it was named The Harris Art College. Then it further developed into The Harris College in 1952, followed by two more developments in 1973 and 1984, to become The Lancashire Polytechnic. In 1992, the latter was granted the full university status, to become The University of Central Lancashire (UCLan) (Pope & Philips, 1995). As a result of evolution and expansion throughout the years, UCLan ended up occupying various separate buildings scattered within the town, most of which are located to the North of Preston's town center, in what is common with many universities in United Kingdom and known as open city campus. In the past few years, UCLan has launched a new evolutionary project, which targets to transform the University campus' public realm; The project has been known as UCLan Masterplan.

This chapter examines the design, development, and delivery of the University of Central Lancashire's Masterplan Programme of Capital Projects in the City of Preston, England, and the impact that shared space streetscape concept design has had on the historical dominance of motor vehicles within a strategic city center interchange. The authors investigate the origins of the shared space design concept, the challenges posed in the context of an urban regeneration scheme, and the reactions of the communities and stakeholders to those opposing this specific type of development, in addition to the predicted urban impact of the project on the city.

This study mainly employed observation, relying on the authors' experience, both in practice and research; bridging across architecture, urban design/planning, and project management. The study introduced here can be perceived as an approach to establish a debate on the usefulness of adopting shared space design tactics within an urban hierarchy. The study also reflects on feedback received from public consultation sessions with different public stakeholders in relation to the UCLan Masterplan, as well as references to consultations with Ben Hamilton-Baillie, conducted back in 2017. The study refers to Preston recently completed street renovation scheme: The Fishergate Project, and explains its relation to the Masterplan. The Authors are aware that there are various similar and/or comparable cases exist in the UK and beyond, but for the scope of this study demonstration, comparisons were not being adopted as a method in this chapter.

The chapter provides a background on Preston, Fishergate Project, and it demonstrates the challenges of employing Shared Space tactics in urban development, with a focus on the Fishergate Project, for implementation in UCLan Masterplan. The chapter concludes by emphasizing how updated, and detailed formal legislative guidance in the United Kingdom (UK) and a better appreciation of cultural context from the perspective of traditional pedestrian and vehicular behaviors is paramount to the successful implementation of shared space projects in the future. It also stresses the importance of planning the shared space strategically as a part of the wider urban/city development scheme towards a 'near' car-free implementation.

BACKGROUND

The University of Central Lancashire (UCLan) Masterplan is not the first application of shared space streetscape design in Preston. The most recent and controversial highway redevelopment schemes Preston City Centre has witnessed was the first phase of the Fishergate project; located along the City's main shopping street, which was opened in October 2014. The redevelopment included the widening of existing pavements and footways to increase pedestrian footfall, removed traffic signal-controlled crossings, and significantly lowered standard kerb heights that traditionally separate and provide a level of protection between vehicular users and pedestrians. The substantial reduction of road width, from a double-lane to a narrow single-lane carriageway, made pedestrian movement from one side of the high-street to the other less challenging for the majority of users, but continues to pose difficulties for some users who have conventionally relied on, and continue to rely on traffic controlled crossings. The completed 2014 scheme removed and replaced traffic-controlled crossings with a series of regular 'courtesy crossings', an informal feature reliant on the reciprocated courtesy of both pedestrian and driver. With the single-lane carriageway still open to public transportation and taxis (not to mention the private car drivers who are either confused of the change of traffic routes, or could not adapt to the new street format) within a major section of the Fishergate street, many locals felt uncomfortable, and sometimes even unsafe, with the new streetscape; particularly pedestrians with special needs and vulnerable groups, who are not trained to share public spaces with vehicles. The achievements and frustrations of the Fishergate scheme have had a significant influence on prejudicial resistance to the UCLan Masterplan Highways design.

There is no official classification as to what constitutes a courtesy crossing, and by their very nature, a precise definition is awkward; given there is no public agreements on the shared format, where an individual can and will choose to cross the highway at any juncture that suits their needs, regardless of the provision of formal crossings, and principally based on a conscious, or subconscious, desire to traverse the shortest conceivable route. Combined with the timeliness of opportunity via visual observation of surrounding vehicular movements, crossing the carriageway at any given point in time to avoid preventable interaction with vehicular and bicycle users, in theory, should provide pedestrians with superior connectivity and accessibility to a greater array of pedestrian-friendly footway areas, relying on drivers to allow pedestrians to cross the road safely, 'out of courtesy' (Jones & Di Guardo, 2019). Kaparias et al. (2015) noted that the shared space concept was received by mixed reactions, and they classified users into opponents, (particularly some elderly and disabled road users) who perceive the space as 'less safe', and proponents, who considered the introduced ambiguity as contributing to the improvement of road safety, as both drivers and pedestrians become more watchful.

Whilst the adoption of courtesy crossings and other shared space concepts along Fishergate made sense from both an aesthetic a socio-economic regeneration perspective, it was obvious to any observer how collective awareness of street users who have been programmed for years to streetscapes that separate between pedestrians and vehicles, made it difficult for some to appreciate how the introduction of informal courtesy crossings along an 800metre stretch of perpendicular highway with no significant urban obstacles or friction could foster an environment conducive to the safe crossing of pedestrians.

Every street represents a balance between movement (the capacity to accommodate through traffic) and a sense of place (the quality which makes a street somewhere to visit and spend time in, rather than to pass through). Shared space is a way of enhancing a street's sense of place while maintaining its ability to accommodate vehicular movement (DfT, 2011)

By observation, the accessibility of the Fishergate scheme appears to work best at peak commuter and shopping times when the carriageway is most congested, with vehicle speeds decreasing to 5mph and under. This level of congestion offers multiple opportunities to cross the carriageway at various marked and unmarked junctures based on natural individual desire lines and 'safe' opportunity. On the other hand, the street occasionally witnesses vehicles travelling in excess of 40-50mph during evenings, which stresses the importance of public awareness and individual responsibility as key elements to support adopting the concept of public shared spaces. There remains contrasting arguments and evidence as to the comparative safety of this stretch of highway as per its current and historical design.

In advance of divulging Masterplan proposals and adopting the same shared space concept design across the Adelphi Quarter, UCLan sought to work in conjunction with the Design Team responsible for the Fishergate Project, including the Local Highway Authority (Lancashire County Council) and their incumbent design consultants, (Planit IE). The theoretical proposals for the surrounding Masterplan highways scheme were already in place and were primarily developed as means of framing a new large, relatively flat, and unencumbered public realm events space to be delivered as part of UCLan's £57m Student Centre & University Squares

(SCUS) Project. The SCUS Project forms an integral component of the University's Masterplan to deliver a one-stop-shop for all frontline student and visitor inquiries and to provide civic outdoor space for the community to enjoy a variety of events of differing scale throughout the year.

With a university city-campus, UCLan aimed, through the new Masterplan, to reduce traffic density crossing through the central city campus area, while enhancing the experiences of both pedestrians and cyclists. The Masterplan Highways proposals adopted similar design strategies of those conducted at the Fishergate scheme, yet accommodating two-way traffic carriageway, rather than one-way; yet, significantly narrowed road, with ultra-low kerbs, tactile paving and a large number of informal 'courtesy crossings' to be identified only by a change in material colour and layout/pattern. Detailed design proposals had not been fully developed and were subject to public consultation, however the scheme had already attracted the attention of national campaign groups whose primary concern was based around the difficulty and inability of blind and visually impaired users to cross the road at any point other than via a traffic signal controlled crossing, a congenital provision across the UK that does not predominantly assist in diminishing traffic congestion within busy city center locations. Despite being a radically different scheme to Fishergate in terms of contextual setting, it was expected that the project to be challenged by the public perception resulted from the community experience of the latter scheme, which has caused public/political controversy.

The introduction of shared space in the Adelphi Quarter aimed to facilitate the retention of key vehicular movements whilst encouraging the greater prominence of non-vehicle use across the central campus area. The assumption being this would reduce vehicular dominance on key streets surrounding campus by introducing extended pedestrian and cycle links, generating a greater sense of place within the areas within and in-between the University campus. To facilitate access between the SCUS scheme and the integrated Highways scheme, this was primarily to be achieved through the introduction of new informal courtesy crossings, supplemented by other design elements aimed at introducing vehicular friction such as roundels, and pedestrian/cyclist pause points e.g. median strips.

Understanding the Challenges of Shared Space Design

Courtesy crossings must take into account the comfort and safety of all road users but especially pedestrians, given that the car driver is under no obligation to stop and give priority to the pedestrian (Jones & Di Guardo, 2019)

Shared spaces are town streets where kerbs, crossings and safe walking routes have been partially or completely removed, requiring pedestrians to share the roadway

Figure 1. Transformation of the streetscape of Fishergate Street in connection to Corporation Street- between 2009 and 2015 (Source: Google Earth Pro)

> *with moving vehicles, and to negotiate their right-of-way with drivers by sight. Blind and certain other vulnerable people cannot do this and are therefore effectively excluded from these streets, which they may have safely walked for many years. (The National Federation of The Blind (UK), n.d.)*

At the turn of the century, Ben Hamilton-Baillie, one of Europe's most prominent and influential advocate of shared space design, travelled across a number of European countries to investigate the impact of urban designed Home Zones where the highway becomes an equally shared asset with improved emphasis on the sense of place and people compared to vehicular movement, concluding that "removing certainty, consistency and clarity for traffic appeared to offer significant benefits" (Hamiltion-Baillie, 2001), which helps improving safety through ambiguity (Hamilton-Baillie, 2004). He observed that the introduction of different materials and patterns across

Figure 2. The completed Fishergate Central Gateway Project, including 'courtesy crossings' (©E. Kamel)

the streetscape had dispelled the notion of the carriageway as a vehicle-only space, creating an enhanced sense of place to an extent where children were comfortable playing in-between moving vehicles. Hamilton-Baillie also highlighted that reducing traffic speeds remains the most critical factor for a successful shared space. Improvements in safety, more efficient use of space, improved traffic flows and huge improvements in the environmental quality of towns and villages were evident from the progress made in the countries concerned, yet, he emphasizes that creativity and development are needed to achieve a perception of safety for visually impaired user groups (Hamilton-Baillie, 2008).

Inclusion

The shared space concepts Hamilton-Baillie observed in Europe slowly gathered momentum within the UK, becoming significantly more recognized, established, and popular amongst the Urban Design community from 2010 onwards. Yet, there is increasing momentum and objection nationally to shared space design, specifically by blind and visually impaired (VI) user groups and supporters. *The National Federation of the Blind UK* (NFBUK) actively campaigns for the retention of traffic-controlled crossings and traditional kerb heights to be available across all towns and cities. It opposes the introduction of shared space design that seeks to erode traditional measures to segregate traffic and pedestrian interactions (NFBUK, n.d.).

The UK Department for Transport (DfT) first formally acknowledged shared space design in their 2007 Manual for Streets (DfT, 2007) noting the key aims as to:

- Encourage low vehicle speeds;
- Create an environment in which pedestrians can walk, or stop and chat, without feeling intimidated by motor traffic;
- Make it easier for people to move around; and
- Promote social interaction.

The manual acknowledged that the absence of a conventional kerb poses problems for blind and partially-sighted pedestrians; hence, it emphasized the importance of providing alternative means for visually-impaired people to navigate by at shared surface schemes (DfT, 2007). Yet, although the problem had been acknowledged, there were no evidence-based design guidelines on such new alternative design elements, were the document concluded by encouraging further research to be carried out by the *Guide Dogs* for the *Blind Association* to consider how the requirements of disabled people can be met. The need for inclusive street layout designs was supported by various bodies, including the House of Commons' Women and Equalities Committee (2017). It is worth mentioning that the UK Highway Code, which at the time of writing was last updated in August 2019, (along with all earlier versions) covers how to cross zebra, pelican, puffin, toucan, equestrian, staggered and human-controlled crossing points, but still does not address newly designed and verbally designated courtesy crossings, many of which have become commonplace through UK towns and cities over the past ten years.

UCLan Masterplan's public consultation process clearly noted opposition from national campaign groups through verbal feedback; the groups consistently complained that every transactional process between vehicle and pedestrian relied on visual eye contact between the driver and the person crossing the road. This point of view was backed up with anecdotal stories of the new Fishergate shared space scheme being a 'no go' area for many, although the Local Highway Authority carried out a visual study within the central shared space area that observed a significant number of the white cane and guide dog users safely utilizing the shared space.

The UK's *Guide Dogs for The Blind Association* have consistently stated that visual contact with drivers is an absolute necessity when crossing the highways, stating: *Pedestrians, motorists, and cyclists have to make eye contact to decide who moves first. This obviously compromises the safety, independence and confidence of people living with a vision impairment* (Guide-Dogs, 2020). Confusion and contradiction still surround this issue with the Department for Transport (DfT, 2011), acknowledging that whilst some drivers and pedestrians may infer things about the intentions of each other. Yet, there is no reliable data that eye contact is used predominantly as a means of communication between drivers and pedestrians.

In general, the achievement of most of these recommendations are perfectly valid and makes sense. Yet, there is a counter-argument, particularly raised by city authorities, that it can be incredibly difficult to fully achieve complete satisfaction, particularly when specific user groups have contrasting and contradictory requirements, and when some have legally recognized and protected characteristics compared to others whose specific and often individual requirements are not yet fully understood or recognized nationally at a constitutional level. In addition, the recommendation that controlled crossings and kerbs are non-optional is somewhat understandably delivered with dictatorial defiance and offers no attempt to open a dialogue with urban designers in trying to understand the wider aims, objectives, and potentially increasingly accessible outcomes of shared space design.

Speed Limit Signage

There is a debate regarding the purposeful removal of vehicular speed limit signage within shared spaces. Those in support of signage removal argue that the confusion and lack of confirmation of the upper-speed limit lead vehicular drivers to air on the side of caution and slow down. On the other hand, the counterargument raises concerns that this encourages drivers to behave irrationally and to drive at greater speed, given there is no visible evidence of an upper-speed limit in operation. If this argument contributes anything to urban planning practice, it mainly highlights the reliability of public space engagement on common trust. Conventionally, the trust evolves as a result of clear regulations and firm monitoring of application, including penalties for traffic violations; while the case of shared spaces, which break down the conventional publicly agreed street settings, this may require a transitional period for comprehensive public training and raising awareness.

It is also important to stress that speed limits can be controlled by several design elements other than signage; for example, the use of narrow roads, varying materials, and colors (e.g., cobblestone), or reducing the straight stretches of the roads where vehicles can build up speed.

Middle-Ground Solution

The concept of 'shared spaces' can be perceived as a middle-ground solution; meaning that it neither improves the roads for vehicle use (therefor drivers are not pleased), not it fully prioritizes pedestrians to provide a safe car-free people-focused public space (therefor pedestrians are not pleased). Although there is no doubt that projects adopted a shift towards shared space designs aimed for enhancing the public realm (and surely it did in the case of Preston), none of the street users' groups feel they have been looked after. This could easily be observed following up comments of local groups on social media; e.g., Blog Preston, where all parties were communicating their anxiety and fear of change, particularly during construction and the first few months following the project opening.

In this chapter, the authors argue that Shared spaces can do the best work within a hierarchical order of urban spaces, where shared space provide a transition between conventional road systems that support vehicles and pedestrianized public/civic urban spaces and paths. This has been confirmed by Kaparias et al. (2013), who observed a decrease in a conflict between pedestrians and cars in shared spaces, due to the decreased number of vehicles over time. Alternatively, authors here also argue that shared spaces, on the long term and if well planned, can perform a soft pressure towards car-free urban zones; for example, slowing down car traffic at shared spaces may, by time, be less appealing for drivers, who may accordingly avoid such spaces unless necessary. When locations of such zones are strategically planned, shared spaces can create poles of attractions for pedestrians, which can develop new pedestrian routes in between to connect. This will be explained further in the following section in the case of UCLan's Masterplan and its relation to Preston's Flag Market square.

UCLan Masterplan Development

Coinciding with several development schemes taking place in Preston City and enthusiasm for improving the public realm and infrastructure of the University of Central Lancashire back in 2015, UCLan launched a £200m project to develop a new masterplan that intends not just to transform its Preston campus, but that will change the urban settings of the City Centre.

It is noteworthy that Preston has always been a city keen to pioneer change and discussions remain ongoing with City partners to partially pedestrianize further areas and key arterial routes through city with a re-focus on public transport, pedestrian and cycle accessibility. The UCLan Masterplan was based on a design philosophy that the public realm had equal importance to the new buildings being constructed between the connecting spaces. The Masterplan has been subject to a Section 247 (Stopping Up) Orders to prohibit vehicular movements along St Peters Street (to become a contemplative garden), Rodney Street (enhanced surface treatment/ landscaping) and part Fylde Street (previously carriageway, now forming part of a significant civic plaza space). The noteworthy drawback in this process is the excessive costs involved in the physical disconnection, removal, and diversion of the Statutory Body apparatus, which has already cost in excess of £3million, and this remains for the known utilities identified in advance of physical excavation. There remain underground services that have not been identified on statutory records and are inconclusive as part of Ground Penetrating Radar (GPR) surveys. In conclusion, creating a new public realm in city center locations can be an expensive business.

The Masterplan is not limited to developing the streetscape, but rather includes the construction of new buildings, green spaces, class-leading technologies, and infrastructure; to showcase UCLan's key principles of openness, transparency, and accessibility, whilst developing a skilled workforce which meets local, national and international needs. This encompassed constructing the University's Engineering and Innovation Centre (EIC), a Student Centre (to perform as the campus' main gateway), students' social spaces, and the transformation and development of the Adelphi Quarter, which consequently requires major reconfiguration to the surrounding highways (Masterplan, 2017).

Figure 3. Image of the completed UCLan Masterplan Central Projects – Engineering Innovation Centre, Student Centre, Adelphi Square, & Highways Project

The Process of Making

On an informal consultation back in 2017, between Ben Hamilton-Baillie and Robert (Bob) Turner, a project manager for UCLan Masterplan and co-author of this chapter, the first stressed that one issue that might be given more emphasis in the Equality Analysis is an audit of the very poor state of the existing public realm of Preston, and its implication for people with disabilities. Although the collection of crossings and facilities allow, in theory, controlled access to/from the university's open campus, the overall effect of the highway measures creates an extremely uncomfortable and discouraging environment. Crossing busy roads will always present concerns and discomfort for pedestrians, especially those with physical or visual disabilities. However, Hamilton-Baillie advised that imposed reductions in car speed, and changes in drivers' expectations, will contribute to a significantly more forgiving and safer environment, and is likely to be a substantial improvement on the inclusiveness and accessibility of this area of Preston.

Given the levels of local criticism received from the Fishergate scheme, it was important the Masterplan Highways design team to review lessons learned from the project; hence The Masterplan Team requested this from Lancashire County Council, who consequently prepared a whitepaper covering three key learnings: Consultation, Design, & Communication, which are explained below.

Consultation

The first and most important lesson highlighted by the Lancashire County Council (LCC)'s whitepaper on the Fishergate project emphasized the importance of public engagement through public consultation sessions with stakeholders. The paper stressed the need for consultations to include as wide a coverage of community groups as possible, especially disability groups. As a result, UCLan arranged a series of consultations, which were extensive and included multiple off-campus sessions held at various locations and times to suit different public stakeholders.

Design

The whitepaper stressed the importance of adopting the New Roads and Street Works Act procedures, including the use of Section 58 and Section 85 notices to predict and co-ordinate planned utility work. The paper stressed that design should intend to minimize disturbance to the newly finished scheme(s). Consequently, UCLan's strategy adopted to address and either to remove or to re-route many statutory utilities and services in advance of the commencement of the main highways' construction contract. The design had to provide a future-proofing consideration for partners' medium and long-term plans; for example, to providing proper infrastructure suitable for future networks and energy expansions, in order to minimize disturbance of finished scheme. UCLan adapted to provide a wide network of Joint Utility Trenching and ducting at the same time as excavation works across the various highways' construction phases.

The County Council's study on Fishergate scheme also recommended attention to be particularly paid to improve the design to better suit visually impaired users; particularly to increase the color contrast between kerb and channel in the Masterplan scheme, especially to avoid ambiguity between surfaces when wet (which is the normal of Preston). LCC ensured increased and acceptable color contrast in final design proposals, but more significantly an increase in kerb height from 30mm (in Fishergate's case) to 60mm – which is criticized by the design team as a move away from the traditional shared space concept.

The specific design responses addressing the lack of connectivity include the introduction of further 3-meter-wide median strips as a resting point for both pedestrians and cyclists, along with the introduction of trees and curves to the highway to provide some visual friction for vehicular users in an attempt to reduce the speed of cars passing through the area regardless of familiarity. Kerb height was originally in line with the Fishergate scheme at 30mm. However, detractors wanted standard kerb in excess of 100mm. In addition, those opposing the scheme wanted every single courtesy crossing eradicated and replaced with a sensible number of signal-controlled crossings. Following an elongated, controversial and confusing planning committee, it was loosely agreed the issues raised should be reflected within planning conditions imposed on the development, the exact wording of which could not be agreed at planning committee and was retrospectively inserted without identifying the exact locations any new controlled crossing points – an underwhelming emotional/political response from all perspectives to a misinterpreted difficulty.

Communication

Communication was the last key highlight of LCC's whitepaper on lessons learned from Preston's Fishergate development project. The paper stressed the importance of podcasting regular updates, especially in areas experiencing disruption from active works. Accordingly, UCLan employed a dedicated Masterplan Communication Officer, who worked with an external PR company to provide written and visual updates on final scheme designs, phasing, disruption, and alternative access routes for pedestrians and vehicle users. In addition, the project team was keen to keep key stakeholders engaged throughout the different project phases; hence, UCLan held multiple face-to-face meetings with wider public stakeholders and local businesses and continues to do so as the Masterplan draws to a conclusion.

For maintaining efficient communication, CGI/3D visuals of designs were extremely useful for engaging and getting press coverage. UCLan commissioned CGI final scheme images and released it to the press in October 2019 to provide

Figure 4. UCLan Masterplan Adelphi Quarter Buildings, Public Realm, and Highways overlapped against original roads layout

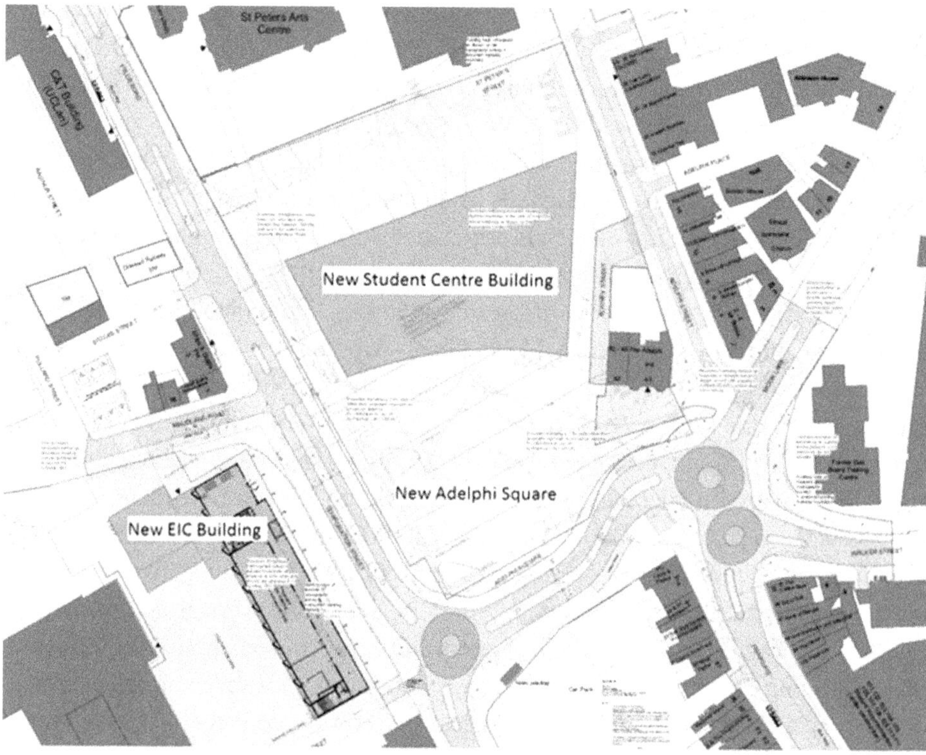

a better visual understanding of the completed scheme. UCLan was also asked to commission and subsequently provide a tactile model of the proposed highways scheme highlighting all newly proposed courtesy crossings within the scheme and all retained controlled pedestrian crossings in the immediate vicinity so blind and visually impaired consultees could kinaesthetically navigate the proposed layout.

It is important to highlight that, in terms of communication and at its most controversial peak, those involved in the scheme have decided to move away from referring to it as ‘Shared Space’ and instead re-branded the updated design as ‘Informal Street Scene’, in an attempt to reduce ongoing controversy.

Wider Urban Considerations

Towards a near Car-Free Zone

As a University, UCLan aims to improve the general settings for its city campus, so students feel safer and to develop a memorable experience throughout their study years. On the other hand, the University also has a responsibility towards Preston, being the largest employer, and occupying a large zone in the City. Accordingly, the Masterplan scheme had to engage with the wider local community to improve the City's public realm.

One of the main challenges for those working on The Masterplan has been attaining the early buy-in to the shared space concept at the highest level within the institution on the understanding the narrowing of carriageways and expanded pedestrianized public realm and footways will discourage the use of cars in this part of the city, without any real empowerment or ability to address specific design and end-user concerns. The institution has not led on any decision surrounding potential amendments to the existing highway design to accommodate specific user groups. The issue has largely been one for the Local Highway Authority (LHA) to deliberate and to make ultimate decisions on. The LHA's Equality Impact Assessment for Highways was included as part of the planning application, and it is the LHA who owns, maintains, and takes full responsibility for the adopted highway upon handover of the works.

Reflecting on the DfT (2011) Local Transport Note on Shared Space, the case for Masterplan Shared Space in the context of creating an approximate additional 7,000sqm of public realm space remains relevant and evident: *Shared space is often applicable where the buildings fronting the street have a strong heritage or cultural significance. It is particularly suitable where the quantity and type of surrounding land-use generates a high level of pedestrian demand for uses other than simply movement through the space. Shared space can also be appropriate at junctions or squares, where pedestrian desire lines are more diverse. Such settings, where streets come together, can provide good opportunities for creating distinct focal points (Ibid).*

This chapter argues that there is a need for a broader debate about the discriminatory effects of conventional road and street design, and especially the limitations placed on vulnerable pedestrians by having to defer to the primacy of traffic flow. The fundamental problem with controlled pedestrian crossings (as with other measures such as guardrails) is that it reinforces such deference. Drivers assume a right-of-way, and each controlled crossing makes the broader street context less forgiving. The green light shown to drivers is particularly damaging in this respect. Any error by either driver or pedestrian can lead to more severe incidents. It is noteworthy that, for the last five years of records, over 80% of pedestrian fatalities and severe injuries

Figure 5. Adelphi Quarter under construction- a sketch of new carriageway layout (©E. Kamel)

were located at controlled crossing points. Lights do not make crossings safer; they create the illusion of safety. This is particularly important in contexts such as the University campus, where multiple crossings will occur and be necessary. It will never be a realistic prospect to install signals at every side road, or every crossing point. There is, therefore, a difficult and important balance to be struck between the installation of crossings and the need to change driver expectations over the broad area, to reduce speeds, and to create a more forgiving and less discriminatory streetscape.

Until the Masterplan Highways scheme is fully complete and in active use (due August 2021 at the time of writing), it is difficult to predict whether or not design compromises offer a better overall solution in comparison to the original scheme with 100% courtesy crossings, or indeed the ideal scenario for blind and VI users with 100% traffic signal-controlled crossings. But Since the construction of the project has started, and for several months (up to the time of writing this chapter), the university campus area of the city has already been witnessing road closure and traffic diversion. The area suffered traffic jams for the first few weeks until drivers started 'avoiding' the route crossing the construction site, and obviously finding alternative routes. Even for university staff and students who regularly drive in, UCLan provided new large parking zones on the campus' boundaries, which helped to reduce their need to drive through. On the other hand, the University has been encouraging green commuting alternatives, such as cycling, public transportation, and shared car parking, through various supporting schemes. This all support the prediction of reduced traffic density when the project is completed.

Power to Reform Street Typology

Preston is a well-known ancient market town in North West England, where its roots can even be traced back to the Romans but were more known for housing a guild merchant, the right that has been granted by King Henry II. The Town was granted city status as part of the Queen's jubilee celebrations in 2002, yet the relatively new city is characterized by keeping its town scale was.

Although Preston is known for being the house of seven historic parks, three of which are Grade II listed, this is in addition to Winckley Square (a central pocket garden), and Preston Docks, which all provide breath for the local community, yet the City has only one civic center; the Flag Market Square, overlooked by Lancashire's most famous building, The Harris (Museum, Library, and Art Gallery), in addition to the City Council building and the Old Post Office (currently being renovated into a hotel). The Flag Market Square, tangent to Fishergate road, the main (and only) Highstreet in Preston, was connected down to the City's train station through the recent Fishergate scheme discussed in this chapter. Historically, the square was connected to the rest of Preston 'Town' via a tramline passing through Friargate road (one of the few remaining roads of the old town's urban grain) (See Figure 6).

Figure 6. Historical map of 1824 Preston, prior the construction of The Harris Museum; showing Fishergate and Friargate connecting at the Market Place (now, Flag Market Square) _ [Source: Preston Digital Archive]

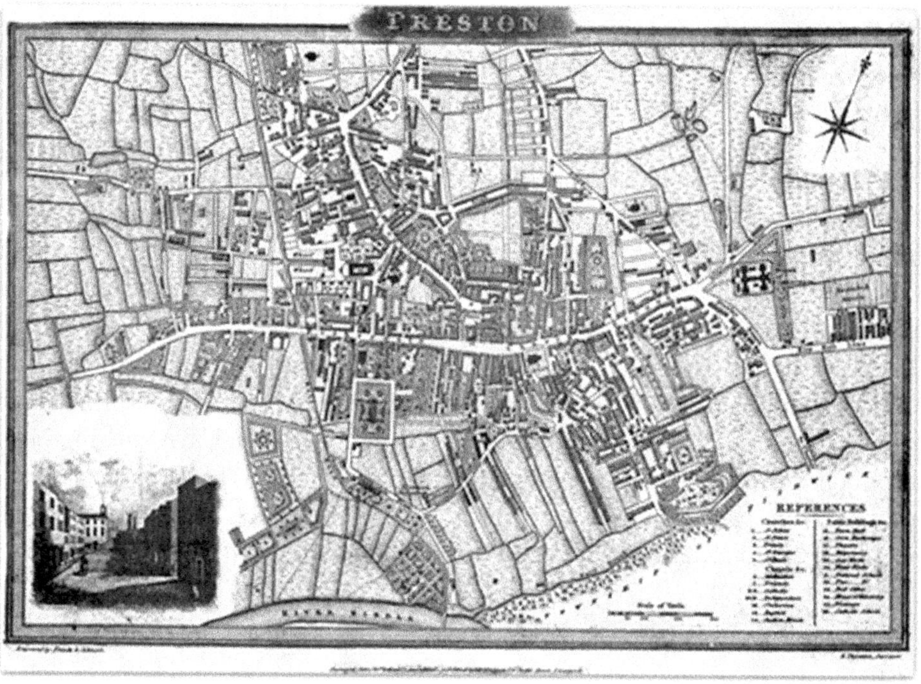

At present, all civic activities and public celebrations in Preston (e.g., Christmas lights, Christmas Market, New Year, Preston Guild, Preston Caribbean festival, etc.) take place at the Market place. On the other hand, a Ringway currently cuts through the Friargate, which has been interrupting the historical connection for several years now.

The significantly wider footway in front of the adjacent Masterplan EIC building combined with the new Adelphi Square opposite two narrow single-lane carriageways with a 3metre wide median strip should improve the current level of pedestrian comfort and allow for ease of movement. Fundamentally, the Masterplan Highways Project, in one sense, is simply an enabling development to facilitate the delivery of one of the UK's biggest new public spaces in the past ten years. The legacy of UCLan's investment in the public realm and adopted highway surrounding the central campus will create a new type of public realm space within a highway-dominated urban setting. The reconfiguration of the road structure will help rebalance the street scene in Preston and will improve accessibility for the vast majority of pedestrian users as well as provide the City and the local community with new multi-purpose events spaces with attractive landscaped areas in which to dwell, interact and enjoy. This fact should not be undervalued, and the newly created opportunity for community, cultural, and artistic interactions and interventions should be celebrated.

This chapter raises the question of whether shared space schemes, when well planned, have the power to reform street typology softly? The UCLan Masterplan development is predicted to reduce the vehicular traffic flow at the new Adelphi Square, while turning it into a civic center. If successful, the two civic centers of the city: The Market Place, connected to The Harris Museum (cultural center), and the Adelphi Square, connected to UClan (knowledge center), would likely attract more pedestrian flow in-between. Currently, while the project is still under construction, the deteriorated Friargate has been witnessing new restaurants and cafes, in addition to students' halls projects, in preparation to receive the new civic center.

It is worth highlighting that Friargate was known for its restaurants and fast food places, yet, some of them were very unstable; simply because of the competition with similar businesses at the near-by city center. In addition, the street itself did not experience high car traffic in the near past. Yet, since the start of the Masterplan project, the road has been temporarily closed down where it connects to the Adelphi Square. This has significantly limited the street use for driving through, where currently (while writing this chapter), no cars other than residents' and shops' users utilize the street.

For the reasons explained above, it can be suggested that Friargate can easily be transformed into a pedestrian route to connect the two civic centers, while the only obstacle remains for planners to solve is where Friargate intersects with the Ringway. It is important to note here that the section of Friargate, extending from Market Place to meet the Ringway, has been used as a shared space for a few years now, where cars are only allowed to access within controlled time intervals during the day (See Figure 7). This has massively affected the traffic density within this section of the road throughout the years.

DISCUSSION

There are multifaceted challenges relating to shared space design within the UCLan Masterplan setting. The project's team must reflect on how the final Highways Project design proposals attempt to address stakeholder attitudes towards various pedestrian crossing types in an attempt to provide equilibrium in the quest for truly accessible

Figure 7. Sketch showing Friargate connecting between the Flag Market Square and Fishergate (right) and the New Adelphi Square (left), with the Ringway cutting through (©E. Kamel)

environments in which everyone feels confident and comfortable to navigate and inhabit. Cultural climates can impact on the pressure for comprehensive design solutions unintentionally affecting access; for example, the UK is seeing an increase in significant hostile vehicle mitigation measures now being formed between the boundary of carriageway and footway in urban city centers following the numerous terrorist attacks utilizing vehicles in Europe during 2016 and 2017. On Fishergate, the LHA has retrospectively installed large, unattractive concrete blocks in a post-project completion response to this threat. It is a shame such reactionary measures are required, particularly when the physical response could have been incorporated into street furniture and/or public art. Indeed, the history and culture of the place itself and the learned behaviors of inhabitants engaging the highway largely dictates the success of introducing changing public realm infrastructure and configuration, which is perhaps one reason why shared space concepts are deemed more successful in Europe than they are in the UK.

It is impossible to ignore the direct impact that Shared Space design and Informal Streetscene concepts have on the accessibility and safety of blind and visually impaired users. Evidence suggests that shared space was introduced in Preston approximately 200 years ago. Winckley Street is located directly off Fishergate, and even today, numerous shoppers and solicitors remain content traversing the cobbled street. Yet, no one appears to have seriously complained about the poor accessibility for wheelchair users nor the fact there exists no formal pavement separating pedestrians from vehicles. Maybe there is a habitual Lancastrian cultural aching for cobbled streets and a historical sagacity that has avoided contemporary resurfacing and remodeling works along Winckley Street and the introduction of structured segregation. The cobbles provide sufficient friction to reduce vehicular speed along such a narrow carriageway and combined with busy pedestrian activity; drivers have a greater sense of awareness of such activity whilst navigating the city center. But mainly, this proves that Shared Space has been functioning satisfactorily in Preston for centuries.

Figure 8. Image of Winckley Street, Preston illustrating level surface between footway and carriageway (©E. Kamel).

We must conclude with some consideration of technical advancements and how the highway of the future might incorporate greater usage of information technology and smart applications for mobile phone users. Having spoken to disabled technology organizations and external companies providing services relating to wayfinding, there are GPS and Bluetooth solutions available to help a plethora of pedestrian users reach their destination safely and efficiently. Whether or not the levels of technological intelligence are currently sophisticated enough to offer an improved level of safety compared to traditional crossing types may take another generation to prove. UCLan is currently in the process of developing a new Mobile Application with voice-enabled GPS wayfinding capability from key City Centre locations to any room on campus on an individual building a floor-by-floor basis. This should eventually have the capability to provide our blind and visually impaired students with enhanced assistance in navigating across campus, through public realm and highways, and include an option in time to choose between navigating via controlled or courtesy crossing routes dependent on sight and confidence levels. Technological developments will no doubt change the way we engage with public spaces.

It appears an impossible task to successfully change the cultural and historical behaviors of pedestrians, cyclists, and vehicular users around the globe within such a short timeframe. Common sense would dictate a natural leaning towards the introduction of an International Highway Crossing Code, yet at its most simplistic level, we cannot even globally agree which side of the road our vehicles should drive on.

Across Europe, certain disabilities appear to have greater status and consideration than others, yet there is no International Hierarchy of Accessibility Requirements. In cities such as Copenhagen in Denmark, where 'green' issues have a higher prominence, and dedicated cycling lanes are respected, the pedestrian movement appears equally subjugated for very different reasons. Why do shared spaces work better in the Netherlands or Denmark? Again, we need to consider the culture of the contextual demographic on a case-by-case basis. In terms of progressing the equality agenda, policymakers in collaboration with urban designers, planners, and general public groups may need to start asking further questions on how to design our public spaces and highways schemes to accommodate the specific needs of those without protected rights or characteristics – e.g., people with restricted growth syndrome, people with dementia, people with learning disabilities, those who have varying degrees of autism, and people with other little-understood sensory impairments.

The Chartered Institution of Highways and Transportation (CIHT) in the UK has made a concerted effort at weighting and prioritizing the requirements of successful shared space schemes. However, unless adopted nationally at a political level, each public sector organization will continue to have their own priorities at a local level, typically with blind/VI accessibility issues being deemed as less vital. In the UK, the term 'accessibility' generally relates to improving access to products and services for those with specific needs and considerations. Yet, in terms of an urban design approach to the public realm, the focus is on 'accessibility for all' in the sense that improving the physical connectivity and aesthetic/functional outcomes in public space mean improving access for all user groups on an equal basis, including non-disabled users. Are Local Highway Authorities through the completion of Equality Impact Assessments and Road Safety Audits best placed to decide the most suitable way to meet the specific needs of individuals with varying degrees of confidence, skills, experience, and disability? There is a duty to consider, but not necessarily to implement design adjustments, and the difficulty remains there is no perfect solution; as an example, in balancing the accessibility needs of a wheelchair user with those of a blind Guide Dog user or a partially sighted cane user.

It is important to consider that many people feel anxious about change, they are unwilling to discuss a compromise, and they believe their specific individual needs have equal if not more important than everyone else's. One must wholeheartedly advocate making a reasonable adjustment in design to address the requirements of those with protected characteristics. However, much like the term 'shared space', the term 'reasonable adjustment' can be misinterpreted and massaged to suit a specific outcome or purpose. In a world where everyone has an increasing sense of entitlement, reactionary outrage can be instantaneously concentrated into a social media soundbite and inaptly authenticated by others, it becomes easy to create a militant approach to quickly and continually deprecate any idea or concept, and then choose to disengage from a progressive dialogue.

Technological advancement in terms of wayfinding continues to develop, however, they can be both an enabler and disabler. Mobile Phone Applications based on GPS technology and Bluetooth beacons can assist with navigation, although informal wayfinding and tactile cues within an uncluttered public realm design should remain intuitive. The emerging development of electric vehicles will make crossing increasingly difficult for the blind where shared space philosophy dominates. The technology question has not gone unnoticed by the WEC, who asks: *How can changes to the way we create and adapt our built environment, such as building information modelling and modern methods of construction, contribute to making environments more accessible and inclusive? (House of Commons: Women and Equalities Committee, 2017).* There remain simple strategies we can consistently adopt to help mitigate these issues, such as locating traffic signal-controlled crossings and public transport interchanges such as bus stops in close proximity to one another.

What is fundamentally clear, and to all those who have been affected both positively and negatively by the redevelopment of the highway and public realm within the Preston City Centre region, is that there is no perfect answer to addressing the complex needs of individual end-users. Only by entering into meaningful, balanced dialogue, embracing compromise, and through the production of composed, objective, evidence-based reports where both designers and users are consulted at the same time, will we stand any chance of influencing policymakers and future legislation to ensure our public realm remains safe, legible and as accessible as possible to all users, regardless of background. When negotiating unfamiliar urban territories, confusion is often temporary, and for urban designers to assume driver behavior, confidence and comfort will remain in equilibrium as familiarity increases, is a dangerous assumption to make.

There is a flawed presumption that making reasonable adjustments does not have any further detrimental impact on general users, or indeed other specific user-groups with protected characteristics. Whether or not the UCLan Masterplan deliverables fully achieve all their original aims and objectives is yet to be realized. In due course and through post-occupancy evaluation, we hope to be in a position to measure this. The Authors remain confident that the Masterplan development scheme will initiate a new healthy public debate and will provide a renewed sense of community reconnecting the City with the University. Having the new civic center UCLan introduces to the City sitting strategically in relation to Preston's only public square, is highly predicted to develop enough attraction between both to improve walkability in-between the culture center and knowledge center; the question here is: will Preston authorities use this development opportunity?

REFERENCES

DfT. (2007). *Manual for Streets. D. f. Transport*. Thomas Telford Publishing.

DfT. (2011). *Local Transport Note 1/11 - Shared Space*. TSO (The Stationery Office). Retrieved from https://assets.publishing.service.gov.uk/government/uploads/system/uploads/attachment_data/file/732739/ltn-1-11.pdf

Guide-Dogs. (2020). *Streets Ahead*. Retrieved 01/01, 2020, from https://www.guidedogs.org.uk/how-you-can-help/campaigning/our-current-campaigns/streets-ahead#shared-surfaces

Hamilton-Baillie, B. (2001). *Home Zones - Reconciling People, Places and Transport Study Tour of Denmark, Germany, Holland and Sweden - July to August 2000*. Academic Press.

Hamilton-Baillie, B. (2004). Urban design: Why don't we do it in the road? Modifying traffic behavior through legible urban design. *Journal of Urban Technology*, *11*(1), 43–62. doi:10.1080/1063073042000341970

Hamilton-Baillie, B. (2008). Shared Space: Reconciling People, Places and Traffic. *Built Environment, 34*(2), 161-181.

Jones, P., & Di Guardo, G. (2019). Analysing pedestrian and vehicle interaction at courtesy crossings. In *The 17th Annual Transport Practitioners' Meeting*. PTRC.

Kaparias, I., Bell, M. G. H., Biagioli, T., Bellezza, L., & Mount, B. (2015). Behavioural analysis of interactions between pedestrians and vehicles in street designs with elements of shared space. *Transportation Research Part F: Traffic Psychology and Behaviour, 30*, 115–127. doi:10.1016/j.trf.2015.02.009

Kaparias, I., Bell, M. G. H., Dong, W., Sastrawinata, A., Singh, A., Wang, X., & Mount, B. (2013). Analysis of pedestrian-vehicle traffic conflicts in street designs with elements of shared space. *Transportation Research Record: Journal of the Transportation Research Board, 2393*(1), 21–30. doi:10.3141/2393-03

Masterplan, U. (2017). *UCLan Masterplan*. Retrieved 02/03, 2020, from https://uclanmasterplan.co.uk/

NFBUK. (n.d.). *Shared Spaces*. Retrieved 04/03, 2020, from https://www.nfbuk.org/campaign/shared-spaces/

Pope, R., & Phillips, K. (1995). *University of Central Lancashire A History of the Development of the Institution since 1828*. University of Central Lancashire.

Women and Equality Committee. (2017). *Building for Equality: Disability and the Built Environment- Ninth Report of Session 2016–17. Online, House of Commons*. Retrieved 03/03, 2020, from https://publications.parliament.uk/pa/cm201617/cmselect/cmwomeq/631/631.pdf

Chapter 7
Roads to Car-Free Cities:
Introducing the Three-Dimensional Balanced Transport Intervention Ladder

Margareta Friman
https://orcid.org/0000-0002-7475-680X
Karlstad University, Sweden

Lars E. Olsson
https://orcid.org/0000-0002-6570-6181
Karlstad University, Sweden

ABSTRACT

Motorized transport has been around for over a century and has benefited people in various ways. As awareness has increased of the negative effects of car use, efforts to reduce pollution, congestion, noise, and accidents have increased. Some cities have taken drastic measures to reduce the number of cars. The starting point of this chapter is a balanced intervention ladder that includes interventions that can either increase or decrease autonomy. The authors introduce the "three-dimensional balanced intervention ladder" as a framework that can be used to describe autonomy relating to reduced car use, balancing this against perceived accessibility and wellbeing. The consequences of travel mode changes have been substantially explored; however, the consequences with respect to accessibility and wellbeing in life have only recently been recognized. By reviewing current research, they identify knowledge gaps in the implementation of balanced interventions and make recommendations regarding the continued development based on autonomy, perceived accessibility, and wellbeing.

DOI: 10.4018/978-1-7998-3507-3.ch007

INTRODUCTION

In several European cities, various interventions to reduce car traffic have either been adopted and implemented or are planned to be implemented in the near future. Stated reasons often relate to environmental and health issues (such as reduced pollution or noise), or to improving safety by decreasing congestion. The types of interventions implemented so far are varied. In Madrid (Spain), petrol vehicles manufactured before 2000, and diesel vehicles manufactured before 2006, have been banned. In Paris (France), vehicles manufactured before 1997 are banned on weekdays and a car-free day, the first Sunday of each month has been introduced. In Brussels (Belgium), a surveillance system has been implemented that enables the identification of diesel vehicles, which are then fined when they enter the city center. Urban road pricing schemes have been designed and implemented in Europe, e.g., London (2003), Stockholm (2007), and Milan (2008), resulting in a modal shift in car drivers toward public transport equalling 10-13% (Croci, 2016). In Oslo (Norway), several interventions have been implemented to restrict car use, including limiting the number of parking spaces in the city center, limiting the road space available to vehicles, and building pedestrian and bicycle lanes. In Copenhagen (Denmark), there has been investment in constructing superbike lanes. Other cities have invested in free public transport (Tallinn in Estonia), with mixed results when it comes to decreasing car use. Cats, Susilo, and Reimal (2017) showed that the proportion of car users decreased by 5% following the implementation of free public transport, but that the average distance travelled by a car simultaneously increased, leading to a 31% increase in total vehicle km. The bus rapid transit (BRT) system is another solution for reducing the need to travel by car. A BRT system is a high-quality bus-based public transport service that complies with an objective quality standard, including off-board payment, segregated bus lanes, platform-level boarding, integration with bicycle-sharing systems, and minimal bus emissions. Implementing BRT systems has, however, shown large variations in car use reduction, ranging from 4% to 50% and indicating that contextual factors are of importance (Ingvardson & Nielsen, 2018).

This chapter aims to discuss interventions aimed at reduced car use in relation to people's autonomy (freedom of choice and self-determination). We also extend our discussion to include potential effects on travellers' perceptions of accessibility, satisfaction, and wellbeing after (voluntary or less voluntary) travel mode changes, the knowledge that is mostly lacking. In doing so, we review current knowledge and elaborate upon balancing the potential consequences of interventions for car-free cities for autonomy, accessibility, and wellbeing. The chapter ends with a discussion on the implications for research and policymaking.

CAR USE INTERVENTIONS AND AUTONOMY

Implementing interventions aimed at restricting car use is not straightforward, something which can be explained by the popularity of the car. The majority of all personal travel is motivated by desires, needs, and obligations as regards participating in out-of-home activities (e.g., Axhausen & Gärling, 1992). Therefore, as has been proposed (Gärling et al., 2002), car use reduction should broadly be viewed as an adaptation on the part of car users that has potential consequences for these users' engagement in different activities, and their satisfaction with this engagement. Out-of-home activities are carried out in a spatially-organized environment which is adapted to the car, and has been so ever since the acceleration of automobile mass production. When moving goods, chauffeuring, protecting against accidents, or participating in different activities during the day, the car is usually the best option. Hence, the car has instrumental value insofar as it is a means toward some other end. The car provides support and is a necessity for mobility when, for example, health is an issue or when people have a busy schedule. For many people, the car also symbolizes freedom and self-determination. People can travel wherever they want, whenever they want, and with whomever they want. In this way, the car is also associated with emotional values related to aesthetics, freedom, and sociability (Jakobsson Bergstad et al., 2011; Steg, 2005).

Decision-makers can implement various interventions to limit or control people's travel behaviour. When choosing an intervention that infringes upon people autonomy in different ways, urban and transport planners should choose the least restrictive alternative. This is justifiable on the basis that people's freedom and autonomy should always be put first by a democratic society. It is in the exercising of autonomy that people maintain some control over, and take responsibility for, their lives. This is also wise considering the fact that reduced car use may restrict access to desirable places and activities, something which, in the longer term, can have negative consequences for people's quality of life and wellbeing.

For many years, a number of different interventions or measures have been available for reducing car use. In the literature, these have come under the heading "Travel Demand Management" (e.g., Kitamura et al., 1997), and have been divided up into more or less coercive groups and labelled as "hard" and "soft" measures. Hard measures include infrastructure improvements, making it costlier to own and drive a car, or restricting cars to certain times of the day or certain zones of a city. These types of interventions are mostly difficult to implement because of public opposition or political infeasibility (Gärling & Schuitema, 2007); generally, however, they have a significant effect on people's travel behaviours. Soft measures, on the other hand, entail a higher degree of voluntarism and have been referred to in the literature as mobility management tools (Cairns et al., 2008). Soft measures are aimed at motivating individuals into voluntarily switching from cars to more sustainable travel modes. Such interventions often include customized information, incentives, or customized feedback. A review of the effects of soft policy measures concludes that there is a great variation in the effects of influencing people's travel behaviours (Richter et al., 2010).

In this chapter, we will apply Griffiths and West's (2015) "balanced intervention ladder" to elaborate on how interventions for reduced car use interfere with the individual's autonomy. It is emphasized that the significance of interventions to the individual's autonomy is highly sophisticated and varies from eliminating his/her travel mode choices to increasing his/her autonomy by enabling more options. Some interventions are "autonomy-neutral" in that they neither increase nor decrease travel mode choices, instead of changing the choice architecture (Thaler & Sunstein, 2009). From an ethical and moral perspective, the choice of intervention should be one that comes with as few restrictions as possible regarding the individual's freedom and right to decide over his/her own mobility. This perspective forms the basis of Griffiths and West's (2015) further development of a popular intervention ladder (focusing on public health, and originally presented in 2007 by an independent body, The Nuffield Council on Bioethics, based in London). "The balanced intervention ladder" puts the individual's autonomy center stage. In the context of travel, above all, a balanced intervention ladder would emphasize that autonomy does not need to be reduced in car-free cities.

On the contrary, this ladder emphasizes, like Griffiths and West (2015), that reflective consent regarding a particular intervention can instead increase the individual's sense of freedom and self-determination. In applying Griffiths and West's (2015) balanced intervention ladder, we aim to use a new lens to view how different interventions in the transport sector can to varying degrees increase or decrease the individual's autonomy, and how this may relate to two important dimensions for the individual traveller, i.e., perceived accessibility and subjective wellbeing. Next, we initially apply the balanced intervention ladder to the context of travel and then introduce "The Three-Dimensional Balanced Transport Intervention Ladder", where perceived accessibility and subjective wellbeing are added as dimensions to balance against.

Applying the Balanced Intervention Ladder to Reduce car Dependency

A key dilemma lies in identifying effective interventions that reduce car use, interventions which are also acceptable to transport professionals, governments, and citizens, and which, in addition, do not lead to negative consequences for the quality of life of those exposed to the intervention. A recent example of what can happen when there is disagreement is the "Yellow Vest movement", which began in France in October 2018 and then spread to other countries and regions (e.g., Canada, Croatia, Egypt, Sweden, and Finland). The Yellow Vests gather hundreds of thousands of citizens who call for lower fuel taxes. The rise of this movement can partly be explained by a decreased level of autonomy in people's everyday lives as fuel prices increase. The effect of decreasing choice becomes particularly severe in regions with few alternatives to the car, such as in northern Sweden. Perhaps transport professionals in particular need to more explicitly consider the ethics and values surrounding interventions, and to develop more sophisticated approaches starting from a balanced intervention ladder.

Normally, interventions for reduced car use have negative or positive consequences for people's freedom to choose how or when to travel. In Griffiths and West's (2015) balanced intervention ladder, it is though argued that, although often being the case, freedom and self-determination do not always need to be reduced in interventions. Their ladder visualizes outcomes from interventions where negative consequences for freedom of choice are at the bottom (to the left of the middle point, denoted with negative signs), neutral consequences at the middle (denoted as 0), and positive consequences located to the right of the middle of the ladder (denoted with positive signs).

If we apply the balanced intervention ladder as an intervention ladder for reduced car dependency in car-free cities (Figure 1), a neutral point of the ladder would thus be in the middle of the ladder with no impact on the individual's freedom. At this center stage, either no action is taken to reduce car use, or nudging techniques may be applied. Nudging is a popular tool for changing people's behaviours by simplifying and changing the choice architecture but does not cause any impact on people's autonomy and self-determination. For example, signs painted on the asphalt showing the way to a bike park, or painting bike lanes to make them more visible, do not impact on the freedom to choose where to park or where to bike. Thus, guiding choices through changes to or reinforcements of the environment is considered to have zero (neutral) impact on the individual's freedom.

On the positive side of the ladder, providing the information is indicated in the first step. One intervention of this type is personalized travel planning (also called "Travel Blending"), where information and support are tailored directly to people's travel needs. With personalized travel planning, it is possible to sway people into thinking about the way they travel while simultaneously informing them of the benefits of other travel modes. This type of intervention is considered to give people greater freedom through increased knowledge.

Educate for autonomy is found in the following step. Education provides opportunities for development by increasing the individual's capabilities and motivation as regards seeing and understanding different alternatives, assessing

Figure 1. The Balanced Intervention Ladder (based upon Griffiths, & West, 2015).

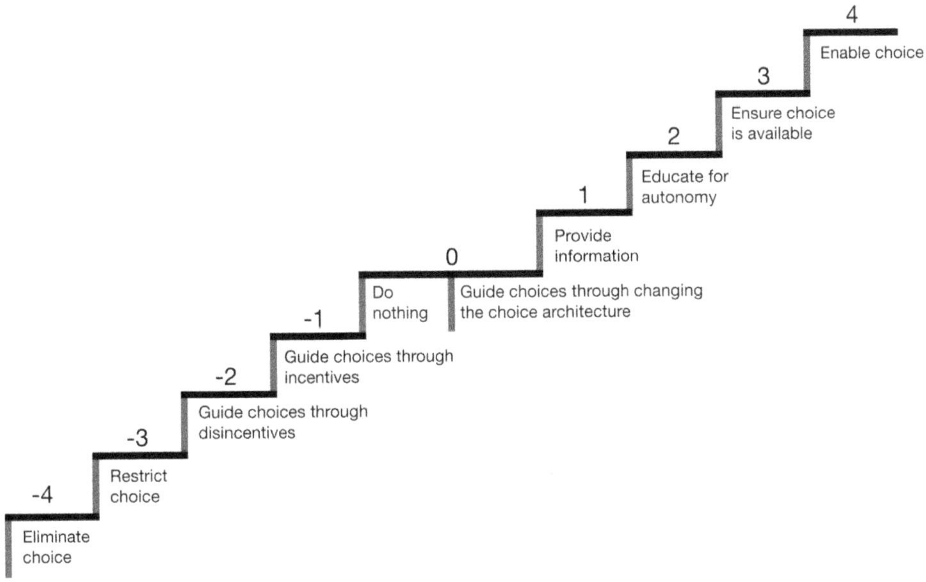

costs and benefits, and becoming more resistant to the manipulation of facts. Traffic schools for children, which include both theoretical knowledge of road signs and practical training in traffic environments, are an example of increasing the individual's possibility to plan and take control over his/her everyday mobility. Another example is coaching the elderly in wayfinding with public transportation, which may be especially important for those no longer capable of using their cars. A third example is providing transport professionals and politicians with training opportunities where they can learn more about the alternatives to the car.

Other interventions aim to ensure accessibility to all possible modes. Developments in digital technology have enabled many such interventions to be implemented in recent years. Since these interventions enable more choices rather than fewer, they are considered to increase the individual's autonomy. An example of this is Lyft (a ride-hailing company in the US), which has expanded an app to guide citizens toward finding a free e-scooter, as well as learning about the public transit system in Los Angeles. Instead of routing a Lyft car to the user's location, typing in a destination will allow users to view nearby transit routes with real-time arrival information and distance information to the nearest e-scooter. Today, trip-planning apps often provide multimodal information.

Another example is various solutions concerning Mobility-as-a-Service, which are now being introduced onto the market. In Sweden, UbiGo has been tested as a fully-integrated mobility service for everyday travel. It aims to make everyday life easier for urban households and to foster sustainable cities by offering a simple, flexible, reliable, and affordable service as an alternative to car ownership. UbiGo functions like a flexible mobile phone account. The service combines public transport, car sharing, a rental car service, taxis, and a bicycle system - all in one smartphone app, all on one invoice, with 24/7 support and bonus points for sustainable transport choices.

At the top of the ladder, some interventions help the individual to achieve personal goals. Such goals become reference values in feedback loops that regulate changes in behaviour. After having set a car use reduction goal, individuals form a plan of how to achieve the goal and make commitments to executing the formulated plan. This process is referred to as the formation of implementation intentions (Gärling & Fujii, 2002) and is influenced by the anticipated consequences of changing current travel behaviours, something which is discussed in the next section. Immediate and clear feedback, as well as motivational support, are essential at this stage. A system for changing current car use, which does not provide such feedback or support, is likely to fail. It is also possible that a goal is reduced or given up if its costs increase too much (Gärling & Friman, 2012). This stage is more demanding and requires the consent and engagement of the participants. The effect of a successful intervention at this stage can, however, be very effective. As an example, Taniguchi et al. (2007) showed that public transport use increased by 76% during interventions in Japan, which directly requested goal setting, compared to 25% with no request for goal setting.

Negative consequences for freedom follow on from interventions located to the left of the starting position on the ladder. On the first steps, incentives and disincentives are used to guide choice. Incentives can be viewed as only slightly restricting the individual's freedom to make travel choices (e.g., electric bike premium) as using the incentive, or not, is still considered a choice. At the same time, such incentives may not be affordable to all. Disincentives (e.g., road charges, kilometre tax) have a stronger effect on freedom and autonomy, but they still enable some freedom to choose, especially so for those able to afford the increased cost. For those with weaker finances, disincentives may severely affect their autonomy if no alternative travel options are available.

Interventions for stronger restrictions to, or the total elimination of, car travel (e.g., road closures, car-free zones), are located at the bottom of the ladder. Laws can also restrict the number of options available to people, for instance, the banning of e-scooters or the regulation of parking norms when building new apartments. A total ban on the car is an intervention that completely eliminates choice as regards owning or driving a car, being the type of intervention that infringes upon autonomy the strongest. Although such interventions are at the bottom of the ladder, they treat all people alike (rich or poor), and thus may sometimes have greater acceptance among the public as they are perceived as fair. Fairness has indeed been found to be positively related to the acceptance of interventions infringing upon freedom when the revenues from such infringements are perceived to be distributed fairly (e.g., Fujii et al., 2004).

When making decisions based on what type(s) of intervention(s) to implement, the emphasis should be on limiting interference with people's autonomy. The balanced intervention stage can be used to support decisions for interventions aimed at reducing car use, bearing people's autonomy in mind. Arguments for a specific intervention should include valid conclusions and reasons based on facts or ethical principles that justify a specific decision. However, a good argument should also include several other reasons, for instance, principles of health, economics, the environment, accessibility, and subjective wellbeing.

Consequences of Reduced car Dependency

People's needs, desires, and obligations to participate in out-of-home daily activities determine car use; therefore, car use reductions should be viewed as changes most likely having consequences for car users' everyday lives. When deciding upon a type of intervention to reduce car use, it is of great importance to assess the potential effects this may have on several parameters simultaneously, for example, the efficiency of the transport system, environmental consequences, financial costs, social effects, accessibility, autonomy, travel satisfaction, and the wellbeing and quality of life of people encountering the intervention. The primary focus of this chapter is aspects related to the perceptions and experiences of individuals undergoing a travel behaviour change, and it will therefore focus on perceived accessibility and wellbeing. In the following sections, we define the concepts of perceived accessibility and subjective wellbeing, then briefly summarizing current knowledge within the transport domain. This is followed by the introduction of a three-dimensional balanced transport intervention ladder that considers people's autonomy, perceived accessibility, and wellbeing.

Perceived Accessibility

In order to take part in family and social activities, to use private and public services, to participate in leisure activities, and to gain access to work or school, a high degree of mobility is required. It is argued that nowadays we live in "a hypermobile society" (Cohen & Gössling, 2015), where accessibility to daily activities is not only important for practical and instrumental reasons, but also important for social interactions that can counteract isolation, loneliness, and exclusion, and promote meaningfulness in life (Stanley et al., 2011). Accessibility has traditionally been viewed as an objective indicator of how various destinations, for example, workplaces, health care, and grocery stores, are reachable within a certain distance or time. Objective indicators have, however, been challenged regarding their validity as they are argued to miss the individual dimension of perceptions and experiences that goes beyond what is captured objectively (Lättman et al., 2018). Perceived accessibility is a complementary measure in the form of a subjective indicator of the individual perspective. It consists of perceptions of how easy it is to access and use the built environment and the transport system, and to access activities of choice. Lättman et al. (2016) define perceived accessibility as "how easy it is to live a satisfactory life with the help of the transport system" (p.36), being based on the individual's own preferences and abilities (rather than objective references); for instance, the choice of which supermarket to go to, what time of day to travel, and options that the individual actually has knowledge of or has the capability to use (Lättman et al., 2018).

Although empirical work on the concept is in its infancy, it was discussed theoretically as far back as 1976 (Burns & Golob), and is currently high on the agenda in the policy debate (e.g., the European Commission, 2015; van der Vlugt et al., 2019). What we know from recent empirical research is that objective, and perceived accessibility differ considerably (Lättman et al., 2018) that perceived accessibility varies with age, that experiences regarding the service quality (e.g., reliability and simplicity) of the chosen transport mode relate to perceived accessibility, that those using active modes perceive accessibility to be greater than the car and public transport users do. That perceived accessibility is positively related to both travel satisfaction and life satisfaction (Lättman et al., 2016, 2020). So far, empirical evidence relates to cross-sectional data on current mode use; an overarching question, however, is whether perceived accessibility would decline with car restrictions and reduced autonomy? Currently, there is only one scenario study that touches upon this question. In the applied scenario, where the car is banned, it was found that frequent car users think they would be more negatively affected (i.e., less perceived accessibility) than less frequent users (Lättman et al., 2019). As no real intervention that restricts car use has been studied thus far, we can only make informed guesses about the possible consequences of perceived accessibility among car users.

Subjective Wellbeing

The concept of subjective wellbeing also referred to as happiness, goes way back to the philosophers of ancient Greece (McMahon, 2008). For several decades, it has been a focus of research in sociology, psychology, and economics (e.g., Diener et al., 1999; Dolan et al., 2008), but has only during the past decade been studied in the travel research domain, initiated by the theoretical model proposed by Ettema et al. (2010). Friman et al. (2018) conclude that there is much research on choice in travel, but less on how to assist people in switching to sustainable alternatives, and the consequences this may have for their wellbeing.

Subjective wellbeing is defined as subjective experiences of positive and negative affects (also referred to as emotional wellbeing), in combination with cognitive judgments of satisfaction with life as a whole (see, for example, Diener et al., 1985). For thorough reviews of the causes and correlates of subjective wellbeing, we refer the reader to Kahneman et al. (1999), and Diener et al. (1999). In general, it has been found that domain-specific satisfaction with leisure, work, and family life correlates positively with overall life satisfaction (Schimmack, 2008). In relation to the transport domain, satisfaction with travel is viewed as domain-specific wellbeing (Ettema et al., 2010) and should hence include both the affective and cognitive components. Self-report scales specifically addressing this for the transport domain have been developed and validated (e.g., Friman et al., 2013, Singleton, 2019a), finding correlations between travel, life satisfaction, and emotional wellbeing (Olsson et al., 2013; Singleton, 2019b). Correlations between travel and other kinds of domain-specific wellbeing, for instance, leisure time, social contacts, work, and work and life balance, have also been found. While we argue that the relationships between travel and life-satisfaction are valid, relevant, and important, others still call for more research before the strength and direction of these relationships can be confirmed (Chatterjee et al., 2019; De Vos, 2019).

Previous studies have found that using active modes (walking and cycling) is more satisfying than going by car that the car is superior to public transport, and that travel satisfaction is beneficial to life satisfaction. For children, the car seems, however, to be the least preferable mode (Westman et al., 2017), and there is evidence indicating that, when followed over a 10-year period of time, adult public transport users even report greater life satisfaction than car users (Martin et al., 2014). Several studies have shown that a number of different travel attributes, e.g., travel time, congestion, reliability, safety, and weather, influences travel satisfaction and that these different attributes may vary in terms of relevance between modes (see, for example, Chatterjee et al., 2019; Ettema et al., 2016). Although attributes may vary in importance for the different modes, the degree of overall travel satisfaction has been found to have a solid and positive relationship with subjective wellbeing (Friman et al., 2018; Waygood et al., 2019). Hence, creating a transport system where people are satisfied with their daily travel may be one key to creating a society where there is wellbeing.

The Three-Dimensional Balanced Transport Intervention Ladder

In order to estimate the potential effects of car use reduction interventions, with varying degrees of autonomy, it is important to understand why people have chosen the mode they currently use. Such choices may depend on a number of motives. As has previously been pointed out, car use can be explained by both instrumental and practical motives, e.g., saving time or increasing reliability, and by affective and symbolic factors, e.g., joy, independence, freedom, mastery, and prestige (Jakobsson et al., 2011; Steg, 2005). Hence, interventions aimed at restricting car use may not only lead to instrumental barriers, they may also impact upon several important psychological motives. For active modes, reasons such as proximity, health, safety, the environment, and freedom play a role (Ettema & Smajic, 2014; Heinen et al., 2010), while for public transport convenience, reliability and comfort are important (Redman et al., 2013). We also need to recognize that some people are to varying degrees captive to their travel mode choice, in that they have chosen the only mode available to them at the time of travel. For some, this may depend on a lack of alternatives in the transport system, and for some health or finances; however, for others, it may be due to (biased) beliefs, attitudes, or a lack of knowledge about alternatives. The effectiveness of interventions may thus depend on several factors.

When introducing interventions to reduce car use, there are trade-offs between different motives and outcomes that need to be considered. The degree of autonomy is at the core of the balanced intervention ladder (Griffiths & West, 2015) that was introduced at the start of the chapter (see Figure 1). We argue that this dimension is not enough to gain a complete understanding of interventions and propose to add perceived accessibility and subjective wellbeing as two important additional layers, thus creating a three-dimensional framework, as visualized in Figure 2; a framework we label *The Three-dimensional Balanced Transport Intervention Ladder*.

In the above figure, combinations of effects on perceived accessibility, wellbeing and autonomy can be placed at any point within the cubicle, giving more or less weight to each of the three layers. A major point with the three-dimensional ladder is that, although autonomy is indeed important, wellbeing and perceived accessibility are also of major importance, but these three layers may not always be easy to maximize simultaneously. To give some examples, four different scenarios are highlighted in Figure 2 using the letters A, B, C, and D. Letters A and B correspond to intervention with very highly restricted autonomy, such as prohibiting cars from city centers. In scenario A, the high negative effect on autonomy is accompanied by high negative effects on both perceived accessibility and wellbeing. This can be the case for a person who is dependent on his/her car to get to important activities, and where there are no alternatives to the car as regards living the life that the person wants to.

Figure 2. The three-dimensional balanced transport intervention ladder.

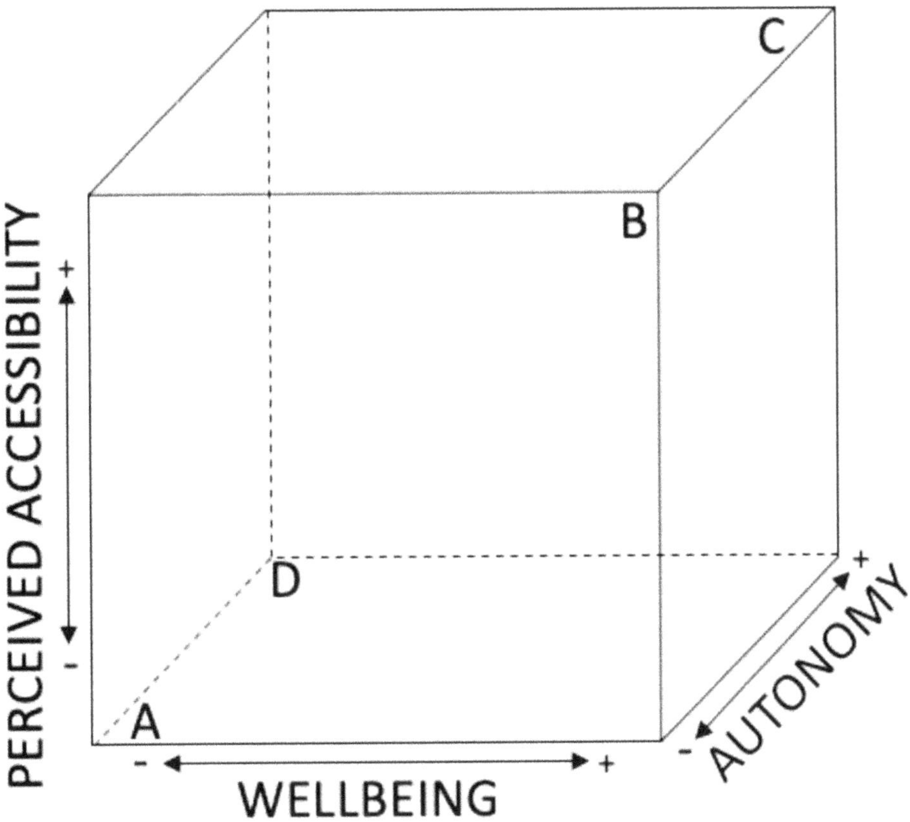

Forcing this person to reduce car use by implementing prohibitions or restrictions of car use would thus have a detrimental effect. The very same outcome may also occur for another person where alternatives objectively exist, e.g., bike lanes or public transportation, but he/she is unaware of the existence of these or has the attitude that they will never fulfil his/her needs. Scenario B can be the result of the very same intervention, in which there are negative effects on autonomy but positive ones on wellbeing and accessibility. In this scenario, the individual may have been forced to use alternative modes of transport but realizes that this works very well and brings positive experiences, leading to a high level of perceived accessibility and wellbeing.

In scenarios C and D, an intervention related to a positive degree autonomy has been introduced, for instance, a marketing campaign featuring a temporary free public transport pass with no restrictions on mode choice. In scenario C, an individual has voluntarily chosen to participate in the free trial period, evaluating it as satisfying since it fulfils the desires, needs, and obligations of that person, and thus contributing to subjective wellbeing. In this scenario, a positive effect on reduced car use may continue even after the trial period has ended. Scenario C thus achieves the most preferable outcome. However, this may not be the case for all those participating in the free test period. Scenario D represents the outcome for an individual who realizes that public transport, in its current form, cannot match his/her desires, needs, and obligations; if he/she continues to use public transport, reduced levels of perceived accessibility and wellbeing will be the outcome.

DISCUSSION, SOLUTIONS, AND RECOMMENDATIONS

There are numerous potential scenarios and outcomes regarding transport interventions that restrict car use, with more or fewer effects on autonomy, perceived accessibility, and wellbeing, that all can be placed within the three-dimensional balanced transport intervention ladder. We have chosen to elaborate on and discuss four different scenarios, visualizing that both high and low levels of autonomy may result in positive and negative outcomes as regards perceived accessibility and wellbeing. When deciding upon and designing interventions, a challenge lies in first to understanding and identifying the different scenarios, and secondly, in introducing actions that counteract potentially negative outcomes, being more or less instrumental or psychological in nature. Hence, it is not only efficiency and environmental measures (e.g., pollution) that need to be in focus, but also autonomy, accessibility, and people's wellbeing.

Based on what is known from previous research, we also conclude that there is scant knowledge of the effects of car use interventions on the degree of autonomy, perceived accessibility, and subjective wellbeing. To our knowledge, there is only one study on children where travel satisfaction, life satisfaction, and the degree of autonomy in daily travel were studied (autonomy was defined as a travel license is given by the parents referring to the child's likelihood to travel without adult supervision). A high level of autonomy gave the children more travel options since they were no longer dependent on their parents' willingness to chauffeur them to their preferred activities. In this specific study, Waygood et al. (2019) showed that children from Canada, Japan, and Sweden, aged 10-12 and with a high level of autonomy, also scored higher on travel satisfaction, which was, furthermore, positively related to life satisfaction.

People with limited travel choices often referred to as captive travellers, score low on autonomy. Although the term "captive" has been included in transport studies, few of these have related captive to perceived accessibility and wellbeing in life. A recent stream of research, however, has acknowledged the concept of "transport disadvantage", denoting the disadvantaging of a specific population group or area that results from difficulty accessing transportation and/or opportunities (Lucas, 2012). How transport disadvantage relates to wellbeing is discussed by Vella-Brodrick and Stanley (2013), who explain that the ability to access opportunities and connect with people, as well as the autonomy to travel to activity sites, enables social interactions, employment, cultural experiences, and/or other goal-fulfilling activities thus boosting people's wellbeing in a positive direction. Research on transport disadvantage is thus highly relevant and can increase our understanding of the relationship between people's (lack of) autonomy, accessibility, and wellbeing. Future studies need to relate the concept of transport disadvantage to car-free cities.

As previously noted, causing as low a negative impact as possible is a challenge, while achieving the greatest effect as regards the aim of the intervention, i.e., reduced environmental pressure or increased public health. Taking a closer look at the four scenarios presented in relation to the three-dimensional balanced transport intervention ladder, an important task is understanding whether or how it is possible to move the outcome of car-prohibition scenario A (reducing wellbeing, accessibility, and autonomy) to the outcome of scenario B (increasing wellbeing, accessibility, and autonomy), and whether or how it is possible to move the outcome of temporary free public transport scenario D to scenario C. Depending on the aim, different measures need to be taken. There may be numerous measures that accomplish this; among these, we have identified three types of actions:

(i) *Improving and offering acceptable alternatives.* Before introducing interventions with strong negative effects on autonomy, sufficient alternatives must be in place. This could, for instance, be new MaaS-solutions, infrastructure investments in public transport, or dedicated bike lanes. If it is not possible to offer sufficient alternatives, there will be two options; (a) cancelling and not restricting car use at this stage, or (b) making trade-offs between autonomy, perceived accessibility, and wellbeing.

(ii) *Increasing knowledge of the existing transport system, changing attitudes, and improving motivation to change.* This could take the form of general information campaigns such as social marketing, or tailoring information on routes and timetables to specific individuals, as well as interventions promoting individuals' capacity for self-determination. Raising awareness, by means of information or free trials, can influence social norms, attitudes, and perceived behavioural control, thus increasing the likelihood of a voluntary change in behaviour.

(iii) *Changing transportation needs and moving away from the hypermobile society.* This type of action emphasizes a reduction in the demand for mobility. It can include diverse actions such as building denser cities, introducing delivery services, or encouraging working from home. With the rapid growth in high-speed Wi-Fi connections, and the soon to be launched 5G mobile connection solutions, the need for physical meetings may decline when both social and work-related activities can be virtually organized on a larger scale.

The above-mentioned types of actions are neither exhaustive nor exclusive, and may indeed overlap each other. Nieuwenhuijsen et al. (2018) elaborate further upon this and propose nine prerequisites for a car-free city, emphasizing the need also to include retailers and the car industry in the equation. In the end, however, no car use reduction measure will work voluntarily if people are not motivated to make a change (Friman et al., 2017). This is one reason for presenting the three-dimensional transportation ladder, where interventions can either enhance or diminish autonomy, perceived accessibility, and wellbeing. More research is, however, warranted on how to enable autonomy through travel interventions. Research has consistently shown the presence of biased predictions in decision-making (e.g., Kahneman et al., 1982). Thus, people may continue to drive their cars, not only due to negative attitudes towards the alternatives but also due to biased negative predictions about future wellbeing when using other travel modes. At some point, unless voluntary measures work, politicians may thus need to implement coercive measures to a larger extent than is already the case; however, the burden of proof will then fall on car-free city advocates to demonstrate that the wellbeing and accessibility benefits of any proposed intervention outweigh the impact on individual autonomy. As current knowledge related to potential trade-offs between the layers in the three-dimensional transport intervention ladder is still in its infancy, it is though difficult to give conclusive recommendations and answers of how policy and interventions should be designed. Based on our research, it is though our conclusion that all three layers need to be considered along the road to car-free cities, and that interventions and supporting actions that increase accessibility, wellbeing and autonomy should be prioritized, taking into consideration that not all individuals are alike.

REFERENCES

Axhausen, K. W., & Gärling, T. (1992). Activity-based approaches to travel analysis: Conceptual frameworks, models, and research problems. *Transport Reviews*, *12*(4), 323–341. doi:10.1080/01441649208716826

Burns, L. D., & Golob, T. F. (1976). The role of accessibility in basic transportation choice behavior. *Transportation*, *5*(2), 175–198. doi:10.1007/BF00167272

Cairns, S., Sloman, L., Newson, C., Anable, J., Kirkbride, A., & Goodwin, P. (2008). Smarter choices: Assessing the potential to achieve traffic reduction using "soft measures". *Transport Reviews*, *28*(5), 593–618. doi:10.1080/01441640801892504

Cats, O., Susilo, Y. O., & Reimal, T. (2017). The prospects of fare-free public transport: Evidence from Tallinn. *Transportation*, *44*(5), 1083–1104. doi:10.100711116-016-9695-5

Chatterjee, K., Chng, S., Clark, B., Davis, A., De Vos, J., Ettema, D., ... Reardon, L. (2019). Commuting and wellbeing: A critical overview of the literature with implications for policy and future research. *Transport Reviews*, •••, 1–30.

Cohen, S. A., & Gössling, S. (2015). A darker side of hypermobility. Environment and Planning A. *Economy and Space*, *47*(8), 166–1679.

Croci, E. (2016). Urban road pricing: A comparative study on the experiences of London, Stockholm and Milan. *Transportation Research Procedia*, *14*, 253–262. doi:10.1016/j.trpro.2016.05.062

De Vos, J. (2019). Analysing the effect of trip satisfaction on satisfaction with the leisure activity at the destination of the trip, in relationship with life satisfaction. *Transportation*, *46*(3), 623–645. doi:10.100711116-017-9812-0

Diener, E., Emmons, R. A., Larsen, R. J., & Griffin, S. (1985). The satisfaction with life scale. *Journal of Personality Assessment*, *49*(1), 71–75. doi:10.120715327752jpa4901_13 PMID:16367493

Diener, E., Suh, E. M., Lucas, R. E., & Smith, H. L. (1999). Subjective wellbeing: Three decades of progress. *Psychological Bulletin*, *125*(2), 276–302. doi:10.1037/0033-2909.125.2.276

Dolan, P., Peasgood, T., & White, M. (2008). Do we really know what makes us happy: A review of the economic literature on the factors associated with subjective wellbeing. *Journal of Economic Psychology*, *29*(1), 94–122. doi:10.1016/j.joep.2007.09.001

Ettema, D., Gärling, T., Friman, M., & Olsson, L. E. (2016). Travel mode use, travel mode shift and subjective wellbeing: Overview of theories, empirical findings and policy implications. In D. Wang & S. He (Eds.), *Mobility, sociability and wellbeing of urban living* (pp. 129–150). Springer. doi:10.1007/978-3-662-48184-4_7

Ettema, D., Gärling, T., Olsson, L. E., & Friman, M. (2010). Out-of-home activities, daily travel, and subjective wellbeing. *Transportation Research Part A, Policy and Practice*, *44*(9), 723–732. doi:10.1016/j.tra.2010.07.005

Ettema, D., & Smajic, I. (2015). Walking, places and wellbeing. *The Geographical Journal*, *181*(2), 102–109. doi:10.1111/geoj.12065

European Commission. (2015). European Accessibility Act. Employment, Social affairs, & Inclusion. Downloaded from: https://ec.europa.eu/social/main.jsp?catId=1202&langId=en&moreDocuments=yes

Friman, M., Fujii, S., Ettema, D., Gärling, T., & Olsson, L. E. (2013). Psychometric analysis of the satisfaction with travel scale. *Transportation Research Part A, Policy and Practice*, *48*, 132–145. doi:10.1016/j.tra.2012.10.012

Friman, M., Huck, J., & Olsson, L. E. (2017). Transtheoretical Model of Change during Travel Behavior Interventions: An Integrative Review. *International Journal of Environmental Research and Public Health*, *14*(6), 581. doi:10.3390/ijerph14060581 PMID:28556810

Friman, M., Olsson, L. E., & Ettema, D. (Eds.). (2018). *Quality of Life and Daily Travel. Springer*. Elsevier. doi:10.1007/978-3-319-76623-2

Fujii, S., Gärling, T., Jakobsson, C., & Jou, R. C. (2004). A cross-country study of fairness and infringement on freedom as determinants of car owners' acceptance of road pricing. *Transportation*, *31*(3), 285–295. doi:10.1023/B:PORT.0000025395.17250.49

Gärling, T., Eek, D., Loukopoulos, P., Fujii, S., Johansson-Stenman, O., Kitamura, R., Pendyala, R., & Vilhelmson, B. (2002). A conceptual analysis of the impact of travel demand management on private car use. *Transport Policy*, *9*(1), 59–70. doi:10.1016/S0967-070X(01)00035-X

Gärling, T., & Friman, M. (2012). A Behavioral Perspective on Voluntary Reduction of Private Car Use. In B. van Wee (Ed.), *Keep moving, towards sustainable mobility* (pp. 109–134). Eleven International Publishing.

Gärling, T., & Fujii, S. (2002). Structural equation modeling of determinants of implementation intentions. *Scandinavian Journal of Psychology*, *43*, 1–8. doi:10.1111/1467-9450.00263 PMID:11885756

Gärling, T., & Schuitema, G. (2007). Travel demand management targeting reduced private car use: Effectiveness, public acceptability and political feasibility. *The Journal of Social Issues*, *63*(1), 139–153. doi:10.1111/j.1540-4560.2007.00500.x

Griffiths, P. E., & West, C. (2015). A balanced intervention ladder: Promoting autonomy through public health action. *Public Health*, *129*(8), 1092–1098. doi:10.1016/j.puhe.2015.08.007 PMID:26330372

Heinen, E., Van Wee, B., & Maat, K. (2010). Commuting by Bicycle: An Overview of the Literature. Transport Reviews: A Transnational Transdisciplinary Journal, 30, 59-96.

Ingvardson, J. B., & Nielsen, O. A. (2018). Effects of new bus and rail rapid transit systems–an international review. *Transport Reviews*, *38*(1), 96–116. doi:10.1080/01441647.2017.1301594

Jakobsson Bergstad, C., Gamble, A., Hagman, O., Polk, M., Gärling, T., Ettema, D., Friman, M., & Olsson, L. E. (2012). Influences on subjective wellbeing of affect associated with routine out-of-home activities. *Applied Research in Quality of Life*, *7*, 49–62. doi:10.100711482-011-9143-9

Jakobsson Bergstad, C., Gamble, A., Hagman, O., Polk, M., Gärling, T., & Olsson, L. E. (2011). Affective-Symbolic and Instrumental-Independence Psychological Motives Mediating Effects of Socio-Demographic Variables on Daily Car Use. *Journal of Transport Geography*, *19*(1), 33–38. doi:10.1016/j.jtrangeo.2009.11.006

Kahneman, D., Diener, E., & Schwartz, N. (1999). *Well-Being: The Foundations of Hedonic Psychology*. Russell Sage.

Kahneman, D., Slovic, P., & Tversky, A. (Eds.). (1982). *Judgment under uncertainty: Heuristics and biases*. Cambridge University Press. doi:10.1017/CBO9780511809477

Kitamura, R., Fujii, S., & Pas, E. I. (1997). Time-use data, analysis and modeling: Toward the next generation of transportation planning methodologies. *Transport Policy*, *4*(4), 225–235. doi:10.1016/S0967-070X(97)00018-8

Lättman, K., Friman, M., & Olsson, L. E. (2016). Perceived accessibility of public transport as a potential indicator of social inclusion. *Social Inclusion (Lisboa)*, *4*(3), 36–45. doi:10.17645i.v4i3.481

Lättman, K., Olsson, L. E., & Friman, M. (2018). A new approach to accessibility–Examining perceived accessibility in contrast to objectively measured accessibility in daily travel. *Research in Transportation Economics*, *69*, 501–511. doi:10.1016/j.retrec.2018.06.002

Lättman, K., Olsson, L. E., & Friman, M. (2020). Restricted car-use and perceived accessibility. *Transportation Research Part D, Transport and Environment, 78*, 102213. doi:10.1016/j.trd.2019.102213

Lättman, K., Olsson, L. E., Friman, M., & Fujii, S. (2019). Perceived Accessibility, Satisfaction with Daily Travel, and Life Satisfaction among the Elderly. *International Journal of Environmental Research and Public Health, 16*(22), 4498. doi:10.3390/ijerph16224498 PMID:31739648

Lucas, K. (2012). Transport and social exclusion: Where are we now? *Transport Policy, 20*, 105–113. doi:10.1016/j.tranpol.2012.01.013

Martin, A., Goryakin, Y., & Suhrcke, M. (2014). Does active commuting improve psychological wellbeing? Longitudinal evidence from eighteen waves of the British Household Panel Survey. *Preventive Medicine, 69*, 296–303. doi:10.1016/j.ypmed.2014.08.023 PMID:25152507

McMahon, D. M. (2008). The pursuit of happiness in history. In M. Eid & R. J. Larsen (Eds.), *The science of subjective wellbeing* (pp. 80–93). Guildford Press.

Nieuwenhuijsen, M. J., Bastiaanssen, J., Sersli, S., Waygood, E. O. D., & Khreis, H. (2018). Implementing car free cities: rationale, requirements, barriers and facilitators. In M. Nieuwenhuijsen & H. Khreis (Eds.), *Integrating Human Health into Urban and Transport Planning*. Springer.

Olsson, L. E., Gärling, T., Ettema, D., Friman, M., & Fujii, S. (2013). Happiness and Satisfaction with Work Commute. *Social Indicators Research, 111*(1), 255–263. doi:10.100711205-012-0003-2 PMID:23378683

Redman, L., Friman, M., Gärling, T., & Hartig, T. (2013). Quality attributes of public transport that attract car users: A research review. *Transport Policy, 25*, 119–127. doi:10.1016/j.tranpol.2012.11.005

Richter, J., Friman, M., & Gärling, T. (2010). Review of Implementations of Soft Transport Policy Measures. Transportation. *Theory & Application, 2*, 5–18.

Schimmack, U. (2008). The structure of subjective wellbeing. In M. Eid & R. J. Larsen (Eds.), *The science of subjective wellbeing* (pp. 97–123). Guilford Press.

Singleton, P. A. (2019a). Validating the Satisfaction with Travel Scale as a measure of hedonic subjective wellbeing for commuting in a US city. *Transportation Research Part F: Traffic Psychology and Behaviour, 60*, 399–414. doi:10.1016/j.trf.2018.10.029

Singleton, P. A. (2019b). Walking (and cycling) to wellbeing: Modal and other determinants of subjective wellbeing during the commute. *Travel Behaviour & Society*, *16*, 249–261. doi:10.1016/j.tbs.2018.02.005

Stanley, J. K., Hensher, D. A., Stanley, J. R., & Vella-Brodrick, D. (2011). Mobility, social exclusion and wellbeing: Exploring the links. *Transportation Research Part A, Policy and Practice*, *45*(8), 789–801. doi:10.1016/j.tra.2011.06.007

Steg, L. (2005). Car use: Lust and must. Instrumental, symbolic and affective motives for car use. *Transportation Research Part A, Policy and Practice*, *39*(2-3), 147–162. doi:10.1016/j.tra.2004.07.001

Taniguchi, A., Suzuki, H., & Fujii, S. (2007). Mobility management in Japan: Its development and meta-analysis of travel feedback programs. *Transportation Research Record: Journal of the Transportation Research Board*, *2021*(1), 100–109. doi:10.3141/2021-12

Thaler, R. H., & Sunstein, C. R. (2009). *Nudge: Improving decisions about health, wealth, and happiness*. Penguin.

Van der Vlugt, A. L., Curl, A., & Wittowsky, D. (2019). What about the people? Developing measures of perceived accessibility from case studies in Germany and the UK. Applied Mobilities, 1-21.

Vella-Brodrick, D. A., & Stanley, J. (2013). The significance of transport mobility in predicting wellbeing. *Transport Policy*, *29*, 236–242. doi:10.1016/j.tranpol.2013.06.005

Waygood, E. O., Friman, M., Olsson, L. E., & Mitra, R. (Eds.). (2019). *Transport and Children's Wellbeing*. Elsevier.

Waygood, E. O. D., Friman, M., Taniguchi, A., & Olsson, L. E. (2019). Children's life satisfaction and travel satisfaction: Evidence from Canada, Japan, and Sweden. *Travel Behaviour & Society*, *16*, 214–223. doi:10.1016/j.tbs.2018.04.004

Westman, J., Olsson, L. E., Gärling, T., & Friman, M. (2017). Children's travel to school: Satisfaction, current mood, and cognitive performance. *Transportation*, *44*(6), 1365–1382. doi:10.100711116-016-9705-7

Section 2
Macro Perspective

Chapter 8
Urban Topology of Car-Free Cities

Sarmada Madhulika Kone
Sri Venkateshwara College of Architecture, Hyderabad, India

ABSTRACT

New technological and automobile discoveries of the 1900s have transformed man's life. Modernism hasn't just brought institutional and sociological evolution in human society but has also brought cultural change. Industrialization and the automobile revolution have made man dependent on machines, and the influence is reflected in his ecosystem. Cities grew to accommodate automobiles, and today, car dominance is affecting the urban environment in terms of health and social interaction in urban open spaces. The car-free city is a new concept that every city has to adopt for a better tomorrow. The study focuses on different topological parameters of car-free cities and identifies different parameters to be considered while developing a conceptual framework towards a car-free urban environment.

INTRODUCTION

Cities grow according to man's needs and wants. Characteristics of a settlement explain the lifestyle of a man. Before automobile invention, settlements were compact, and everything is accessible to the mode of transportation of those days. Communities were built where walking and horse carts provide the limits to a settlement extent. A settlement pattern is influenced by the mode of transportation, which also explains its influence on cultural practices and social relations. In the late 19th century with the discovery of the petrol-fueled engine and Henry Ford's contribution to the automobile industry with his first moving assembly line for mass

DOI: 10.4018/978-1-7998-3507-3.ch008

production of an automobile has made private cars available to people in affordable ranges. A car made farther destinations accessible in a short period of time and with less human effort. After World war II, development was oriented more towards a new lifestyle guided by modernism and cities were planned to accommodate this change, the new road network was designed to connect far destinations and allow speed. This evolution of modes of transportation and its effect on expansion cities and communities has its influence on social relations and the human lifestyle as a whole (Claude Comtois, 2006).

As per the 2018 Traffic Index by TomTom across 403 cities globally in 56 countries, 75% of the cities reported an increase in traffic congestion between 2017 and 2018. Mumbai recorded 65% of extra travel time is spent stuck in traffic, Bogota with 63% extra travel time, Lima and New Delhi record 58%. Worldwide car sale in 1990 is 3.92 million; today, the car sale in 2019 is 77 million. According to WHO, the transport sector a major cause for urban air pollution, with an estimated 3.7 million premature deaths are attributed to outdoor air pollution. Poor land use and transport policies are the primary cause of climate impacts, air pollution, traffic congestion, traffic injuries, lack of physical activity, which leads to non-communicable diseases, and Noise. Not just environmental and health issues, cars do have huge social costs. J.H. Crawford says, "No technical improvement to cars can restore the vital function of streets as community social spaces: when something as dangerous and intrusive as the car rules our streets, civic life vanishes from them." Dr. Ian Walker, a specialist in traffic psychology, says that people who cannot afford a car or will not drive a car struggle to accesses the city when it is designed for motor cars as this cut off people who cannot drive particularly low-income group and older people. The objective of this chapter is to understand the topological difference between an automobile city and a car-free city; and rethink our planning models for tomorrow (TomTom, 2019).

EVOLUTION OF URBAN PLANNING MODELS

Jean-Paul Rodrigue, in his book "The geography of transport systems", explains the historical evolution of transportation in five phases. Transportation in the Pre-Industrial era, pre-1800, can be characterized as limited transport technology with non-mechanized modes of transportation, animal labor, and limited long-distance trade. In the second phase, from 1800 to 1870, the Industrial Revolution started the mechanization of transportation with a steam engine and railways. The emergence of modern transport systems from 1870 to 1920 noted the growth in international transportation and the dominance of rail transport. Introduction to tramways in urban areas and bicycles provided cheap mobility for masses. Fordist era from 1920 to 1970, the era knows for the internal combustion engine and mass production system, which have changed the mobility habits of human beings. The last phase of evolution in transportation as per Jean-Paul Rodrigue is the post-Fordist era after 1970, where information technologies are merged with transportation systems, and IT has become a tool to manage transportation systems. With reference to the evolution of transportation by Jean-Paul Rodrigue, urban planning concepts are compared in terms of the communication system, and mobility patterns worked out. Urban models explaining different concepts of a settlement, the transition in these models along the timeline are correlated with the evolution of transportation (Jean-Paul Rodrigue, 2013).

1870 – 1920: Emergence of Modern Transport System:

During the third evolution phase of transportation, railways were dominating mode of transportation. Tramways were introduced into urban cities, and bicycles provided cheap trips for the common person. The concept of Garden cities by Ebenezer Howard in 1898 is proposed for overcrowded industrial cities for a population of 32,000 in a permanent green belt, linked by canals and transit. The roads are radial with a width ranging from 120 to 420 feet for the Grand Avenue. The central park containing public buildings is surrounded by commercial land use, which is further surrounded by residential land use, and the outer circle contains industries and factories. Railroad bypasses the town, meeting the town at tangent. The access system of this concept is clearly dependent on railways. Railways provided intercity accessibility connecting the industrial zones at the periphery of the city. Central boulevards served horse carts, pedestrians and other manual mode of transit. Though the land use zoning (as shown in figure 1) in Garden cities is primarily based on industrial pollution levels, the transit system in late 19the and early 20th century i.e. railways, gave intercity accessibility and improved regional network, whereas the urban pattern was simple and land use were clearly defined in late 19th century (Howard, 2007).

1920 – 1970: Fordist Era:

Advancements in technology and invention of the internal combustion engine gave the automobile to man. Imprints of Industrialization and World War II are seen in mid-20th century cities with overpopulation and poor urban management. Great devolvement in urban planning techniques and urban planning concepts of the 20th century. Le Corbusier's Radiant cities in 1924 are proposed for industrial towns addressing the same issues Howard did with his Garden cities. His city zoning majorly focused on the zoning of industrial & residential land use.

Burgess' Concentric Model in 1925 talks about the zoning of workplace and housing based on socio-economic status. His model explains that a city grows

Table 1. Evolution of Urban planning concepts along with Evolution of transportation.

<table>
<tr><th>Urban Model</th><th>Proposed By</th><th>Year</th><th colspan="2">Evolution phase of Transportation by Jean-Paul Rodrigue</th></tr>
<tr><td>Garden Cities</td><td>Ebenezer Howard</td><td>1898</td><td>Emergence of modern transportation system</td><td>· International transportation
· Dominance in railways
· Introduction to tramways
· Bicycle provided a cheap mobility
· Increase in urban population
· Urban sprawl and specialization of economic functions</td></tr>
<tr><td>Radiant cities</td><td>Le Corbusier</td><td>1924</td><td rowspan="12">Transportation in Fordist era</td><td rowspan="12">· Invention of internal combustion engine
· Extended flexibility of movements
· Fast and inexpensive transport modes
· Mass production system
· Mass consumption of resources
· Propelled flight
· Sub-urbanization and expansion of cities</td></tr>
<tr><td>Concentric model</td><td>Burgess</td><td>1925</td></tr>
<tr><td>Broadacre city</td><td>Frank Lloyd Wright</td><td>1932</td></tr>
<tr><td>Central place theory</td><td>Walter Christaller</td><td>1933</td></tr>
<tr><td>Sector model</td><td>Hoyt</td><td>1939</td></tr>
<tr><td>Bid rent theory</td><td></td><td>1939</td></tr>
<tr><td>Multiple nuclei</td><td>Harris & Ulman</td><td>1945</td></tr>
<tr><td>The Losh model</td><td>Losh</td><td>1954</td></tr>
<tr><td>Mann model</td><td></td><td>1965</td></tr>
<tr><td>Ecological model</td><td>Murdie</td><td>1969</td></tr>
<tr><td>Mercantile model</td><td>Vance</td><td>1970</td></tr>
<tr><td></td><td></td><td></td></tr>
<tr><td>· Sustainable towns
· Compact cities
· Smart cities
· TOD</td><td colspan="2">-</td><td>Post-Fordist era</td><td>· Transportation systems merging with Information Technology
· Internet, satellite communications and wireless networks</td></tr>
</table>

outward from the central area, and upper classes live farther from the city center as they can afford to commute. Homer Hoyt modifies the Concentric Model in 1939 to account for major transportation routes. Bid rent theory explains the change in land values form the center of the city and how transit nodes, at the intersection of main roads and ring roads, increase the land value. The Central Place Theory by Walter Christaller in and Losh model explain marketing strategies with minimum centers to serve the entire population; these models explain in the regional scenario.

Figure 1. Garden city, Grand Avenue, (Howard, 2007)

MODERN TRANSPORTATION PLANNING GOALS

Cities are constantly changing with human needs, resources available, and time. Rapid urbanization has increased stress on the infrastructure system of a city. An existing access system of a city though caters a certain range of traffic volume for a certain population capacity, change in the mode of transportation has also had its influence on its capacity. The carbon emission from a unit volume of an urban area is increasing with an increase in private automobile usage. The capacity of the road network system can be increased with the help of technology to the cost of various social, environmental, and health issues. Modern transportation planning and policy aim at reducing the negative impact of the urban transport system on the ecosystem. Car dependency is one of the major problems of the urban transport system as it leads to environmental problems, traffic congestion, and the social-economic impact of the automobile on a city. City development and economic growth are directly proportional to its transit network and its efficiency to carry goods and information and people. Towards sustainable transportation planning, transport policies aim at addressing the socio-economic and environmental problems while improving the efficiency of the transport system.

TOPOLOGY OF CAR-FREE CITIES

Urban topology explains the characteristics and arrangement of urban spaces as nodes at different hierarchy and transport networks as links. An urban setup where nodes as traffic generating points connected with different transport linkages based on the mode of transit and type of traffic the node generates, for example, people, freight, and information. This section aims at understanding the urban setup, characteristics of nodes, and different parameters of links in a car-free city. Topology is a function of geographic, socio-economic and many other technological variables as they determine the urban context in terms of its geography, economic links, modes of transport, and technological advancements in transport. Every city has its topology. General characteristics of nodes, links, and network geometry of a city can be considered in understanding the topological characteristics of car-free cities.

Urban Topology

Urban space is composed of two major components, which are places and an access system connecting these places. Urban topology is about understanding the composition of urban places and the characteristics of an access system or the network connecting urban places. As nodes in a network system are referred to traffic generating parcels, the important parameters of these parcels are (i) land use of the parcel and (ii) its accessibility. The composition of the parcels refers to land use patterns and how different characteristics of linkages to these parcels depend on land use. The topological characteristics of an urban access system include various elements of urban street organization and different parameters influencing the urban access system, as given in figure 2. Technological advancements and infrastructure that enable transportation also have an impact on urban topology (Christa, Taylor, Joe, & Bettencourt, 2015).

Cities experience biological growth. Though cities are planned, change can only be controlled. A planned city after a certain period of time becomes biological, experiencing natural forces of urban growth. No city can be purely planned, and every city has a certain extent of biological character to its structure. Considering every tree structure is a simple semi-lattice structure, an urban structure resembles a semi-lattice. Various elements and hierarchy of elements in the urban transport network make it follow a complex natural order, where a simple tree structure fails to explain the complexity of an urban structure.

A topological space in the field of Geoinformatics talks about the structure of space. The topology of an urban access system is to understand the topological characteristics of urban geometry. These geometrical parameters of the urban access system explain the street spatial structure and its characteristics. The topological parameters of a surface geometry as per classification theorem for surfaces are connectedness, compactness, orientability, 2D, and boundaries of the surface. Parameters correlated to urban topology, where the above geometric topological parameters of an urban network correlated to functioning parameters are vehicular speed limits, mode of transportation, range of traffic volume capacities, surface area (Christa, Taylor, Joe, & Bettencourt, 2015).

Today, Information Technology has become a tool for everything we deal with in everyday life. We manage, run, and develop systems with their performance data. Similarly, the urban road network system does record its performance data, which explains how the network has performed in the past. The metadata of the urban road network system explains the character of the network. The metadata of the access system is correlated to its topological character.

Deconstructing Topology of Car-free Cities:

Urban topological parameters are projected to a car-free scenario where the mode of transit, speed limits, and other parameters determine the topological parameters of a car-free city. This section attempts to understand the topological aspects of a car-free city. Today's modern cities are designed for automobiles accommodating the ability to travel longer distances and greater speed. The proximity range of particular land use in terms of travel time gave a larger radius covered under distance. Man has his work, home, and play network established across cities designed for automobile-dependent life. A care-free city is where equal travel comfort levels are achieved in work, home, and play network after deserting automobiles. Based on the access network parameters connected to the topology of an urban structure, the topology of a car-free structure is deconstructed. The table below compares the parameters connected to the topology of urban structure in automobile cities and car-free cities.

As the topological characteristics of a network are general parameters that describe the geometry of a surface and topology of every city is different, these topological parameters are comparative scales that measure the geometry of a network in comparison to others.

Figure 2. Urban topology and parameters connected to topology (Christa, Taylor, Joe, & Bettencourt, 2015).

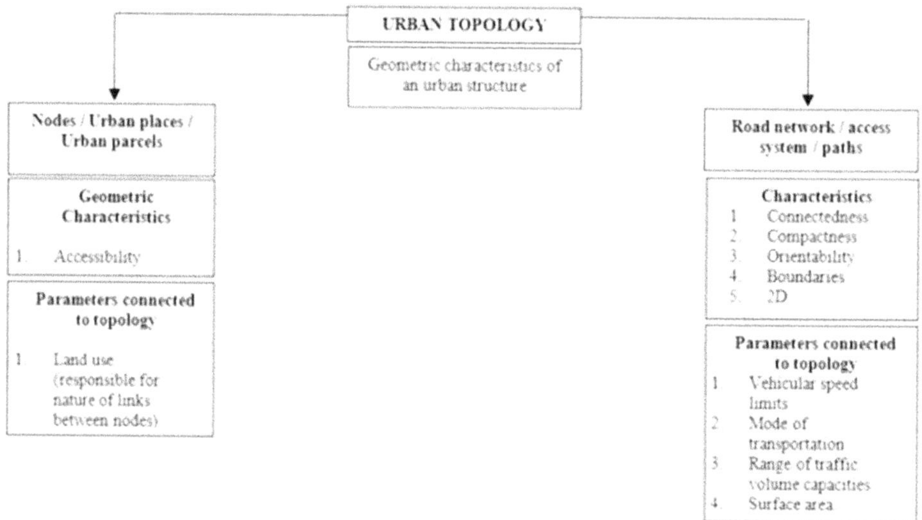

Table 2. Comparison of parameters related to urban topology between automobile and car-free cities

Parameters connected to the topology	AUTOMOBILE CITIES	CAR-FREE CITIES
Mode of transportation	Private Cars & Automobile, Public transportation	Walking, Bicycling, Public transportation
Vehicular speed limits	High – as designed for vehicular flow.	Low – as designed for pedestrians
% of vehicular roads	Every node or urban parcel has vehicular accessibility.	Urban nodes of greater hierarchy in the network are vehicular accessible through public transportation and routs of freight.
Traffic volume capacities	Roads are designed for higher capacity of traffic volume	Roads are designed for pedestrians, bicycles and public transportation.
Development	Transit-oriented	Community-oriented
Land use pattern	A complicated work, play, and home relationship	A compact network of work, play and home

Connected: A network is said to be connected when any point on the surface is connected to any other point by a path on the surface. Since every node on a transport network is accessible in an access system of a city, the degree of connectedness is compared here. The speed that linkages are designed for automobiles cannot take any path to connect the points; while in car-free cities, access system is designed for pedestrians and bicyclists, the degree of connectedness is high in car-free cities than in automobile cities.

Compactness: Compactness of a structure explains the ability to triangulate the surface geometry. The degree of compactness depends on the number of triangles in a unit surface area. Depending upon the travel time, automobile cities connect farther nodes. As a result, there are larger triangles are framed, while in car-free cities comparatively smaller triangles are formed. Thus, car-free urban structure has higher degree of compactness when compared to automobile cities. The high degree of connectedness also indicates the complex surface triangulation in a car-free network structure.

Orientability: A 2D surface is orientable if the surface has a consistent normal vector. As we are discussing the road network system on the earth's surface, the urban structure of both the automobile cities and car-free cities is orientable.

2-Dimensionality: As both the network structures have 'x' and 'y' directions on the surface and have zero movements in 'z' directions, the structures have 2-dimensionality, though a 3-dimensional space is required to frame the network structure.

The topology of the urban structure of a car-free city has a higher degree of complexity, a higher degree of connectivity, and greater in terms of details. It is a semi-lattice structure with overlapping loops of the pedestrian network, bicyclist's network, and public transit network (Zeeman, 1996).

Components of a transport system, as shown in figure 3, are Nodes, Networks, Demand, Mode of transport, Infrastructure, and Flows. Nodes constitute origins, destinations, and intermediacy. Networks can be deconstructed to links between nodes and their capacity, while demand determines the volume that the system needs to take care of and the frequency it has to adopt. Modes of transport, Infrastructure facilities available with technological advancements and flows also determine the structure of a network system. Transforming cities to car-free is not just a physical transformation; it is a socio-cultural change to lifestyle, a design that initiates a change in cultural habits and lifestyle of people using the space.

Figure 3. Components of transport system, (CE 261, 2017)

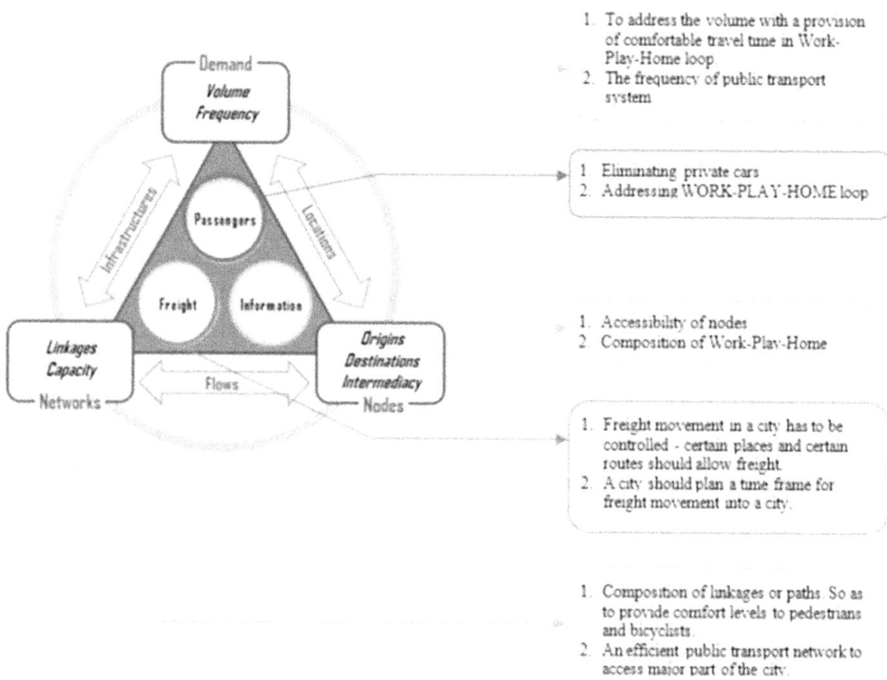

CASE STUDIES

Humanizing urban environment, where cities are revitalized with a priority to reduce private car usage and encourage public transportation, has become one of the major urban movements today in addressing urban sustainability. Car-free strategies are being implemented at different degrees in response to different urban contexts. The case study discussed below looks into different strategies the city employed to create an urban car-free environment.

Masdar City is planned as a smart eco-city for tomorrow. A sustainable community planned in the desert of Abu Dhabi (as shown in figure 4, the location of the planned city) is designed to harvest renewable energy sources. Traditional Arabian design or the vernacular style is combined with unconventional technological solutions. The city is planned for a population of 90,000, where 40,000 residents and 50.000 commuters (Goldenberg, 2016). The urban development concept of this green-tech neighborhood with is economical, social, and environmental sustainability aims to be a commercially viable and energy conscious. “The principles of one planet”, the ten-guiding principles on which the masterplan of the city is laid include zero

Figure 4. Location map of the city Masdar (Goldenberg, 2016).
Masdar, Abu Dhabi, UAE:

carbon, zero waste, sustainable transport, sustainable materials, sustainable water, sustainable culture, and heritage. The concept targets rapidly increasing population and economic growth, i.e., the project opens a wide opportunity to business from real estate to free market and covering the housing shortage of Abu Dhabi. However, the project is implemented more as a development project attracting global level investments with its clean technology and sustainable approach to living.

The land-use plan and its access system in figure 5, explain the topology of the city. With its smart urban management and algorithmized transit system, Masdar could provide a wide range of alternatives to private cars. Located 17km east of Abu Dhabi's urban center and within the proximity of Abu Dhabi's infrastructure, the city has a great potential to develop as a business center. The central business district with offices and retail is accessible through the metro connecting Downtown Abu Dhabi and Abu Dhabi international airport. Research & Development and Institutional are zoned in the North. Industrial in the Southeast and utilities on the periphery. The Personal Rapid Transit that circulates underground in electric pods helps the residents of Masdar to travel free connecting major nodes in the central business district and the metro stations. The access system connects the urban nodes at different hierarchy with a complex network of public transit across the city. Masdar provides a variety of transit systems to its citizens across the city, connecting different zones. Based on the zoning of the city with respect to its regional context, the travel demand, trip purpose, and movement pattern across the city zones are identified. Based on the trip purpose, a different hierarchy of public transit is designed connecting different zones of the city. The metro line connecting Downtown Abu Dhabi, central business zone, and Abu Dhabi international airport, while the light rail transit (LRT) runs diagonally across the city from North to the south connecting Masdar's business district to Downtown Abu Dhabi. A city with a population of 40,000 residents and

50,000 commuters is designed with a complex public transport network connecting different zones with a different mode of transportation as an alternative to private cars (Kherdeen, 2016).

Figure 5. Land use plan of the city Masdar (Kherdeen, 2016).

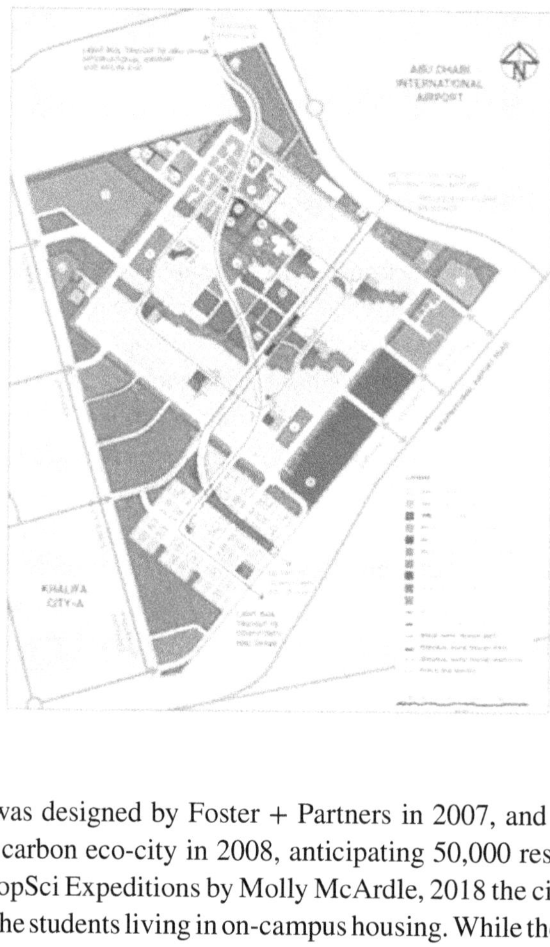

The project was designed by Foster + Partners in 2007, and UAE announced its first net-zero carbon eco-city in 2008, anticipating 50,000 residents within the decade. As per PopSci Expeditions by Molly McArdle, 2018 the city's population is 1300, including the students living in on-campus housing. While the PRT has opened two stations out of planned 100. The project of smart eco-city is more of a business experiment to attract international business with its smart urban technology. The project that has to be completed by 2016 is extended until 2030, due to the global financial crisis in the late 2000s. Today, with the pace of its development and its occupancy parentage, the city is slowly turning into the green ghost town. The 22 billion-dollar billion project has abandoned its initial idea of net-zero carbon city and is a 50% neutral carbon-free city, as the eco-city slowly aiming to become an economic project. Chris Wan, the design manager of the Masdar city projects, says that Masdar's development in the bigger picture is doing better for planet Earth and it is much more efficient", in a report by Digital trends by Lulu Chang (Kherdeen, 2016).

The 6sqkm eco-city, which employed a green transport system to reduce carbon emission, has multiple public transit options like PRT, Metro rail, GRT, LRT, Public bus routes, for private cars, shown in figure 6 how land use and public transit are planned. Today with technological developments in the auto sector and widely available of zero-emission electric cars, the use of these electric cars has been increased, and these electric cars are slowly replacing the public transit system in Masdar. The global financial crisis has disturbed the implementation of Masdar's development plan, and Masdar had to redistribute its weightage more towards economic development and social progress. Though the plan failed to design with respect to demand and feasibility of the regional context, there are certain takeaways in how zoning and transport networks are laid across the city. The geometry of the city's access system is designed around public transit is as shown in the zoning.

A SMART APPROACH TO CAR-FREE CITIES

Today rapid urbanization and growing urban regions have made people rely on motorized transit to cover a larger distance in speed. The technical improvements made car easily available to people, where the existing public transport system designed for certain population capacity fails to address the demand of increasing population and the degree of accessibility across the city. Therefore, the key to achieving a car-free urban environment is by providing people with other alternatives to cars, which can be more affordable at the same time give the same level of comfort to the commuter with the same travel time. Public transportation is the alternative to a car, designed for the following measures;

1. Larger capacity and increasing population
2. Speed and frequency
3. Comfort and degree of accessibility across the city

Cities need to scale up their infrastructure to cater to the ever-growing urban population for tomorrow. A rejuvenated public transport system for tomorrow, which can replace car and other private motorized transit and can provide access to any place in the city. A complex system that connects different urban zones and urban nodes at different hierarchy to enable access across the city and smart management to achieve the efficiency of the system. Today the technological developments like

Figure 6. Non-residential zones (work zones) and topology of public transportation connecting these zones.

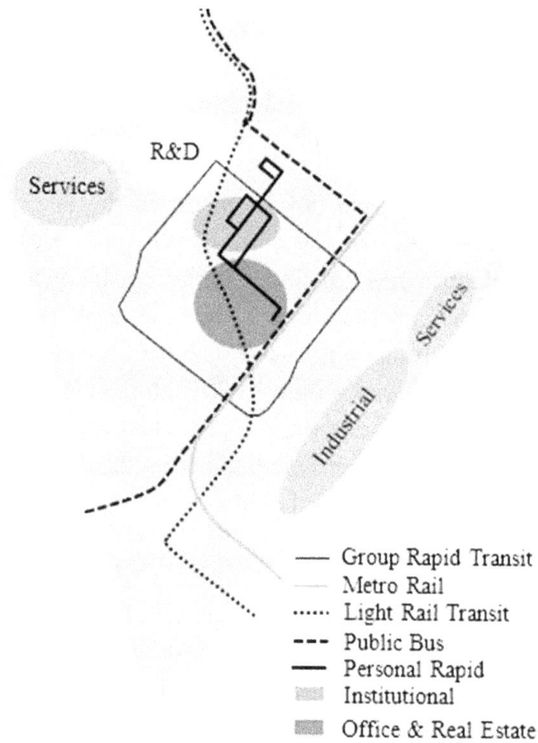

Figure 7. Classification of urban transport system.

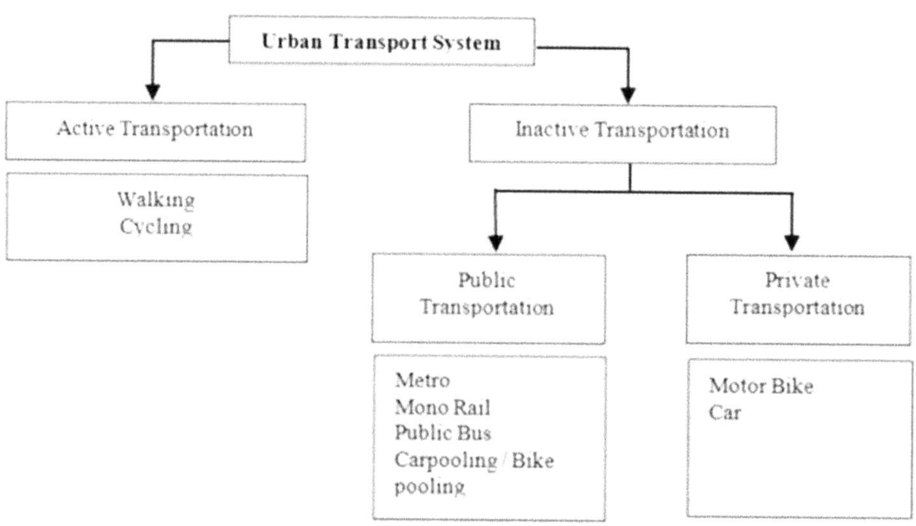

the Internet of Things and Artificial developments have made system management easy and gave cognitive abilities to systems. Real-time data can be obtained from the system, analyzed, and the system can be managed. In the era of Algorithmic city, a smart public transport system is one of the better alternatives to the car-free urban environment. As we design for tomorrow, it is important to project what commutators would want so as not to disturb the social harmony.

A CONCEPTUAL FRAMEWORK TOWARDS A CAR-FREE CITY

The conceptual framework to a car-free city outlines the vision and strategies with a sustainable approach, not just focusing on environmental and economic goals but also social cohesion. With 70% percent population estimated to live in urban areas by 2050, existing cities need a management plan to guide urban transport system towards a car-free environment. Growing cities do not just require the addition of different modes of public transportation with an increasing population but also the improvement of existing infrastructure. Today, we have different strategies that cities have been implementing to reduce the number of cars. These strategies either focus on particular zones within the city or particular days of a week, i.e., these strategies regulate the traffic towards the car-free urban environment.

Change is inevitable, and every system needs the flexibility to scale up to external forces and require a transformation plan where the existing pattern is disturbed to the least. This conceptual framework for existing cities focuses more on important aspects to be considered while designing a transportation masterplan towards a car-free urban environment. The methodology to draft the conceptual framework for existing cities has three stages, as shown in figure 8, the first stage diagnosis the city's economic, social, and environmental performance and the urban transport system. In the second stage, vision and goals for the city and urban transport system are set up, while in the third, the vision and goals set up are deconstructed to performance indicators, and they are measured to monitor the urban transport system.

Stage - I:

The stage one, diagnosis of the city to understand the urban character and how the city functions in terms of economic, environmental, and social indicators. The urban transport system is diagnosed to study its topology and its management in correlation with the urban performance indicators. Based on the urban topology and its management, an alternative to the car is identified. As the vision of this conceptual framework is a car-free urban environment, an alternative to the car is identified from either Active transportation or Public transportation. An automobile

Figure 8. Different objectives of the conceptual framework towards a Car-free city

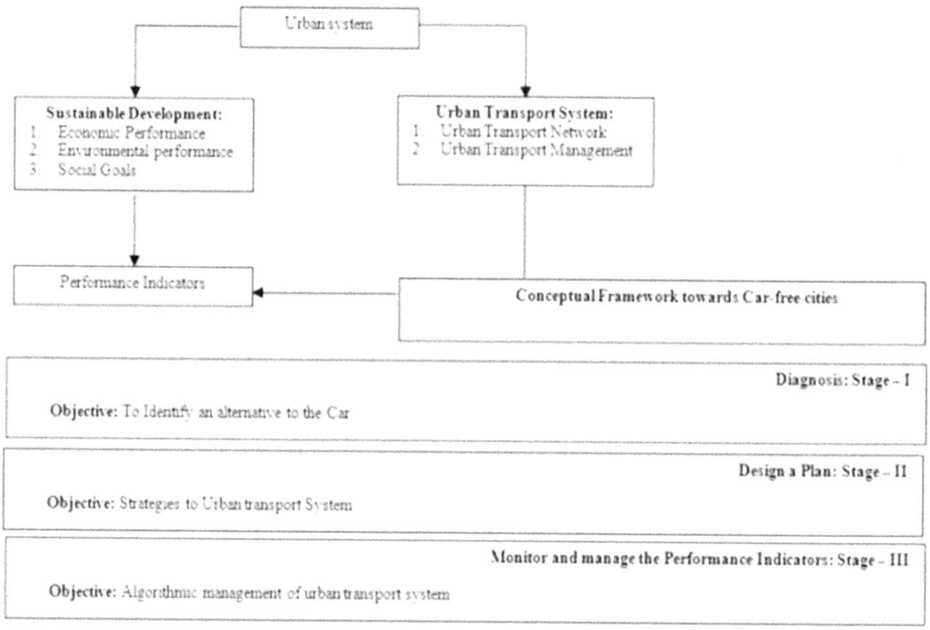

can travel a greater distance in less time with comfort. Today in cities, we see a car is used to travel long destinations as well as short destinations. Public transportation can help man travel far destination with speed, but to travel destinations within the neighborhood, we need cycling or walking. Thus, an alternative to the car is a complex multi-modal transit system of public transportation, walking, cycling, and other pooling transits. Cities need to come up with a plan for public transportation with a great degree of coverage across the city and another means of transit connecting these major nodes to other nodes at the different hierarchy.

Stage - II:

The objective of stage two is to outline different strategies to the given urban transport system to rediscover the public transport system and an implementation plan to least disturb the social fabric of the city. The design methodology of the public transport system depends on the existing topology of the city. The methodology identifies different measures to be considered while developing a public transportation plan. Figure 9 outlines a methodology to design a topological concept for public transit in the existing transit system of a city.

Stage - III:

Figure 9. Methodology to Design a conceptual topology for public transport system

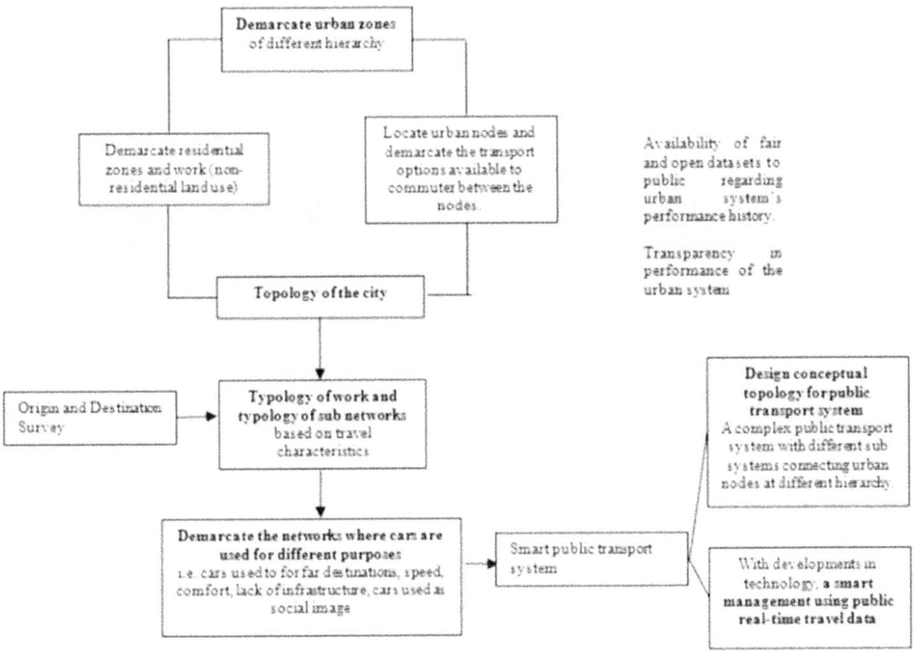

Figure 10. Components of Public transport system

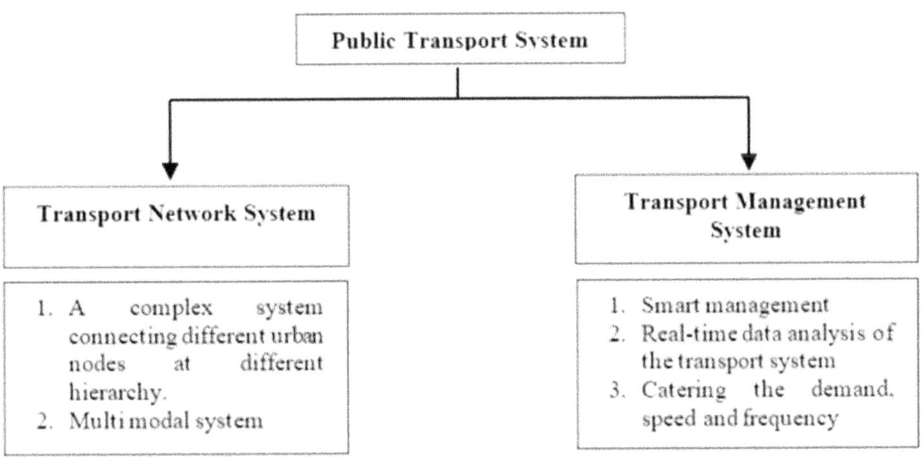

Developments in technology has made man's life easier. Real-time data collection has helped to monitor systems and manage them. Internet of things and Algorithms has made easy to collect commuter's information and analyze the data to identify the demand. At the same time, Information technology can make demand responsive decision making possible while managing systems. Since public transport system connects the major urban nodes. A network system connecting these nodes to lower hierarchy, where electric car-pooling from public transport stations to other nodes (those without public transport accessibility). Demand responsive relocatable electric car-pooling network based on the real time is collected about the commuter's destination based on tickets of public transit generated. (The ticketing system can be designed in such a way where the commuters pick up a drop off data is collected to feed data to real time systems in assessing demand.) Use of IoT and artificial intelligence in real-time data collection of commuters and traffic, public transport system is managed to address the demand. Communities are to designed with cycling lanes, pedestrian trails and a smart management plan where citizens are informed about the schedule of public transport system within the city. Innovative mobility options available in cities need to me revitalized and PPP model for public transport system where innovative mobility options provide accessibility to those nodes where public transport system cannot.

REFERENCES

CE 261. (2017, September). Retrieved from http://athena.ecs.csus.edu/: http://athena.ecs.csus.edu/~yaoz/CE261_class3.html

Christa, B., Taylor, M., Joe, H., & Bettencourt, L. M. (2015). *The Topology of Cities*. Santa Fe Institute.

Claude Comtois, B. S. (2006). *The Geography of Transport Systems*. Routledge.

Goldenberg, S. (2016). Masdar's zero-carbon dream could become world's first green ghost town. *The Guardian*.

Howard, E. (2007). *Garden Cities of To-Morrow*. Routledge.

Jean-Paul Rodrigue, C. C. (2013). *The Geography of Transport Systems*. Routledge. doi:10.4324/9780203371183

Kherdeen, R. M. (2016). Masdar City: Oriental City of the Twenty-First Century. *Academia (Caracas)*.

TomTom. (2019). Retrieved from TomTom: https://www.tomtom.com/en_gb/traffic-index/

Chapter 9
Highway Robbery:
Measuring the Land Use Priorities of a British Town

James Dyson
University of Central Lancashire, UK

ABSTRACT

Just how much land do we devote to highways, not just road surfaces, but verges, embankments, cuttings, car parks, and other related infrastructure byproducts? Do the professionals planning our towns and cities know? This chapter introduces an analytical tool to estimate the land-take of highways using mapping techniques to produce a plan or percentage figure and makes a comparative analysis across other urban centres. The outcome might be linked to quite different measures of civic success such as urban happiness, pedestrian safety, or retail occupancy.

INTRODUCING AN ANALYTICAL TOOL

I think you perhaps over-estimate the extent to which planning and transport policy are integrated. The borough's transport priorities have been set outside of the context of the development plan.[1]

Prompted initially by public consultation response to a council allocation of £70m (+) for town center highways, and latterly the same council's plan to allocate £500m (+) to a proposed relief road,[2] the methodology outlined below has been trialed through discussions with the same council planning office. An introductory text provides an overview of planning and policy, interrogating the orthodoxies accepted

DOI: 10.4018/978-1-7998-3507-3.ch009

in committee, and introduces the case study of Stockport, a British town in Northern England. A series of analytical drawings that map the "land grab" from pedestrian and public realm to vehicle and infrastructure follows, revealing the extent to which this attitude continues to shape our urban space; an excursus based on these drawings exposes the issues within a suite of paintings that provided a platform for public discussion. Together this commentary and analysis expose the quotidian dedication of contemporary planning in our towns to vehicle traffic at the expense of human comfort, with a conclusion offering a reflection on the future of public policy.

This chapter is written from the perspective of practice-based experience founded on the authors' close engagement with design and planning policy and "on the ground" research in favor of literature review. The research and proposition here as stated in the course of the text is based primarily on recording and knowledge of case studies, in tandem with analytical design tools, the authors' correspondence with policy decision-makers, the exhibition as a discursive platform, the architect as an activist, and the use of some archive material. First-hand groundwork rather than secondary reference material has formed a primary source, useful though the latter might be in understanding some of the background context and wider issues.

The study takes as its starting point Stockport, a former cotton mill town now absorbed into the Greater Manchester conurbation, Figure 1. The town lies at the confluence of the Goyt and Tame, which combine to form the River Mersey, culverted below a 1960's shopping center. The Romans found a convenient crossing point here, and although its defensive position was equipped in medieval times with a castle and market, the town never rose to prominence, and remained throughout the industrial revolution and beyond a poor cousin of its near neighbor Manchester. The changes of level evident within the historic town center have been rightly praised,[3] though today rendered incidental by two-dimensional town planning and invasive highways networks.

A. River Tame. B. River Goyt. C. River Mersey. The confluence of the Tame and Goyt is now beneath the M60. D. Merseyway shopping center. E. Ring road formed by Tiviot Way, St Mary's Way, Shaw Heath, King Street West, Wood Street, George's Road, and Belmont Way. F. A6. North to Manchester, South to London. G. M60 Greater Manchester orbital. H. A6-M60 SEMMMS (South East Manchester Multi-Modal Strategy). (Author).

Planning for the Future

In a permanent campaign of congestion relief, and against all contemporary evidence, many provincial local authorities in the UK are pressing ahead with highway building on an epic scale. Whether in Norwich or Northampton, Shrewsbury or Skegness, relief roads and town center access works are at full throttle. Official statements

Figure 1. Map of Stockport, its dual carriageway and motorway systems

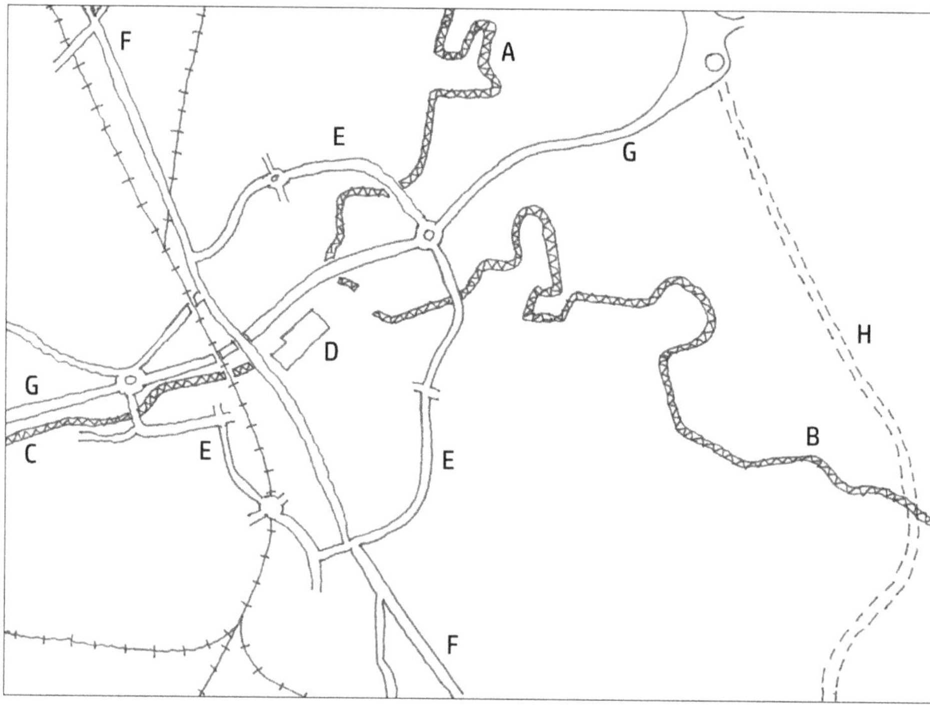

in support of these projects address the relief of congestion and thereby pollution, civic pride, and economic growth.[4] An inveterate alliance of engineers, planners, and councilors, interdependent local economies now reliant on the motor trade is pushing back against environmental policies that promote parallel green measures in their boroughs. Sajid Javid, the current (October 2019) Chancellor of the Exchequer recently announced:

We will soon launch the new Roads Investment Strategy with £29 billion committed for strategic and local roads over the next five years.[5]

Any country building roads on this scale has distanced itself from a Union of Europe where there is a consensus of doubt concerning the benefits of motor vehicle proliferation. How does this marriage of convenience of Local Authority and Highways England protocols trump planning for the 21st century? Forward-looking European cities such as Barcelona and Stockholm are progressing towards more sustainable and cleaner modes of public transport, leaving Britain well behind in terms of future economic resilience.[6] It is pedestrian, not vehicle traffic, which brings prosperity to local business and which, therefore, argues for the adaptation of congested town route grids as reconditioned and safe spaces for people.

Figure 2. The Urban Form of Highways.

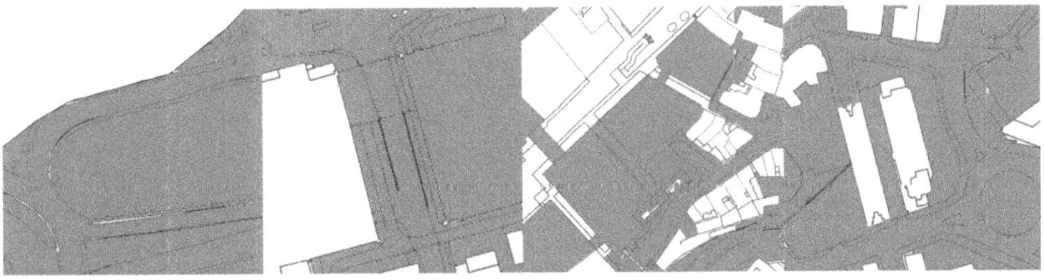

Where red = land-take of the highway, white = land for non-highway use. (1-4 from left) 1. The curves of a motorway junction leave unusable spaces suitable only for vestigial scrub growth. 2. Town center supermarket enveloped in roads, junctions, ramps, and car parking. 3. Town center retail stores used as the plinth for multi-story car parking. 4. Office buildings as islands in a sea of traffic planning.

Fear of Failure: Marketing Road Dependency

Planning for towns in the UK is directed primarily at the need for vehicle access in the belief it will sustain a community. An irrational fear of the future, it is fueled by the propaganda of the highway lobby, seeking support from those with insufficient understanding of the outcomes and alternatives to believe their claims about congestion relief, sustainability, and economic prosperity. "Access" in this context means door-to-door vehicle access, from home to work, with a stated parking provision enshrined in planning policy and supplemented by the tolerance of fly parking on every available and accessible space, Figure 2. At what stage will there be enough vehicle access?

The careful lexicon of signage, "Relief Road" or "Town Center Access", reinforces the public relations message that a project will cure a problem or enable an equality provision for local communities - an increasingly common feature of road constructors' tropes, when the undeclared target is simply to widen the path for commercial haulage. Some local communities, or stakeholders in contemporary corporate-speak, are more equal than others, and perhaps Tesco, Asda, Amazon, and DPD are more equal than the residents of the Stockport borough wards of Hazel Grove, Offerton, Bredbury, and Brinnington.

Inducing Demand

A vote-winning claim for the local political campaign is that a project will relieve congestion. More roads equal less congestion is a logic that has been successfully sold by the highway lobby to direction-less councils for almost a century. Evidence rather confirms that road building simply leads to increased traffic flow and further congestion:

If you increase the number of highways in a city by 1 percent, it causes driving to also increase by 1 percent.[7]

This phenomenon is known as induced demand, where the standard response of road widening results in increased traffic flow, causing congestion to occur elsewhere:

Widening roads to ease congestion is like trying to cure obesity by loosening your belt.[8]

Statements such as "improving the flow of traffic" are entirely misleading, as these projects do not improve flow but rather increase the quantity.[9]

The proposition that the current conditions of slow-moving traffic contribute to air pollution and by implication, this will be remedied by more highway is particularly specious – increasing capacity will increase car usage and therefore pollution. Atmospheric and particulates pollution from highways is now in the public consciousness, water and noise pollution are also significant.[10] The electric car promoted as a solution by the highways lobby remains a niche statistic complicit in traffic volume calculations.[11]

A 20th Century Bypass Case Study

Claims are made that highway building results in economic prosperity. Stockport has a 6-lane motorway through the town center. Prior to its construction in the 1980s, it was heralded as a fount of riches, both locally:

As the arrival of the M63 brought investment, so will this (M62). The town's accessibility will be increased, and its catchment area widened. Stockport is indeed set to become one of the "boom towns" of the 80's and 90's.[12] *And nationally:*

I believe that if we can keep the roads schemes moving forward in the way that is proposed in the Stockport and Hazel Grove areas we shall do our duty by the people and provide a wealth-creating infrastructure.[13]

Figures for the economic prosperity of the town and its inhabitants contradict this outcome, indicating that prosperity was improving in the 1970s before the motorway, but that the town has been in a steady economic decline ever since its construction.[14] The motorway stands as a monument to the continuing malaise afflicting those who are and have been responsible for decision making. At what point did the hypothesis that is building roads in a town will somehow make it prosper become an accepted norm?

With economic prosperity comes civic pride, which is again promoted as a benefit of colossal highway earthworks. Cars and the illusion of convenience represent aspiration and status, while pedestrians and public transport road users are seen as economically irrelevant to competing for road vehicle hierarchies. The town itself adopts this mantra. Approaching its construction in the 1980s, the M60 through Stockport was eagerly awaited by politicians:

I am strongly of the opinion that when these roads are built, a new sense of civic pride will be engendered.[15]

Less than twenty years on, Stockport was voted 12th "the worst place to live in Britain" by a national open access survey.[16]

Competing Priorities

Optimistic cost planning at the business case stage tends to be overtaken by the realities of construction later, for example, the miscalculation of excavated ground during construction of the recently completed A6MARR in Stockport.[17] Undeterred by this oversight, the council has built-in only 1.2% inflation cost into the Business Case for its proposed A6-M60 link; this appears optimistic given that many specialist construction skills will need to be bought in from the EU (where for instance tunneling is commonplace rather than exceptionally rare which it is in the UK). Further, under this Business Case, the council "owns" all risks associated with the project rather than the central government via Highways England.

Funds are, therefore, sequestered by highway projects at the expense of investment in town center re-civilization. £500m, the reported cost of Stockport's proposed A6-M60 link, would go a long way in providing public transport, affordable housing, and community facilities to provide footfall and encourage business development. The town center is at risk of long term decay. Planned population evictions from Portwood and the town center in the 1950s were followed by the insertion of the colossal retail-only, car park serviced, Merseyway Shopping Center in 1968, factory and railway closures, competing Greater Manchester motorway retail parks and now a commercial airport enterprise park 7 miles away. The decisions made by the planning authority prioritizing vehicle access over pedestrian access have created an inhospitable place, where land is grabbed for vehicle use at every opportunity, a demonstrably ill-adapted valley town with a near 20% commercial vacant footprint.

Planning by Accident

Greenbelt policy has delivered to the 21st century continuous lines of undeveloped land, a gift to highway planners who can instantly identify new routes with little need for politically difficult compulsory purchase or resident removal. The highways unintentionally become boundaries, so fields become separated from their parent use; although still notionally designated as greenbelt, they are blighted by noise, pollution, and disuse. Land becomes attractive as prospective housing development, or as out-of-town commercial or leisure sheds. Planning, that is, suburbanization, takes place by accident. Case studies confirm this phenomenon.[18] Where a bypass skirts a single settlement bordered with farmland, green space enclosed by the bypass is suburbanized. Where the bypass navigates green gap spaces between two existing suburban communities, both occupy any remaining green space. What therefore are the prospects for adjacent communities? What will be the impact on the town center of building low-density housing and retail sheds on the outskirts? These questions are not addressed by relief road business cases or planning inquiries, because these reports are framed simply for the roads in isolation.

Chicken or egg: highways at first appear, in the case of Stockport, to be an assumed by-product of urban planning policies, which in turn assume all journeys will be made by private vehicles. Public transport infrastructure, by contrast, has not kept up with supermarket car parks, new housing developments, new urban priorities, or the M60 motorway and its slip roads. Ring road projects are synonymous with a zoned approach to planning, which segregates functions rather than encourages diversity, thereby accelerating the decline of urban centers. Why is each major project in Stockport town center now accompanied by a multi-story car park?

Ticking the Consultation Box

Authorities often claim that full and inclusive consultations have been undertaken. However, consultation questions are carefully worded to elicit the desired reaction,[19] and the millimeter-designed final drawings accompanying the text are for the public an unreadable contract document designed to outwit any proper response. Rather than a consultation in the sense of an exchange or discussion, these documents are a statement of intent. Stockport's SEMMMS A6-M60 Relief Road Feasibility Study includes detailed technical plans but with no simple perspective renders of the projected works.

When produced, images of highway proposals are invariably massaged. The promotional video for the A6-MARR renders very few vehicles, nor the ubiquitous highway paraphernalia of fencing, signage, CCTV and streetlights, and markets the ridiculous impression of clutter- and traffic-free environment. To understand what the SEMMMS proposals would actually look like, the author compiled a 3D model, used by local community groups in opposition to this highway project.[20] In parallel by contrast, local councils do pay attention to their contractor development partners and battery of engineering and planning consultants, who stand to gain a great deal from these projects. Are they fit and proper agencies to determine decision-making in regard to development and infrastructure in our towns and cities? So, projects scythe through landscapes, which once built on, are lost as green spaces forever, with no remediation possible. These places are not ours to destroy.

We are only trustees for those that come after us.[21]

Figure 3. The Urban Form of Highways

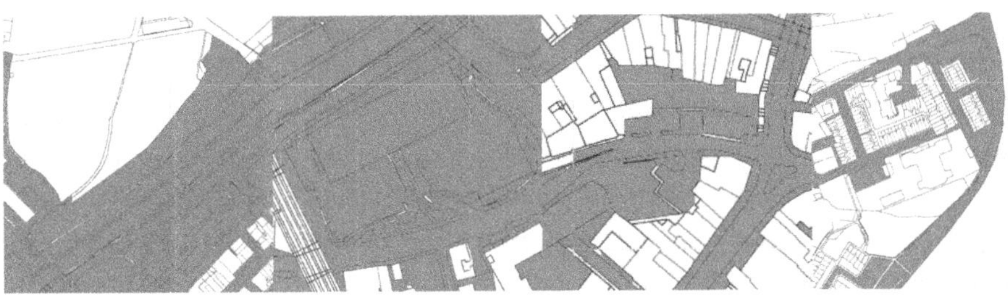

Red = land-take of the highway, white = land for non-highway use. (1-4 from left) 1. An urban motorway cuts swathes of unproductive space. 2. The whole areas of the town center are now a sea of traffic. 3. Fly parking on gap sites renders Conservation Areas meaningless. 4. Residential areas isolated from their neighbors by an unintended separation enforced by ring road planning.

Mapping a Land-Grab

Planning is thus redefined by councils like Stockport as the joining up of roads whose curves are the product of a designed speed in mph, in the process creating multiple vestigial spaces with no purpose, Figure 3. Any functional use simply occupies the space left over after the highway has gone in. So how much of the surface area of a town center is now devoted to highways? How does this compare with other urban centers, places where we might feel more comfortable as a human being? Moreover, at what percentage coverage in tarmac will planning priorities get refocused from vehicles to people? As congestion will never be solved by more highway construction, what exactly is the end game for the highway engineer-planner?

To understand the scale of this problem, detailed consideration can be given to urban centers, with the express aim of categorizing land either as the product of highway planning, or not. A simple black or white representation of the predominant use of land for clarity and communication, no intervening grey shades. In order to explain this process, several assumptions need to be clarified.

Firstly, an intimate knowledge of the physical context is necessary. The study of mapping and web-based satellite data must be supported by personal knowledge of the urban situation. In order to verify digital data, the streets need to be walked. What exists on the ground is measured, irrespective of land ownership, pending redevelopment, planning permissions, or council feasibility plans. As a designer of spaces (internal, external, and urban), the author has taken a professionally informed position on the original design intent. Normally top surface use can be taken as highway, which turns out to be a robust assumption. As far as Stockport is concerned, there are several instances of multiple levels of parking above retail below, so parking, which appears on the top surface, is the primary functional use; the negative effects of noise, pollution, and storm water runoff accompany vehicle movement regardless of mixed-use.

So the land-take of highways includes roads and car parking, vestigial space left over by highways planning, grassed and planted areas around car parks, pavements and verges alongside roads resulting from highways planning, mono-functional car park buildings / multi-story car parks, surface car parking, paths giving access to car parks, traffic islands, roundabouts, car showroom forecourts, lorry parks, and bus stations. It excludes all buildings and adjacent private enclosed spaces, private gardens, allotments, builders' yards, enclosed heritage sites, military compounds, graveyards, cemeteries, shopping malls, private sports pitches, open garden space in housing, back yards to terrace housing, domestic garages, waterways, and railways.

If this mapping were to be developed further, iterations could involve percentage marks for surface areas, different representational shading, and so on, but these would give a less direct round figure percentage expression, which was the objective here.

It remains an inexact science, and to be representative, a balanced view sometimes has to be taken, which the instances in Table 1 below should help negotiate.

Table 1. Mapping Criteria for Highways (Satellite view from Google Maps, Mapping by Author, where red = land-take of the highway, white = land for non-highway use)

Satellite view (map)	Mapping decision (map)	Typology. Logic.
	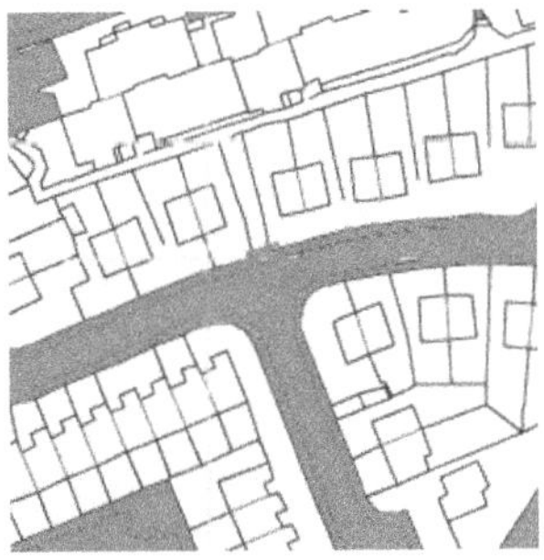	**A normal street** Highway. Pavements alongside trafficked highways are included – they only exist to provide a modicum of safety to pedestrians along the flow designed for vehicles. Residential areas are squeezed between routes, disconnected from each other by the highway rather than forming connected neighborhoods.
	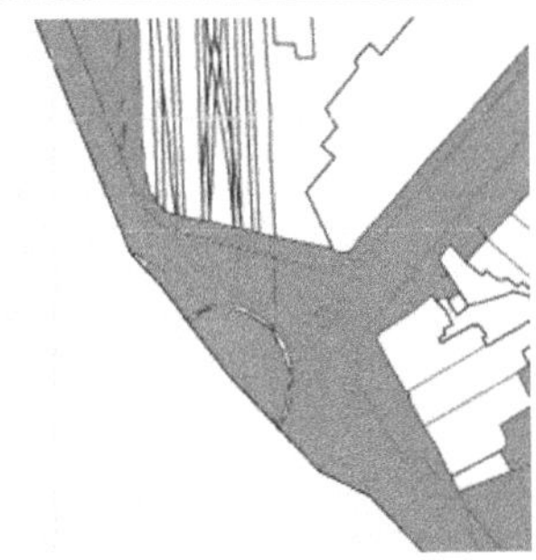	**Urban perimeter ring road** Highway. Taken to the centerline of the highway and junctions, this avoids potential ambiguities about the opposite side of the road and where the urban boundary might be.
	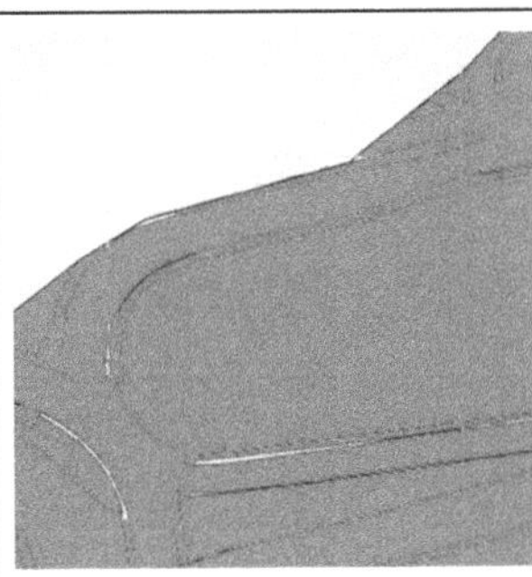	**Vestigial space** Highway. SLOOP (space leftover on plan) after the 30mph and 70mph curves, and the various roundabout/slip road junction curves, which they link, have been laid out by a highways engineer.

continued on following page

Table 1. Continued

Satellite view (map)	Mapping decision (map)	Typology. Logic.
		**Buffer area topography** Highway. Where green space is provided, it is largely unusable, its topography has been rendered so by the engineered road levels, and its buffer areas are only necessary to buffer something that is desirable (domestic space) from something that is not (highways). St Mary's Way in Stockport shown here has recently (2018) been further widened to 6 lanes in places.
	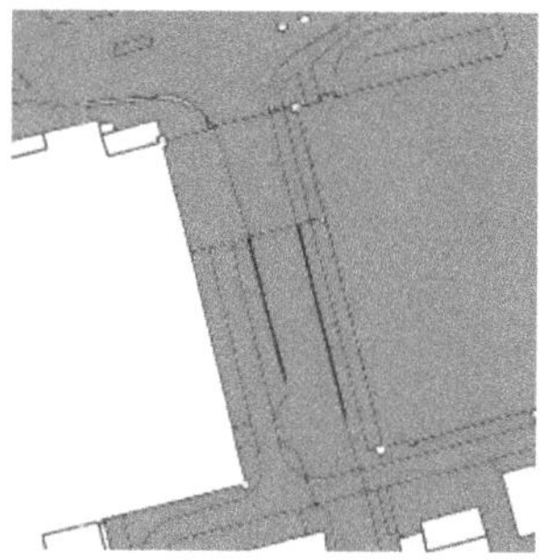	**Roads below buildings (Asda, Stockport)** Highway. The prime use of the space is the highway, the bridge exists to connect an inconvenient site, a superstore articulated from its multi-story car park (A1 Shops, Food Retail, 1 space per 14 sq m).[22]
		**Yards (e.g., Builders)** Not highway. Productive, functional space. Vehicles have now replaced industry as the main source of air pollution.[23]
	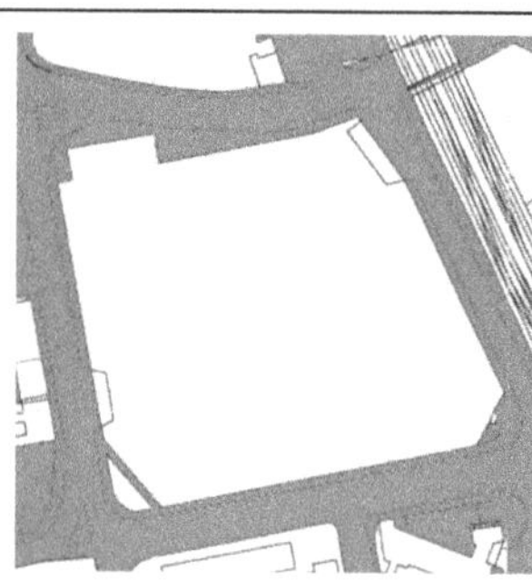	**Bus depot interior** Not highway. Firstly the internal plan of the building might not be known. Secondly, it fits a similar category to buildings and builders yards, being for productive maintenance and repair activity as much as parking. Thirdly it represents a positive investment in public transport.

continued on following page

Table 1. Continued

Satellite view (map)	Mapping decision (map)	Typology. Logic.
	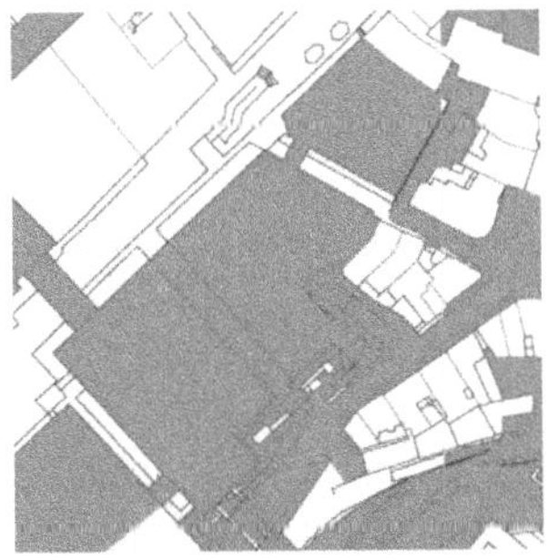	**Parking above retail below (Merseyway, Stockport)** Highway. Multiple levels of parking above retail below, so parking is the primary functional use. (A1 Shops, Non-Food Retail, 1 space per 20 sq m).[24]
	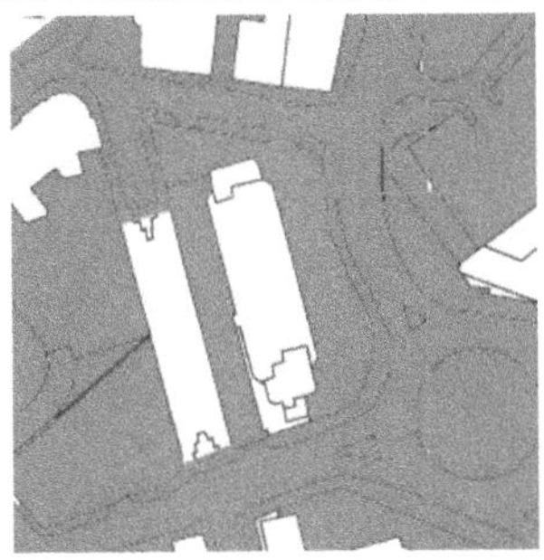	**Sub-level parking below buildings (Heron House, Stockport)** Highway up to the building perimeter, not highway within. The prime use of the site is offices. Space around is parking and therefore highway. (B1 Offices, 1 space per 30 sq m).[25]
		**Surface parking** Highway. Fly-parking, inappropriate land use, especially in Conservation Areas, where a more urban character is required.
	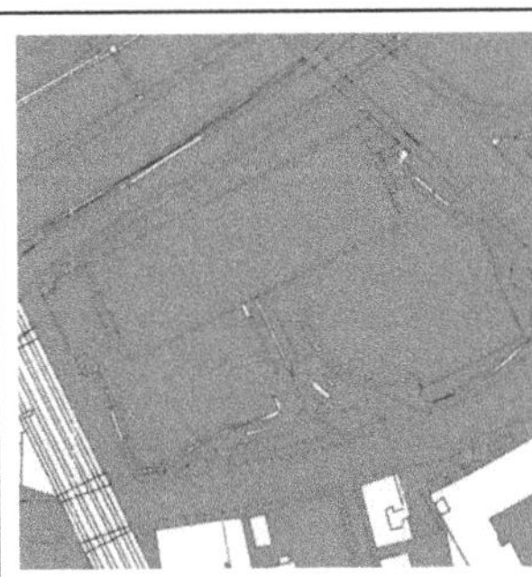	**Multi-story car parks** Highway. But counted once only; this study is concerned with land use, not the total car space area, which is much greater. Stockport now has 7 multi-stories in the town center.

continued on following page

Table 1. Continued

Satellite view (map)	Mapping decision (map)	Typology. Logic.
	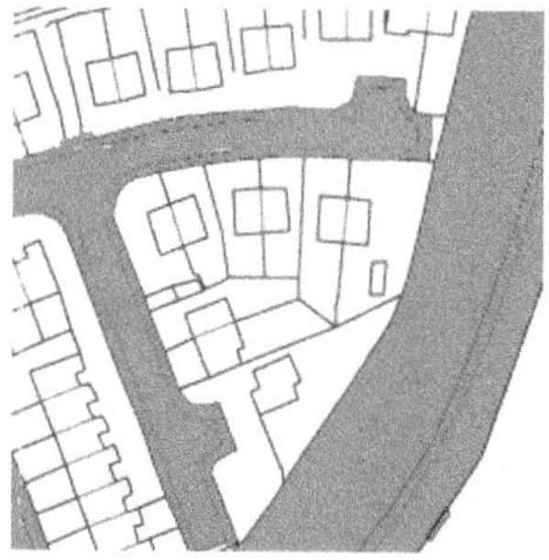	**Domestic garages** Not highway. Although obviously a product of highway planning, they are now rarely used for car parking, being mostly storage or workshop.
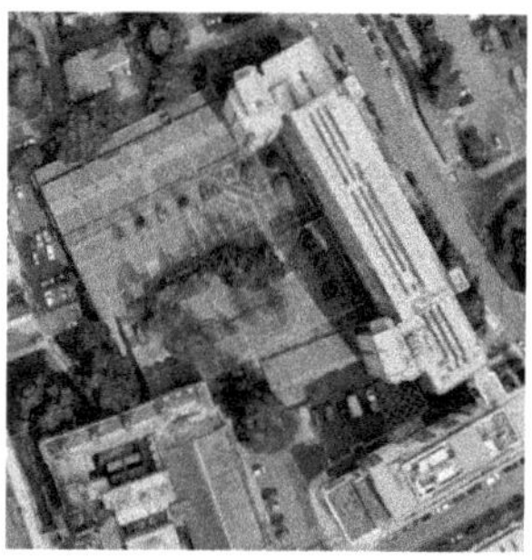	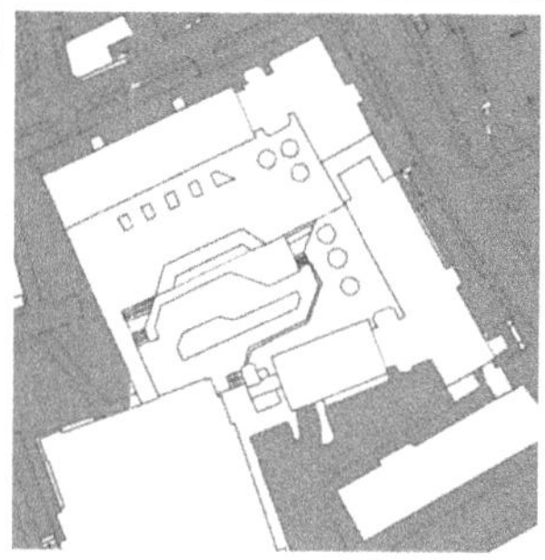	**Original design intent (Stopford House piazza)** Not highway. This piazza, though nameless, and neglected by the very council offices it complements, is primarily a generous public space, part of the 1970 civic offices (space designed for public use), not simply a decorative cover to car parking (space designed for highways).
	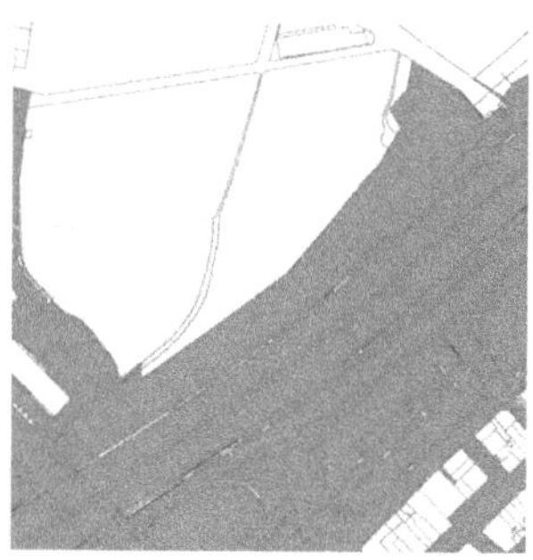	**Land around motorway (example 1, between M60 and Heaton Norris Park, Stockport)** Highway. Land sterilized by motorway route. Had the M60 taken a different route or design, the former railway land, and its escarpment alongside might then be available perhaps as an extension of the park down into the town center, perhaps as a transverse landscape link through the town connecting accessible routes alongside the Mersey and Tame. How much better an environment and connectivity with its surrounding context could Stockport have if this parkland flowed downwards to meet the town center? How much better a resource could Heaton Norris Park be if town center users could get to it? The M60 denies all this and instead creates vestigial space.
		**Land around motorway (example 2, Marsland Street and Water Street to the River Tame, Stockport)** Part highway, part not. Where land-take is less defined, then whole sites against the M60 are assumed highway resultant, but not those further away. Highway could strictly be limited to the M60 itself, or a margin of land adjacent; equally the entire multiple sites which have been vacant since the motorway was first built could be deemed highway resultant. The motorway is undoubtedly the primary reason this land remains vacant, but there may be other factors such as land remediation, access, servicing. Here is a good example of taking a balanced view.

The Walking City

As a parallel but subsidiary exercise, pedestrian space has also been mapped, but is only counted as public space where wholly traffic-free, including buses during normal working hours; Stonegate, Davygate, Swinegate and Petergate in York are examples. This is not the case in Little Underbank in Stockport, where a full-size council-promoted bus continues to squeeze past parked cars and damage the heritage cobbles on a heritage street. Pedestrian space includes pedestrian-only streets and squares, parks, informal accessible games areas, planned urban spaces, footpaths, public playgrounds, accessible (not fenced) sports pitches, churchyards, not graveyards, Table 2. Space not for either highway or pedestrian use and buildings can be described as neutral in the discussion below.

Table 2. Mapping Criteria for Pedestrian Space (Satellite view from Google Maps, Mapping by Author, where purple = land dedicated to pedestrian space, white = land for non-pedestrian use or vehicle-populated.)

Satellite view (map)	Mapping decision (map)	Typology. Pedestrian or Not. Logic.
	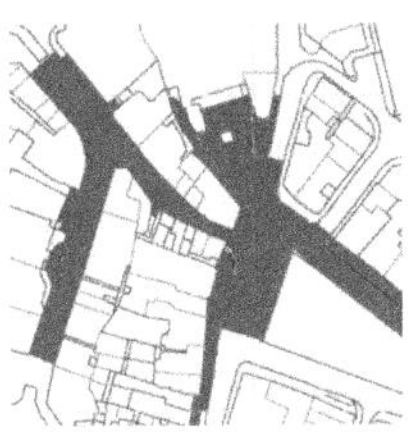	**Pedestrian only paths, streets, and squares** Pedestrian. Not where vehicles are admitted during normal working hours. (But this is Stockport; since the mapping was done, vehicles are now admitted to the pedestrian space around the market, visible in the more contemporary satellite view).
	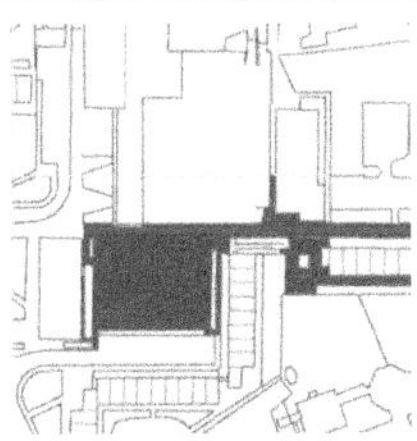	**Parks, playgrounds, accessible games areas** Pedestrian. Not access-controlled security fenced sports pitches to schools and leisure centers.
	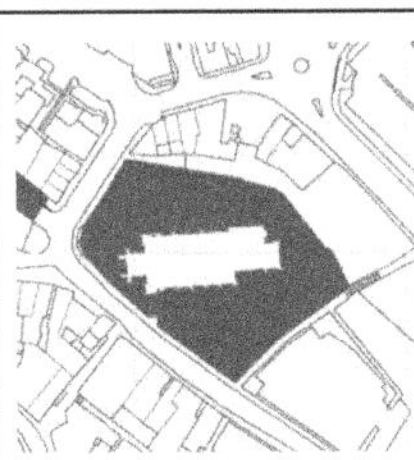	**Churchyards** Pedestrian. Accessible churchyards now used as public relaxation space or parks. Not graveyards and cemeteries in current use.

Mapping Results and Discussion

Using the above methodology, an analysis is made of four urban centers with similar populations: Stockport, York, Lincoln, and Preston.[26] This selection of comparator urban centers is deliberate, based initially on the author's personal knowledge to be confident about mapping, while investigating any correlations between spatial provision for highways and perceived sense of comfort and happiness as a pedestrian. Stockport is compared with two cities that are relaxing to walk around, Lincoln and York. That Stockport is uniquely part of a large conurbation is of no relevance, the similarly populated urban areas of Nuneaton and West Bromwich are no more pleasant to the senses than Stockport, but do these places constitute an aspirational goal? Preston is included again due to personal knowledge, and because like Stockport has recently bid for city status.

The urban center is here defined as land within an inner ring road. Therefore, for example, Stockport is defined by Tiviot Way, St Mary's Way, Shaw Heath, King Street West, Wood Street, George's Road, and Belmont Way. Clearly visible are the M60, Mersey way, and on a cross-axis the A6. Preston has no ring road to the South, but the River Ribble forms a similar boundary to the town. The mapping of urban centers in Table 3 below is arranged in ascending order of percentage area for highways.

Table 3. Mapping of Urban Centers (Mapping arranged in ascending order of percentage area for highways. Highway land red, public space purple. Commercial and retail vacancy rate statistic is also given.)[27]

Mapping	Statistics
	**LINCOLN** **Town center area 256.4 ha** **Area for highways 62.9 ha** **% area for highways 24.5%** Area for public space 9.3 ha % for public space 3.6% Commercial and retail vacancy 7%

Table 3. Continued

Mapping	Statistics
 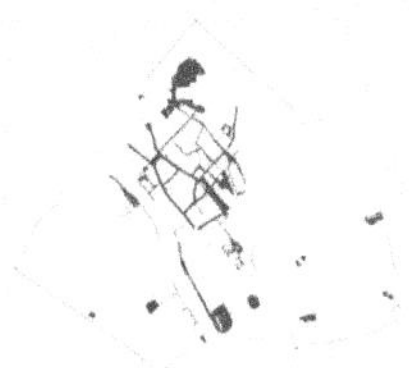	**YORK** **Town center area 117.1 ha** **Area for highways 33.9 ha** **% area for highways 29%**
 	**YORK cont.** Area for public space 7.9 ha % for public space 6.8% Commercial and retail vacancy 6%

continued on following page

Table 3. Continued

Mapping	Statistics
	**PRESTON** **Town center area 275.1 ha** **Area for highways 92.3 ha** **% area for highways 33%**
	**PRESTON cont.** Area for public space 26.5 ha % for public space 9.7% Commercial and retail vacancy 6%

continued on following page

Table 3. Continued

Mapping	Statistics
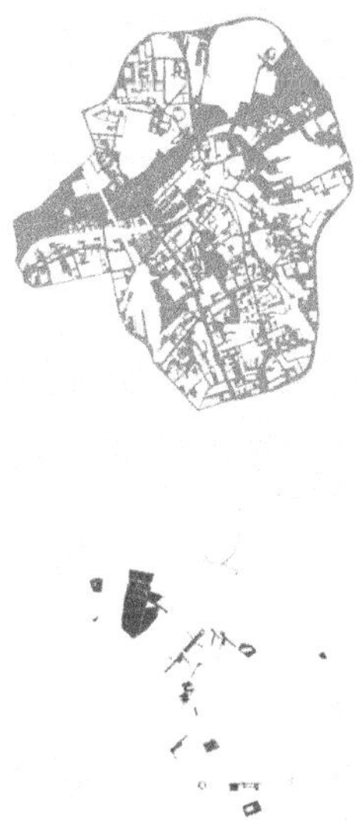	**STOCKPORT** **Town center area 250.2 ha** **Area for highways 128.0 ha** **% area for highways 51%** Area for public space 10.1 ha % for public space 4% Commercial and retail vacancy 13%

Public space: User-friendly spaces in Lincoln and York comprise safe linked pedestrian streets and squares, which make these cities pleasant for human occupation and enjoyment. Preston benefits from the large area of Avenham Park within its center. Stockport's Heaton Norris Park improves the town's ratio of public space, but is a neglected resource, being walled-off from the center by both the M60 and the recent Redrock development. The largest paved traffic-free urban spaces in Stockport are Merseyway, closed down outside retail shopping hours, and at the council's offices at Stopford House, an expansive piazza under-utilized and disconnected from the general public realm.

Neutral space: Many of the spaces in Lincoln are open and green but private; sports clubs, fenced school grounds, walled heritage sites, private gardens. Rivers form a focus for commercial and leisure activity in York, Lincoln, and Preston. The Mersey at Stockport is culverted. Railways are evident in Stockport and Preston, yet occupy surprisingly little space; the former sidings of both are now mostly car parks. Although well known as a transport hub, the railway infrastructure of York lies outside the city walls and ring road. Housing is plentiful within Lincoln, York, and Preston, contributing if not to public space, then to the civic sense of community and commercial footfall. Stockport has housing in small isolated pockets only, disconnected by highway boundaries, plus the estate towers of Lancashire Hill and Hillgate.

Highway space: Ring roads largely define the boundary of the contemporary city center. Lincoln's inner city is less defined to the North and East, where urbanity eases gently into suburbia with fewer barriers. York's extra-muros urbanity, particularly along with Clifton, Blossom Street, and Bishopthorpe, make the inner ring road less of a boundary and obstacle. Preston's Southern boundary is the Ribble, beyond which is further open space: the flood plain playing and agricultural fields of Penwortham. North of the Fishergate axis, highways-led planning dominates. Over half of Stockport's town center is subsumed in highways engineering.[28] The A6, M60, and ring road together contrive to sever urban connectivity and render all walking routes unpleasant and high-risk. Ironically this network of roads also enables the town center to be easily bypassed by those who might contribute to its wealth.

Figure 4. The Urban Form of Highways

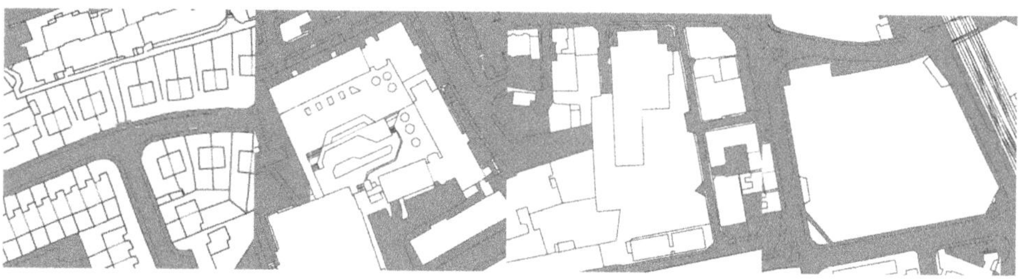

Red = land-take of the highway, white = land for non-highway use. (1-4 from left) 1. Sub-urban residential estate, a vehicle corridor with plots either side. 2. A town hall piazza marooned. 3. Industrial yards, now less polluting than the highways that serve them. 4. The bus depot – a public transport oasis!

Black and white; the land is pilfered from already deprived communities or retained by those with the means to resist. The result of mapping highway land occupation like this is, at first sight, a curious perversion of a figure-ground or Nolli[29] plan, in which space for vehicles takes the place of building form, Figure 4.

Excursus: Improvisations on a Masterplan

This figure-ground for Stockport is also oddly anthropomorphic, inviting the image to be considered simultaneously as landscape, abstract, and portrait. This has formed the basis for a series of artworks, and gallery exhibition has provided a useful public platform for discourse on the wider issues, Figure 5.

Figure 5. Gallery Paintings

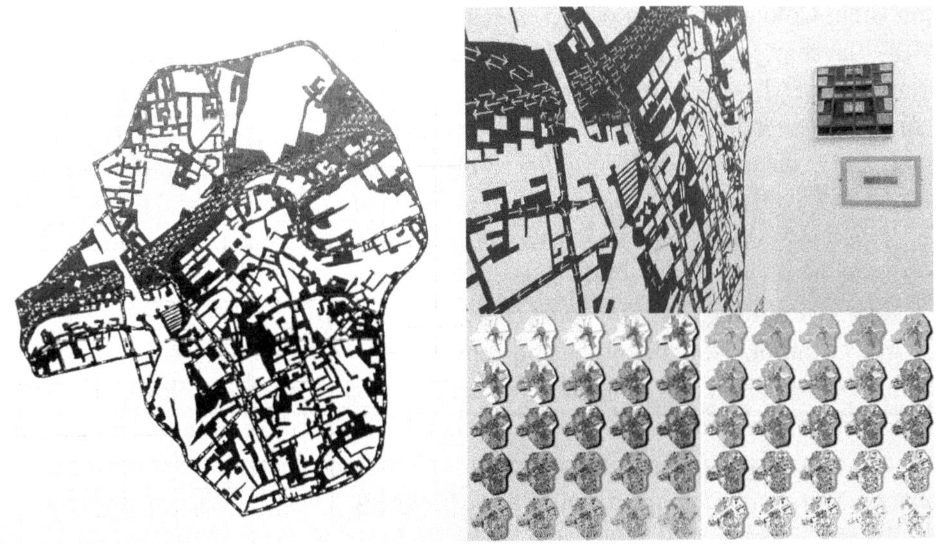

Left: The Stockport Planner No1. Top right: The Stockport Planner No3. Bottom right: The Stockport Diptych. (Author).

Gallery Painting, *The Stockport Planner No1*: an initial work exhibited in Stockport, made of authentic materials. The plain white canvas represents the *tabula rasa* created firstly by sweeping away the existing urban/landscape context; the very idea of a blank canvas here is somehow appropriate. This is drawn using a technical drafting pencil with engineering blue lead, firstly a Cartesian grid representing the carving-up or acquisition of landscape, and secondly, the plan of Stockport's road system. To this latter is applied black bitumen tar, the blacktop area corresponding to land used for, or sterilized by, highways. The shape is that of the town center plan within a central ring road. It is painted with an old brush, the broad brush of top-down desk-based planning. The top layer is a series of directional arrows in white road paint, gooey and difficult to apply, but ethically correct. If the ring road appears one-way, this is because it is drawn to its centerline. The arrows are a nod to Louis Kahn's drawings for Philadelphia,[30] and perhaps direction and flow arrows are all the traffic planner sees. The temptation to cover the whole thing with a final coat of whitewash to represent the consultation process has been resisted. 2' x 2' canvas exhibited at Stockport Art Gallery in 2016. The Stockport Planner No2: a "mini-me" executed in the same materials, this time without a definition of internal non-highway space. Perhaps the entire town is now under tarmac and is little more than a roundabout. 5" x 5" canvas exhibited at Stockport Art Gallery in 2016.

Gallery Painting, *The Stockport Planner No3*: A mural executed in vinyl matt emulsion. Colour sampling of the extensive highways depicted on five Stockport MBC flagship projects found RAL 7012 to be the closest match for the emulsion work. Part of "*The New Townscape*", 2016, an exploration into a new art movement, *"In no sense a planned agenda but a type of synergy; some emerging artists shared concern for the promise of urban life as the country experienced a polarization of wealth and poverty. These artists like a deliberately subversive act, found beauty in "subtopia"; the urban fringe where the rot had set in".*[31] 3m x 3m wall painting exhibited by invitation at EPOK Gallery, 2016.

Gallery Painting, *The Stockport Diptych*: an urban take on The Marilyn Diptych by Andy Warhol, using historical and contemporary OS mapping to document the change in fortunes of a British town. Warhol's original depicts the colorful life of the star on the left, contrasted with her fading demise on the right. Here the purposeful use and form of pre-planning Stockport morph into the contemporary condition of no use and no form planned by our highway robbers. 2 x 80cm x 80cm hinged print exhibited at Stockport Art Gallery in 2019.

CONCLUSION: THE PLANNER MINDSET

Figure 6. Urban Spaces Compared

Left, Stockport (51% of land due to highways) - highway clutter, prohibitive signage, vans, cowering pedestrians, cctv, empty shops. Right, York (29% of land due to highways) - vehicle-free lively relaxing pedestrian space with a successful business. (Author).

Looking at the urban scenes of Figure 6, we know where we would rather be, and why. One space is comfortable, the other not, but is this measurable, to a scale? The analysis table above is populated with a percentage figure for a commercial and retail vacancy, but further research might correlate the percentage highway figure from the above methodology to physical and psychological measures of comfortable outdoor space. These measures would include access to nature, space for exercise and social interaction, safety, sleep patterns, connectivity, affordability, and pollution, all of which can be shown to affect mental health and wellbeing.[32] A wealth of background research is now available to understand the background issues, such as green space as a psychological buffer from noise pollution, but as far as the author is aware, it remains to be statistically connected to the physicality of our urban space as it exists.

Stockport is currently spending and planning £982 million[33] on highways projects at the same time as it is cutting important services:

We are faced with a huge challenge in meeting the level of savings required.[34]

Perhaps the council has more than enough funding at its disposal but uses it irresponsibly. What UK local authority faced with a central government grant of this magnitude would say, well, thanks guys, but we would rather improve our provision for pedestrians and cyclists instead? So council desks bid for and take the cash, and town centers become inhospitable places characterized by roads, car parks, and vacancy. A pattern of motor vehicle space invasion has emerged steadily over the past 50 years, not unique to this proud mill town, where “no entry” is both a road sign and an assumed status for the entire town.

Approaches have been made to local politicians, councilors, and MP about moving away from infrastructure-led planning to plan for people and communities because this requires a political shift, not something local council desks are tasked with. The politicians were asked where they thought the end game of continuing to follow highways-led planning at the current rate would be, and by extension, what their vision of the physicality of a future Stockport was; will all town center land eventually be ambushed for “access”? These are difficult questions for people preoccupied with retaining their seats, elections, and Brexit, and have met with no reaction.

A dialogue was, however, established with the Head of Policy and Planning at Stockport MBC. Initially, the response was quite dismissive, but much discussion followed, both about the specifics of mapping, which helped form and modify the above methodology and about the wider context of planning for vehicles and planning for people. An agreement was reached that the analysis, however broad brush, stood up as a fair representation of what exists on the ground.

Following this discussion, the Head of Policy and Planning at Stockport Council concluded:[35]

- *"I would suggest that the conclusions you reach – and certainly your vision for what Stockport might alternatively look like – are very hard to argue with."*
- *"I think you perhaps over-estimate the extent to which planning and transport policy are integrated... The borough's transport priorities have been set outside of the context of the development plan."*
- *"I do not think we have yet changed the mindset anywhere near enough – and by that I mean the mindset within the Council as well as developers and the public at large – I think there is a gradual change of direction... Moving forward the Council is going to have to evolve far further in this respect, and at a far quicker rate."*
- *"It is important to ensure that any consultation undertaken actually has the potential to result in a proposal being amended."*
- *"I appreciate your input because, being as well considered and supported by evidence as it is, it provides helpful reinforcement to pushing for change to happen faster."*

These comments suggest that the mapping approach outlined here is amongst other things innovative, and no similar work is known to the author from elsewhere. As discussed, further development might be to correlate this to emerging indicators of happiness or comfort in cities, and/or explore the relationship between urban planning and transport policy, and/or the validity of consultation, and/or the development potential of vestigial land. Furthermore, as suggested by the above response, a long-term aim might be to contribute to a critical re-evaluation of planning priorities; in favor of creating useful, intelligent, and beautiful places.

ADDITIONAL READING

Borsi, S. (1993). *Roma di Benedetto XIV: La Pianta di Giovan Battista Nolli, 1748*. Officina Edizione.

Chandler, D. (2015). *Stockport: Bricolage Town* (MA Thesis). Manchester School of Architecture.

Gehl, J. (2011). *Life Between Buildings*. Island Press / Center for Resource Economics.

Jacobs, J. (1961). *The Death and Life of Great American Cities*. Vintage.

Rogers, R. (1997). *Cities for a Small Planet*. Faber and Faber.

Urry, J. (2017). The Little Book of Car Free Cities. Imagination Lancaster, Lancaster University.

ENDNOTES

1 Email from Head of Policy and Planning at Stockport Council to author, December 2015.

2 The two projects comprise Stockport TCAP (Town Center Access Plan), and the A6-M60 Relief Road, aka SEMMMS (South East Manchester Multi-Modal Strategy). The TCAP includes highway widening, replacement bridges, key junctions, new signal junctions, new one way systems, increased capacity at roundabouts, parking bays, bus lanes, new link roads and roundabouts. Budget figures are quoted, not final cost.

3 For example Pevsner, "Architecturally Stockport is rarely praised, but it has one tremendous asset, the difference of level between the older and newer town." in Pevsner, N. and Hubbard, E. The Buildings of England, Cheshire (1971 edition). Also see Ian Nairn. Nairn Across Britain, 1 – from London to Lancashire. (1972). Recovered from: https://www.bbc.co.uk/iplayer/episode/p01rn270/nairn-across-britain-1-from-london-to-lancashire. (From 00:28:20).

4 For example, "I support construction of this relief road. For that is what it will bring, relief." "We need faster through roads and for traffic to bypass us, moving fumes away from our homes, school, hospital and high street." "If we act to invest in infrastructure we stand a better chance of growth and improvement." "The A6 is the 10th worst road in the UK for lost time and congestion. The "think of the children" brigade should actively consider this lost time." All these statements are from one councilor, Cllr Julian Lewis-Booth of Stockport (Con), letter to Stockport Express 21.2.2018. The councillor in question is director of a bulk food packaging and catering equipment distribution business.

5 Sajid Javid. Address to Conservative Party Conference, September 2019. Recovered from: https://www.conservatives.com/sharethefacts/2019/09/chancellor-conference-speech

6 For Barcelona Superblock programme, see: https://www.youtube.com/watch?v=ZORzsubQA_M. For Hammarby Sjostad (Stockholm) see: https://www.bbc.co.uk/news/business-45564514.

7 From Duranton, G and Turner, M, The Fundamental Law of Road Congestion: Evidence from Us Cities. (2009).

8 Roy Kienitz, executive director, US Surface Transportation Policy Project, quoted in Litman, T. *Generated Traffic and Induced Travel Implications for Transport Planning*. (Victoria Transport Policy Institute, 2011).

9 Quote from *SEMMMS A6-M60 Relief Road, Feasibility Report, Stage 1*. (Transport for Greater Manchester and Stockport Metropolitan Borough Council, WSP Parsons-Brinckerhoff, 2016).

10 For noise pollution, see: https://www.theguardian.com/lifeandstyle/2018/jul/03/sonic-doom-noise-pollution-kills-heart-disease-diabetes. For water runoff pollution note: *"Between 75% to 90% of the de-icing salt applied ends up within 10m of the road, although, elevated concentrations of Na+ and Cl– have been observed tens and even hundreds of metres away from roads. Once in the soil, salts can drain to ground waters or through surface soils, and from there enter streams and rivers. Some salts are transported directly to surface waters through drainage systems. The detrimental impacts of elevated concentrations of Na+ and Cl– on roadside soils, vegetation, and ground- and surface-waters (as well as on cars and road surfaces), have been well-documented.... Many questions about the long-term effects of road salting currently remain unanswered."* From: Recovered 2017 from: https://www.york.ac.uk/environment/postgraduate/msc-dip-environmental-science-and-management/road-salting/

11 "The electric car is one of our biggest concerns. It will make you feel like you've done something positive without having to change your behaviour. Most of the pollution's still there - from the particles produced by tyres and brakes - and it doesn't address obesity and health, it doesn't address congestion, yet it will make people feel like they now have the moral high ground. Potentially, electric cars are a real problem."Chris Boardman, the Cycling and Walking Commissioner for Greater Manchester, quoted in Cycling Plus Magazine, March 2019:

12 From Dick Hargreave, Director of Development and Town Planning, quoted in "*Economic Enterprise – Stockport's Way*". (Undated c1980 document).

13 Peter Bottomley, MP for Worthing West, Parliamentary Under-Secretary (Department for Transport), quoted in *Hansard,* 26 January 1988.

14 Figures for Percentage Breadline Poor in Stockport are as follows: 1970: core poor percentage 16% Breadline poor percentage 25.1% (41.1% total). 1980: core poor percentage 9.9% Breadline poor percentage 17.1% (26% total). 1990: core poor percentage 14.6% Breadline poor percentage 21.7% (36.3% total). 2000: core poor percentage 11% Breadline poor percentage 26.8% (37.8% total). (Source: *Poverty, Wealth and Place in Britain 1968 to 2005: Understanding the Transformation of the Prospects of Places*. SASI, University of Sheffield).

15 (Sir) Tom Arnold, Conservative MP for Hazel Grove 1974-97.Quoted in *Hansard*, 26 January 1988.

16 Jordison S. and Kieran D. *Crap Towns*. (2003).

17 Reported in *Manchester Evening News*, 28.1.2018, to be miscalculated by 340,000 tonnes. Recovered from: https://www.manchestereveningnews.co.uk/news/greater-manchester-news/manchester-airport-bypass-when-complete-14210969

18 See *Prospects for Stockport*. Author (2017) at: https://goo.gl/kMz3kp.

19 "Framing the Congestion Question: If you ask people, "Do you think that traffic congestion is a serious problem?" they frequently answer yes. If you ask, "Would you rather solve congestion problems by improving roads or by using alternatives such as congestion tolls and other TDM strategies?" a smaller majority would probably choose the road improvement option. This is how transport choices are generally framed. But if you present the choices more realistically by asking, "Would you rather spend a lot of money to increase road capacity to achieve moderate and temporary congestion reductions and bear higher future costs from increased motor vehicle traffic, or implement other types of transportation improvements?" the preference for road building might disappear." From Generated Traffic and Induced Travel Implications for Transport Planning. Victoria Transport Policy Institute. 2014.

20 See *Prospects for Stockport 2*. Author (2017) at: https://goo.gl/6fjvHN.

21 William Morris. Address at the Twelfth Annual Meeting, Society for the Protection of Ancient Buildings. (1889).

22 Stockport Unitary Development Plan Review – Adopted 31st May 2006. APPENDIX 9 PARKING STANDARDS.

23 *"In towns and cities, the main source of air pollution is road transport."* The British Lung Foundation. (2019). https://www.blf.org.uk/support-for-you/air-pollution/where-does-it-come-from

24 Stockport Unitary Development Plan Review – Adopted 31st May 2006. APPENDIX 9 PARKING STANDARDS.

25 Stockport Unitary Development Plan Review – Adopted 31st May 2006. APPENDIX 9 PARKING STANDARDS.

26 Populations 2001; Lincoln 104,221, York 137,505, Preston 184,836, Stockport 136,082. Source: Office for National Statistics. (2015). https://www.ons.gov.uk/ons/index.html

27 Source: Office for National Statistics (2015). http://www.neighbourhood.statistics.gov.uk/dissemination/. (The retail vacancy for Stockport in 2015 was reported to be 26.6% by The Local Data Company (LDC) report "Cautious Optimism").

28 The mapping analysis for Stockport was carried out prior to implementation of the Town Center Access Plan, a £73m programme of highway works in the town center.

29 Roma di Benedetto XIV, La Pianta di Giovan Battista Nolli, 1748. Nolli gave a hatch fill to private building space and the solid parts of public building space, so defining the accessible city.

30 Plan of Proposed Traffic Movement Pattern, Philadelphia. Louis Kahn. (1952).

31 Introduction to: *The New Townscape*, EPOK Gallery, Stockport. David Chandler. (2016).

32 This list, from The Center for Urban Design and Mental Health, recovered from: https://www.urbandesignmentalhealth.com/how-urban-design-can-impact-mental-health.html . See also: McCay, L. *Designing Good Mental Health into Cities: the Next Frontier for Urban Design.* (Design Council, 2019). *Cities Safer by Design: Urban Design Recommendations for Healthier Cities, Fewer Traffic Fatalities.* Welle, B. et al (World Resources Institute 2015). Also Gehl, J. *Life Between Buildings*. (2011).

33 Included in this figure are: A6-MARR (Manchester Airport Relief Road) £290m, ref: https://www.stockport.gov.uk/a6-to-manchester-airport-relief-road. Stockport TCAP (town Center Access Plan) £73m, ref: https://www.stockport.gov.uk/town-center-access-plan. A6-M60 SEMMMS (South East Manchester Multi-Modal Strategy) £477m, ref: https://www.manchestereveningnews.co.uk/news/greater-manchester-news/controversial-477m-m60-link-road-14832120. Stockport Interchange £120m, ref: https://www.manchestereveningnews.co.uk/news/greater-manchester-news/how-stockports-huge-new-bus-15362717. A560 Cheadle Corridor £4m, ref: https://www.gov.uk/government/uploads/system/uploads/attachment_data/file/653071/npif-regional.csv/preview. Stockport Exchange Multi-Storey Car Park £8m, ref: https://www.ncpsolutions.co.uk/our-markets/case-studies/rail/a-trusted-partner-for-stockport-exchange/. Redrock Multi-Storey Car Park, estimated £10m of £45m, ref: https://www.bdonline.co.uk/carbuncle-cup/carbuncle-cup-redrock-stockport-by-bdp/5094676.article. Total £982m. All references recovered October 2019. Also, A34 Phase 1, one of 16 projects bidding for £700m in 2019.

34 Keith Holloway, Executive Councillor for Supporting Adults, on explaining why Stockport is cutting £4.7m from care services to vulnerable adults, reported in Manchester Evening News, 22.10.2015. https://www.manchestereveningnews.co.uk/news/greater-manchester-news/stockport-town-hall-bosses-plan-10314050. Stockport is currently (October 2019) conducting a public consultation exercise for closing museums in the town to save £150,000.

35 Email from Head of Policy and Planning at Stockport Council to author, December 2015.

Chapter 10
Success Measures for Transforming Into Car-Free Cities:
Recommendations for Implementation

Rahma M. Doheim
https://orcid.org/0000-0001-5443-4451
University of Business and Technology, Saudi Arabia, & Assiut University, Egypt

Alshimaa Aboelmakarem Farag
https://orcid.org/0000-0003-1539-6928
Zagazig University, Egypt

Samaa Badawi
Mansoura University, Egypt

ABSTRACT

Private cars contribute heavily to air pollution and significantly lower air quality in cities. The number of deaths because of pollution and car accidents is increasing on a global level; therefore, achieving sustainable mobility in urban areas is essential. Hence, the transformation into a car-free model is not a marginal issue but rather a crucial need that should be a global trend. The biggest challenge in this transforming process is to minimize the dependency on private cars. This chapter reviews thoroughly some global practices of inspiring models of transforming into car-free cities around the world. This review aims to identify the success measures for the transformation of a car-free city through investigating the challenges that affected the adoption of the transformation process. This would potentially guide governments and policymakers to select the approach that copes effectively with the cultural, social, geographical, and economic characteristics of their countries.

DOI: 10.4018/978-1-7998-3507-3.ch010

INTRODUCTION

In the 21st century, the private car mode becomes an important part of the transportation system, since it is an affordable, and comfortable transport alternative. In fact, private cars have become a dominating force on the streets of cities after the Industrial Revolution. The dependence on private cars as an essential transportation mode has grown rapidly during the last decades, which caused the past century to be defined as *"A Century Of The Car"* (Gilroy, 2000). The number of motorized vehicles in the world grew from about 75 million to about 675 million between1950 and 1990. Around 80% of these vehicles were primarily private cars (Linda, 2003). The number of motorized vehicles is still increasing; in the middle of the year 2010, there were more than 1 billion cars registered worldwide. At the beginning of 2019, they were about a quarter of a billion more. Furthermore, it is expected that by the year 2035, over 1,8 billion automobiles will be in use worldwide (Worldwide Automobile Productions, 2020).

The heavy dependence on private cars has resulted in many environmental and socio-economic problems. It contributes heavily to air pollution and significantly lowers air quality in urban cities. Cars emit CO2, black carbon and other greenhouse gases and air pollutants. Additionally, they cause heat, road damage, congestion, and oil dependence (Santos et al., 2010(a)). According to the European Environment Agency, automobiles and airplanes are considered to be more damaging to the environment than other modes of transport. The Co2 emission for a car is 124.5 gm/km compared to 130.2 for the airplane and coming in second place (Adbelhamid et al., 2018). Thus, car use is one of the most significant contributors to climate change, accounting for 27% of global CO2 emissions, and is the fastest-growing source of greenhouse gas emissions (Zarrilli, 2019). Car dependence also affects the social aspect as it threatens the quality of life due to noise pollution. Moreover, the cars dominate the city's streets leaving less space available for walking due to the car lanes and parking spaces, which occupies most of the urban spaces in cities that could be used for other purposes such as pedestrian walkways and parks which are more beneficial to public health and wellbeing (Amoly et al., 2014). Another critical problem is safety. Motor vehicle crashes are a direct cause of a large number of deaths and injuries. According to the World Health Organization, in the Global status report on road safety in 2018, the number of road traffic deaths is getting worse and continues to climb, reaching 1.35 million in 2016. Nearly 3700 people dying on the world's roads every day, and tens of millions more are injured or disabled every year. These losses severely impact families and communities (WHO, 2018). Car dependence also causes economic problems, which mainly represent a decrease in accessibility to economically important destinations due to congestion. Congestion in European cities costs have increased from about 100 billion euros in 2003 to 270 billion euros per year in 2020, and are projected to increase in the future (Boffey, 2020).

Megacities all over the world have high rates of car use, and their citizens spend hours in traffic suffering from the previously mentioned problems every day. Therefore, there is a real need for careful planning and long term strategies in the local and international levels to cope with the emerging crises due to the increasing demand for urban mobility. A prominent example of international action towards the deteriorating environmental status is Cities Climate Leadership Group (C40) which represents a successful model for global collaboration between the world's megacities to exchange policies, programs, and solutions that focus on tackling climate change and driving urban action that reduces climate risks while enhancing the wellbeing and economic opportunities for citizens. C40 is a global community that its members could share ideas and solutions through 16 thematic networks; transportation actions is one of them. Those actions include using innovative financing approaches for transportation, encouraging Transit-Oriented Development, and dependence on Low Emission Vehicles (Zarrilli, 2019). In this regard, many cities worldwide have taken serious steps towards the car-free zone model.

Freeing zones or cities from cars is likely to free up urban spaces, which can be filled up with attractive alternatives such as green spaces, public squares, markets, shops, or other facilities (Nieuwenhuijsen & Khreis, 2016). In particular, replacing roads and parking with green environments can be one step forward to change the environment from detrimental to beneficial. It provides opportunities for physical activity and active transport mode shifts while mitigating air pollution, heat, and noise levels (Wahlgren and Schantz, 2014). In fact, the car-free concept is not a new idea that until the late 19th century, all cities were car-free. However, there were attempts to limit the types of transport modes in cities dated back to the Roman Empire when wheeled traffic was not allowed to exist in the streets during the daytime by law. Moving much forward to 1930, "The Automobile City" started to appear with low density and long distances between work and residence (Wright, 2005). In the 1950s and 1960s, most of the cities' public spaces were dedicated to cars. Then, the modern concept of car-free cities has been revived in the late 1980s as an opposite trend (Medina, 2018).

GLOBAL PRACTICES OF TRANSFORMING INTO CAR-FREE CITIES

The early 1990s have seen a growing interest among many cities in transforming into car-free cities and applying sustainable transportation. However, this does not mean they prevent driving cars completely, as this idea is currently out of reach due to city life complexities. It is hard to imagine a world without cars. However, many capitals and megacities around the world are beginning to shift their mobility solutions away from the private car and towards more environment- friendly and people-focused mobility means. Accordingly, the term "Car-Free" is applied to the efforts by city governments to increase the attractiveness and economic vitality of their cities through reducing the existence of parked and moving vehicles and encouraging access by sustainable means of travel, hoping to reach the minimum and only essential needs of car use (Topp and Pharoah, 1994).

Transforming into car-free zones or cities becomes essential, and it is the trend that many cities around the world have approached, and many others should follow. This chapter reviews thoroughly global practices and policies of leading models of transforming into car-free cities around the world, in which the experience of each city represents a unique case and special characteristics that do not precisely match other experiences. In the following part, several examples of inspiring car-free practices are illustrated and discussed for nine cities from around the world, which are Paris, Mexico City, London, Bologna, Johannesburg, Addis Ababa, Madrid, Oslo, and Copenhagen. The selection of the cities aimed to explore different models with different cultural, social, geographical, and economic characteristics, in which each tackle different challenges and various approaches to overcome those challenges. In this regard, this review aims to identify and to highlight the challenges and success measures that affected the adoption of the transforming process as a sustainable and less polluting transportation practice. This would potentially help governments, urban planners and policymakers to adopt the concept of sustainability when approaching urban transport policies in metropolitan areas.

Paris- France

Paris, the city of beauty and charm, has been categorized among the worst in Europe in terms of air quality. This problem has resulted from motor vehicles, which are the primary source of GHGs (30%) and is responsible for almost 40% of CO_2 emissions. The levels of nitrogen dioxide (NO_2) and particulate matter (PM_{10}) concentrations in the Paris region significantly exceed the limit values (Bulteau, et al., 2019). Paris City has been reshaped like other cities during the 1960s and 1970s, by car-oriented planning. After the Second World War, the city avenues were dominated by car use, which turned them into gasoline-filled arteries and throttled the city core with congestion and pollution (O'Sullivan, 2016), which costs France more than €100 billion each year (Rodriguez, 2015).

Paris is leading an international political movement to control the excessive use of cars and to limit the accelerating number of used vehicles hoping to cut down on air pollution that in recent days has reached dangerous levels. In order to achieve the previously mentioned goal, the French government has made lots of efforts, including redesigning urban areas, encouraging bike use, encouraging electric-car use, and change people's travel behavior. On the urban design level, the French have worked on a comprehensive development plan that includes the entire city, and all neighborhoods will be turned into car-calmed, pedestrian- and bike-dominated zones. The major seven squares in Paris have been redesigned to reduce vehicle lanes and parking while increasing pedestrian space and greenery (O'Sullivan, 2016), figure 1. Redesigning the seven squares would result in transforming the vehicles' congestion, noise, and pollution into open places where people could breathe.

Figure 1. Plans for Paris's Place Gabetta, with shaded areas showing more space for trees and pedestrians. (O'Sullivan, 2016)

French have extended their efforts to afford sustainable transport facilities. Paris doubled the total length of bicycle lanes (from 700km to 1.400km) and created a cycling express network to cross Paris from north to south and from east to west. To ensure the successful operating of the cycling service, the city added more facilities such as a new secured offer for bike parking (Velobox), and bike stations in railway stations (Veligo). In order to promote the bike culture, the city established cyclist-friendly circulation rules and programs, support for repair-workshops, and « bike schools » to make bikes easily accessible to all and stimulate bicycle touring for tourists (Najdovski, 2017).

Paris has had previous efforts in encouraging riding bikes. In 2007, Paris rolled out a bicycle-sharing program (Velib), which introduced bicycle locking stations around the entire city and more than 20,600 bicycles. The schemes are run by private companies, which introduced annual subscriptions, smart-card payment, and reservation technologies to promote the use of the bicycle-sharing program (Santos et al., 2010(b)). Electro-mobility has also emerged in Paris as a clean sustainable alternative for transportation. To encourage the new technology, the electric vehicles have been entitled to free parking in addition to public charge points since 1993. Then, in 2011 (Autolib), the world's first electrical car-sharing rental service brought electric vehicles into the mainstream. The new service got a great acceptance quickly among its users as it is silent, instant torque, zero pollution from the exhaust while driving, and allows users to access a vehicle whenever they want while saving the cost of car ownership. The cars are exclusively recharged using energy from renewable sources, which makes the whole process ultimately clean (Dusart, 2015). Moreover, the Parisian plan to meet its targets under the Paris climate accord is to ban all petrol and diesel-fuelled cars from the city center in addition to end sales of petrol and diesel vehicles by 2040. Simultaneously, Volvo started to make fully electric or hybrid cars from 2019 onwards. Thus, the car making industry has to be well equipped to participate in the governmental plan (Chrisafis & Vaughan, 2017).

Parallel to all previously mentioned efforts in redesigning public spaces and affording sustainable alternatives for transportation, Paris has also made many efforts to change people's travel behaviors. An example of the efforts made to raise the people awareness and encourage them to change their travel behavior even for one day is turning on Paris's city center into car-free for the first Sunday of every month from 10 a.m. to 6 p.m., to improve air quality and share public spaces by pedestrians and cyclists. Residents, delivery vehicles, public transport, and taxis are permitted to enter and leave through designated access points with a limited movement speed of 20 km/h. The initiative is rolled out to several streets of the city, except major roads such as Boulevard Sébastopol (Coffey, 2018). The car-free day initiative allows all Paris residents to live differently and experience moving without a car for one day and let them see what it would be like if the city was designed entirely for them. Parisians and tourists on this day would enjoy the city without noise, pollution and therefore without stress and will discover a new Paris.

Nevertheless, the French Government has faced many challenges after imposing strict policies that have always been met with a lot of criticism and public resistance. In the past, the French tried emergency measures to reduce air pollution. The vehicles registered before 1997 and motorcycles before 1999 were banned from driving within city limits during the weekdays, and limiting the moving speed was a common measure to get people out of their cars. In March 2016, a new applied rule allowed only cars with odd-numbered license plates to drive on Monday. On Tuesday, the city's streets were to be open only to vehicles with even-numbered plates. Although this rule was followed by 90 percent of Parisian drivers and resulted in reducing air pollution by about 30%, it was criticized by a number of Parisian citizens. For example, some people were not aware of the ban and got fined. Others could not let cars as they have to drive their kids to school every day with no exception. Thus, they did not have another choice than paying the fine (Memmott, 2014).

Moreover, Parisian governance has taken many decisive actions, such as banning the most heavily polluting vehicles from the city. There is a system of shields detailing that record each vehicle's age and emissions, and all cars must display for examination or face a fine. Meanwhile, it offers drivers financial incentives to buy cleaner new ones (O'Sullivan, 2016)

Despite Paris' reputation for the strict penalty, the financial incentives it bestowed on its citizens cannot be denied. Parisian citizens can get the benefits of following governmental projects when they get rid of their cars. The Parisian citizen can get around $700 to buy a bike, sign up for a car-sharing service, or buy a public transit pass. In the case of having a small business, the citizen can get around $10,000 from the city to buy an electric truck or bus (Peters, 2018).

However, all of these concerted efforts to limit the control of cars as the primary cause of environmental pollution and constant congestion have coincided with some objections from the public side, in which some decisions were not effective that caused more problems rather than solving them. An example is banning motorized vehicles, and limiting the transport to specific lanes resulted in cases of congestion in those lanes, while the paths that were reserved for bikes and pedestrians are almost empty. Additionally, most alternatives that are introduced for people are still unfinished, but solutions focus on forcing people to move away from cars. Another criticism has been directed to the French government, such as the ignorance of transport difficulties that are happening outside the borders of Paris (O'Sullivan, 2016). In addition, the decision of replacing diesel-fuelled cars with new electric cars has raised the people's concerns that such a decision would cost households a lot of money to buy a new car and may be considered an intrusion into individual behavior (Badger, 2016).

The mayor of Paris, Anne Hidalgo, was criticized for her inapplicable ambitious plans, in which she wanted to make all public transport in Paris City for free. The opposition had another opinion that exempting people from paying the fares may motivate them to use public transport, however, it will have economic effects on taxpayers, who will pay the unpaid tickets. Additionally, they declared it is unfair to make public transport free for Parisians but not for people living in the suburbs (Osborne, 2018).

Despite all the criticism, a study that has been conducted by Eurostat, the EU's statistics agency, in 2015, showed that public transport in Paris City is one of the most used in Europe. More than 60% of Parisian people are using public transport means such as the metro, buses, and train system in return for 25 percent who said they use their car to commute (Osborne, 2018). One of the most relevant results that Paris has been shifting away from driving. The car use inside city borders has dropped 45% since 1990, while the percentage of people riding bikes grew ten times.

It seems that the city of Paris is on track to achieve its goals, however, there are more efforts that need to be done as Paris still has a problem with dangerous levels of smog, despite the accelerated efforts to cut pollution from cars in the last few years (Peters, 2018). In addition, Paris should provide significant efforts to involve Parisian society into its programs and policies to limit cars' use. This would potentially reduce the continuous criticism and be replaced by fruitful cooperation with the community.

Mexico City- Mexico

Mexico City has been exposed for a long period of time to environmental threats to its survival as a viable city. Mexico City was declared the "most polluted city in the world" in 1992, according to the United Nations (WHO/UNEP, 1992), as it is considered a massive source of greenhouse gas. It is also thought to represent the ecological disaster wrought by rapid urbanization in developing countries. Additionally, in 2016, Mexico City was recognized as the most congested city among the 390 cities, according to the Tom-Tom Index. The main reason for the pollution and congestion in Mexico City is the heavy reliance on automobility and car ownership that significantly increased during the last decade (Avila-Forcada & Medina, 2019; Dieleman, 2013). The high dependency on private cars with high dense neighborhoods and widespread informal public transit are contributing to worsening environmental problems, congestion, infrastructure, and public health. This dilemma forced the Mexican people, policymakers, and advocacy groups to give more attention to containing and slowing the rapid rise in private car use (Guerra, 2014).

In response to this deteriorating situation, the precautionary measures adopted by Mexico were similar to those of Paris, which also met public resistance. The initiation of the "Day Without a Car" was in place in 1989, when the Mexican Authorities banned each car from driving a specific day of the week. The regulation specified that cars with license plate numbers ending with digits 0 or 1 do not drive on Monday, 2 or 3 do not drive on Tuesday, and so on, however, no restriction was on weekends. An exception was given to those of the fire department, firms, and households. Surprisingly, car use increased as well as gasoline consumption. The people were able to circumvent the regulations by purchasing additional cars since each car was permitted to drive different four workdays (Eskeland & Feyzioglu, 1997).

In 2016, Mexican Authorities decided to expand its initiation to mitigate the city's high smog levels by prohibiting cars from driving into the city center two days every work week and two Sundays per month. People can drive on a given day using a rotating system based on license plate numbers (Garfield, 2018). In 2017, Mexico City choked in an extended pollution alert and worst smog for two decades that Ozone levels exceeded the accepted threshold. This resulted in putting the Mexican residents under high health risk to the extent that the government warned people from going outside, especially children, older people, pregnant women, and people with respiratory and cardiovascular problems. Besides, the Mexican Authorities banned cars off the streets, especially the less efficient vehicles (Baynes, 2017). Because of the seriousness of the situation, it has emerged that the measures have been taken as a feature of strictness. However, Mexico was bound to find other solutions that would improve the deteriorating situation.

It seemed a promising initiative, in April 2007, when Mexican Authorities decided to implement the program "Sundays without a car", close 10 km (6 miles) of Mexico City streets each Sundays for car use, and opened those streets for people to practice many activities such as dancing, jogging, biking, connecting community in many ways. Those activities have expanded over the years to salsa lessons, urban theater and music, healthcare and social services, political campaigns, specialized stores, and cycling mechanics (Mendez, 2014). This initiative was well received by the citizens and encouraged them to let the car for a day in exchange for fun and enjoyment.

In 2009, the Mexican government launched new projects to encourage sustainable mobility. A bike-share program called (ECOBICI), has been created by the Ministry of Environment and the Non-Motorized Mobility Strategy Office. They aimed to build robust bike infrastructure, creating a cycling culture, and integrating cycling into the primary transport system. To achieve that, they built a strong bike network with 85 stations, 1,200 bikes for rent, 48.6km (30 miles) bike lane network throughout the city, and bike parking at transit centers. In 2018, the system had 452 stations and 6,500 bikes. Accordingly, (ECOBICI) became alternative mobility for the middle classes of the center of Mexico City. The lower classes were already bike users. To foster sustainable mobility behavior and a healthy lifestyle from an early age, they created "bike schools" for young students for training on biking. A massive change in the mobility patterns of the city has been achieved. In 2012, 40% of (ECOBICI) participants stopped using cars for short-distance trips. Another reason that caused the increase in using bikes and motorcycles is the high congestion levels that resulted in reducing car speed and thus made the bike use more convenient. This change saves lots of time in traffic and reduces tons of CO2 emissions. This achievement was a source of hope to promote healthier, faster, and more efficient commutes. The Mexican Authorities aspire that people can take back the public realm from cars and make sustainable mobility a reality (Avila-Forcada & Medina, 2019; Mendez, 2014).

Several strategies were adapted to influence the behavior of commuters and reduce car use, such as new traffic regulation that speed limits were set at more stringent levels, parking meters were installed, and a system of sanctions. In July 2017, a new building regulation was issued to limit the number of parking spaces. In addition to rehabilitating many sidewalks and streets in the downtown area of Mexico City and some other neighborhoods in the core. Two streets were closed to revamp the historic downtown and promote a walking environment. Several environmental organizations have joined these efforts with the Ministry of Environment and the Government of Mexico City (Avila-Forcada & Medina, 2019).

Although the efforts spent by the Mexican government to reduce the pollution and congestion in Mexico City, there are serious challenges that need to be solved. For example, there is still a significant increase in the number of registered vehicles in Mexico City that was 5.6 million in 2007 and reached 12.3 in 2016. The growth rate of vehicles is almost two times the rate of population growth, which means drivers did not shift to other modes of transportation, and there is no reduction in emission levels and more congested streets.

The reasons beyond the delay in achieving sustainable goals include the elimination of cars' sales tax in 2012 and the increase of auto-credits in the market. Another reason is the income growth and households' preferences, in which people tend to buy a car when their income increases to a specific limit (Avila-Forcada & Medina, 2019). Additionally, there are millions of Mexican people who did not get any benefits from the governmental development projects. For example, many neighborhoods are not connected to the new mobility networks; thus, people rely massively on the microbuses, which contribute considerably to air pollution as well as the low quality of life (Dieleman, 2013). Despite that the public metro service increased, it did not reach a bigger fraction of residents who lived far from metro stations (Guerra, 2014). Therefore, additional improvements are needed in the area of sustainable mobility, community involvement, and effective urban planning.

London, UK

In the UK, London has an inspiring model in which traffic was reduced in the city center by a third (Garfield, 2018). However, London is still the most polluted by Nitrogen Dioxide among the European capitals. This poor air quality poses a significant environmental risk to public health in the UK, resulting in a loss of £2.7bn in productivity in one year (Asthana and Taylor, 2017). The British government found that raising the charge for diesel cars that drive into Central London is not enough to reduce air pollution and protect the health, and stricter rules are needed to be imposed. Therefore, the British capital decided to follow Paris and ban diesel cars. Along with that, it was announced in July 2017 that the country would ban sales of diesel and gas cars by 2040 as a part of the government's clean air plan (Asthana and Taylor, 2017).

To reduce the major risk on public health caused by traffic emissions, particularly from diesel vehicles, comprehensive political action is required to minimize the over-dependence on motorized private transport and invest in public transport, and ensure that everyone has access to safe convenient networks of walking and cycling routes (Graham, 2014). Therefore, the UK government is determined to invest, take strong action, and accelerate the development of local plans to improve air quality in the areas where emissions have breached EU thresholds (Asthana and Taylor, 2017). In order to reduce pollution and road congestion, the UK government encourages more sustainable modes of travel and highlights the benefits of car-free housing. In the car-free housing schemes, car parking is not allowed within site but only for disabled people (Morris, et al., 2009). London has a responsible procurement policy in which all new double-deck buses became a hybrid, electric, or hydrogen since 2018, and all-new single deck buses will be electric or hydrogen from 2020 so that all busses in 2037 will be zero-emission (C40,2019).

Oxford Street in London is a good example of a comprehensive model for transforming into a car-free zone. It was known as the most polluted street in the world with the highest levels of Nitrogen Dioxide (Griffiths, 2014). It has been pedestrianized after being a fume trap for years and causing a huge pollution problem because of the buses that pass through this vital, busy, and central shopping place. It is planned to be pedestrianized gradually through banning cars from 800m of the street in a year and the remaining parts of the street in two stages in the following years (O'Sullivan, 2017). The plan includes raising the streets to the pavement level to be more accessible and adding new seating areas, new public spaces, cycle lanes, and wider pavement.

The importance of the Oxford Street pedestrianization case is due to its complexity, where the places surrounding the street exist a long time before cars exist, almost two millennia old. In other words, although Oxford street is wide enough according to London standards, the surrounding streets are narrow and maze-like that buses had no alternative route but through Oxford street (O'Sullivan, 2017). Therefore, residents are concerned about forcing the traffic into the surrounding residential streets, and some of them see this as shifting the problem to somewhere else instead of solving it (Crerar, 2017). The answer for residents' concerns lies in the Crossrail, a huge underground east-west heavy rail project that will release the massive congestion from the central tube line underneath Oxford Street. The Crossrail project - (high-frequency service is due to open in 2021) - should take most commuter traffic and provide two major new stations on Oxford Street (O'Sullivan, 2017).

What distinguishes the British experience is that the government is keen to engage the people in decision making. The British government has made lots of efforts in engaging the public community, in which it does public consultation and listens to the people and asks them what they think about the detailed plan before taking final decisions. In the first public consultation, more than 60% of the 12,000 people who responded have supported the plan (Crerar, 2017). To avoid public resistance and prior to introducing the London congestion charge, information campaigns were held to provide knowledge about expected results and to explain in a straightforward manner the policy with details, and to inform the public about feasibility studies and assessments (Bongardt, et al., 2013).

Bologna- Italy

The city of Bologna's efforts for adopting car-free strategies dated back to 1989. Over the years, the city of Bologna has had steady steps contributing to its transition towards a car-free city, and the city government has taken several actions to achieve its aim. However, all actions have been taken with respect to the containment of the community and its needs. Unlike Paris and Mexico, and the strict policy they have adapted with citizens, Bologna understands that if citizens are well informed and engaged in urban mobility, the transformation will be successful. In this respect, the city invited citizens in June 2006 from different categories to share ideas with municipal technicians about new regulations and traffic restrictions. Based on their inputs, the Urban Traffic Master Plan was approved in 2007. Additionally, the city planned a campaign to raise awareness on the impacts of traffic on health, energy efficiency, and safety with the goal to change citizens' travel choices. The campaign aimed to reach all citizens including pupils and university students. A website was also planned to foster participation, exchange of experiences and information on existing cycling routes and make suggestions for improvements (CIVITAS, 2013).

Simultaneously, the city has applied smart technology to facilitate the transformation process and fit the community needs. It installed electronic pillars around semi-pedestrian areas within the limited traffic zone (LTZ) in the historical area. The pillars are active 24 hours a day and are automatically lowered from 6:00 to 10:30 and from 15:30 to 16:30 for freight delivery vehicles. The time slots can be adjusted, for example, when there are events or construction works or to allow the access of deliveries and transit of emergency vehicles. Besides, Bologna developed new software and updated the mechanical features of the system. It is connected to the database for the vehicle pass that grants access to the LTZ. Cameras record license plates of the vehicles that enter the LTZ and fine car owners that are not on the authorized car list. The city provides free software to manage the data of employees' trips. It also provides online questionnaires to collect information on employees' mobility habits. Based on this data, companies can develop better mobility plans (CIVITAS, 2013).

In addition to those actions, the city of Bologna applies a service to provide parking lots with public transport connections that allow people to leave their cars and transfer to public transportation, this service is known as Park-and-Ride system. It is worth mentioning that Bologna has used clean alternative fuels on public buses since 2002, and encouraged private car owners to shift from petrol to liquefied petroleum gas (LPG) and compressed natural gas (CNG) (CIVITAS, 2013).

Along with public transportation, Bologna provides an innovative mobility service that complements the local public transport and, at the same time, avoids the rigidity of traditional public transport. This mobility service is car-sharing. It is suitable for people who want to rent cars for a few hours and pay based on how long they have a car and the distance traveled. Each shared car can replace between 6 to 20 cars. The user can choose the operator near to his place, then register on the website, reserve a car located in a specific parking lot either online or by phone. The operator sends a membership card that can be used to unlock the cars (Richard, 2010). Accordingly, the Ministry and the Municipalities of Bologna signed with other Italian cities the first agreement to lay the basis to establish a national car-sharing network and establish an authority to coordinate the overall project. As a result, the "Initiative of Car Sharing" (ICS) was established, and it is responsible for the spread of car sharing in Italy (Bernardi & Rupi, 2015).

Moreover, Bologna developed a system for mobility eco-savings with credits to encourage sustainable mobility modes, which can be quantified and converted into credits. It is a reward mechanism for citizens granting free bus tickets or a free parking bonus (CIVITAS, 2013). Another motivational policy is that the city initiated a project to encourage citizens to use bikes in their daily travel. The project-based on providing different incentives for citizens that started shifting from cars to bicycles, such as a folding bike for users who successfully achieved the assigned targets. A bicycle for the participant who had covered the highest share in 'virtuous' kilometers (Di Bartolo, 2015).

On the other hand, some restricted rules were imposed by the city to reduce undesirable behavior such as illegal on-street parking, including the introduction of differentiated parking fees depending on the environmental characteristics of vehicles, and introducing paid parking permissions for residents for the second car per family. Additionally, Bologna implemented a road pricing policy based on an intelligent transport system (ITS) that can distinguish vehicles according to vehicle emissions, models, and size. In 2003, to minimize private car dependency and to promote walkability, Bologna adopted an Urban Traffic Safety Plan based on the

Italian road safety plan and the EU's White Paper on Road Safety plan. Both aim to minimize accidents through the focus on the introduction of a 30 km/h speed limit, and creating detailed plans for individual cross-roads or street networks (CIVITAS, 2013).

The city also promoted strong Internal and external cooperation, which involved different aspects of management and practical organization provided by the ATC, which contains the parking management department, and the technological department which plan call center and administration center. Marketing offices organize marketing campaigns to increase awareness among the people of this new means of transport (Thøgersen, 2009). In addition, a new traffic control center named "Cisium" that will eventually manage traffic in the entire metropolitan area. Cisium will facilitate sending messages on the traffic situation in real-time via car radios, onboard GPS navigation devices, and to other traffic operators like the motorway control center (CIVITAS, 2013).

All of those actions yielded positive results, such as the reduction of pollution emissions and a shift from private vehicles to public transport. In 2010, about 4,000 users refrained from using their own private cars, which resulted in reducing the harmful emissions by 35% compared to 2008 (CIVITAS, 2019). There has been a strong increase in the use of public transport following an agreement with the local public transport operator (ATC) and the national railway operator (Trenitalia) that made it possible to cut the costs for annual transport passes (CIVITAS, 2013). Furthermore, in 2009, Bologna was considered as the most sustainable city in Italy because the city has applied many innovative actions such as depending on cars with low environmental impact, car sharing, bike sharing, mobility manager, carpooling, etc.), and it has a strong public transportation network (Eltis, 2014). In 2011, Bologna won the European Mobility Week "EMW" Award for its efforts towards sustainable mobility modes. In addition, Bologna won the CIVITAS "City Vitality and Sustainability" Award for Technological Innovation in creating a transport system that integrates traffic monitoring and rule enforcement (Porta, et al., 2009).

In addition to the long experience of Bologna in the process of transforming into a car-free city, the incentive policy that it adopted to mainly work on the people's comfort and involves their ideas and needs, have helped Bologna to achieve its intended goals and exceed.

Johannesburg, South Africa

Johannesburg, the economic hub of South Africa, is one of the highest emitters on the African continent of carbon dioxide (CO2) emissions. Therefore, it has committed to international partners and governments around the world to ensure that the city works on reducing greenhouse gas emissions. Accordingly, it has to reduce the dependence on fossil fuel vehicles and to take steps towards more sustainable mobility options (Maota, 2015).

Similar to Bologna, the city that put its people first, the experience of Johannesburg City is considered less stringent than other experiences, in which the local officials aim to persuade people to opt for public transit rather than imposing strict laws. In Sandton, the most congested area in Johannesburg City, 70% of travel trips were undertaken in private cars and more than 85,000 vehicles, carrying close to 120,000 people, move in and out during peak hour traffic each day, and these numbers are increasing at more than 3% per year. With such an increase, the city would not find enough space that accommodates those numbers in addition to congestion and pollution risks. This problem probably generates the need to move to more sustainable solutions. In 2015, in the Sandton business district, the government held an event called "EcoMobility Challenge". It was a month-long celebration for sustainable urban transport that involves featuring activities, exhibitions, and dialogues. For one month, the people were encouraged to move by public transport, bike, on foot, or any other sustainable transport option (Joburg, 2015). The goal of the event of "EcoMobility Challenge" is to allow people to visualize their city without cars within a sustainable applied approach and to experience how walking and cycling can be accessible, safe, and more attractive. To achieve this goal, the city residents ought to ditch their cars and use greener, healthier modes of transport. In order to prepare the city and community for the EcoMobility Challenge, the local officials installed new bike lanes and bus routes, wider sidewalks, park-and-rides, light rail cars, and bike rental stations, in addition to closing down certain streets and some car lanes in the Sandton center to promote bus use. People were not prevented from driving cars (Garfield, 2015(b)). However, it was more comfortable and more attractive to use eco-mobile modes of transport. The city designated the closed streets for walking and cycling, for leisure, sport, contemplation, and any other activities for public participation. The people also had the chance to participate in a series of public dialogues with international mayors, and local stakeholders. Throughout the dialogues, the different partners with citizens discussed possible solutions for Johannesburg City to improve sustainable and low carbon urban mobility, reduce the dominance of private vehicles in cities, and build an urban transport system that meets the needs of citizens while minimizing energy consumption, emissions, and space requirements. Besides, there was the EcoMobility Exhibition in which the eco-vehicle manufacturers, service providers, and innovators displayed their products and obtained live feedback from visitors who had the chance to test eco-vehicles on a specially-designed track. The EcoMobility Challenge event was socially successful, in which people showed positive responses and experienced cleaner, less stressful, and more joyful urban spaces with fewer cars and less CO_2 emissions (Joburg, 2015).

The city aimed to get 30% of drivers to use public transport (Maota, 2015). The majority of people were not willing to give up their cars. However, the number of car commuters dropped by 22% during the month of the event, and the city hopes that people change their commuting routine. In such a transformation to a new form of the city, the people feel anxious to change a behavior that they are used to. This is why the public should be involved in making future urban development decisions (Garfield, 2015(b)).

However, the city is still facing real changes. There is still a lack of reliable public transport that could accommodate the current population, and people rely on unregulated minibus taxis, which handle last-mile connectivity but increase congestion. Additionally, the aging traffic flow control systems contribute to the congestion where a large number of traffic lights in Johannesburg do not work. New pedestrian and cycle bridges are being built to promote bicycle use, but few are using them due to many reasons including theft, affordability issues, and the social stigma associated with using a bicycle. The accessibility and affordability of public transport for disabled and low-income groups are challenges. The rail system (Gautrain), Metrobus, and Rea-Vaya Bus Rapid Transport (BRT) systems need to improve stations and expand services for disabled people. Many people use private transportation to avoid vandalism and the theft of train cables, both of which are prevalent on trains. Minibus taxis help with first- and last-mile transport problems in Johannesburg but are mainly unregulated, poorly maintained, and overcrowded.

Therefore, the city should work on a comprehensive framework that tackles the different challenges and not rely only on promoting people's awareness. However, no one can ignore the efforts that have been made. The city of Johannesburg is taking steady steps forward to promote active modes of transportation, as the government is expanding a BRT system, and improving the city roads. To reduce the air pollution in Johannesburg, the city has introduced eco-friendly bus fleets, which are expected to save 1.6 million tons of carbon dioxide emissions by 2020. The city is creating well-planned 'Corridors of Freedom' with accessible public transport options and safe neighborhoods to promote walking and cycling. Electric vehicle infrastructure projects are expected to become an innovation hub in Africa, testing mobility solutions in emerging markets (Deloitte City Mobility Index, 2018).

It is worth mentioning that Johannesburg is recognized as one of the best three leading cities in Africa according to its environmental progress and extraordinary commitment to taking urgent action and building a greener future over the year (C40, 2019).

Addis Ababa, Ethiopia

Addis Ababa is another African example of cities that aims to solve mobility problems while simultaneously building the community's culture. Addis Ababa, the capital city of Ethiopia, is an exceptional example to transform into a car-free city since it is still among the least motorized countries in the world, and car ownership among residents is very low due to the high import and ownership tax of a combined 300% (Gerth-Niculescu, 2019). Investigating on the transportation means used in Addis Ababa reveals that the majority of the population in Addis Ababa have low-income levels, and most citizens are used to walk or take public transport such as minibuses, buses, and taxis, as they are relatively affordable means of transportation (Abubaker, 2019).

However, Addis Ababa is the fourth most polluted city in Africa (Gerth-Niculescu, 2019). The rapid urbanization in Addis Ababa has led to the urban crisis such as informal housing, poverty, and social inequality, which in turn led to intense traffic congestion due to the heavy reliance on poor minibuses, as demand for modern transport has increased faster than the city can provide it. In turn, this situation creates health and safety risks by producing more greenhouse gas emissions (Oranga, 2015).

In response to these challenges, the city offered a flexible and affordable option to solve the mobility problems, while building a community, which is skating. For low-income residents, skating seems to be an affordable option, especially for children who cannot afford bicycles to access education and job opportunities. Such a transport option that promotes a healthy, active lifestyle, demands practice and learning. The role of local organizations was effective in this regard; as Ethiopia Skate, a grassroots organization contributes to teaching low-income youth how to skateboard. Moreover, the NGO Ethiopia Skate organization takes this urban sport to underprivileged children, providing them with skateboards and regular training. In turn, all of these efforts contributed to transforming skating into a legitimate transport option that is affordable, sustainable, and promotes a sense of community. An extended effort to activate skating as a new transport option was suggested, such as building a new ramp around the city to teach lessons and get children excited about skating. Another suggestion was to create a skate park. The cooperation between transport and public parks into the city could potentially contribute to mitigating climate change, reduce congestion, and improve mobility (Abubaker, 2014).

Additionally, Addis Ababa has joined other African cities such as Cape Town to shift towards more pedestrian-oriented urban development by holding a regular car-free day "one Sunday every month". On this day, the city closes the roads at the heart of the city and encourages pedestrians to go out to public spaces and streets. Similar to other cities, Addis Ababa encourages citizens to walk, socialize, play sports, and do activities with colorful gear and proudly lead group exercises in a festival atmosphere that reflects the diverse social and cultural fabric of the Ethiopian community, figure 2. It became a familiar scene seeing people dance in public spaces. Tents were also set to offer free health checks to those who were walking and exercising. Instead of accumulating the whole activities in a specific area, the decentralization of this event to six districts within the city contributed to making the event more accessible and far-reaching, and activities were more likely to be within walking distance for residents (Abubaker, 2019; BBC, 2018).

Figure 2. Residents in a workout led by local sports associations at Addis Ababa's 4th car-free day(Abubaker, 2019)

The increased number of skating communities and people who join car-free days inspired the city's administration that holding this event more regularly can make a clear difference.

Cycling was not a common commuting option because of the topographic difficulties and the relatively high prices of bicycles. However, the city's administration aspired to promote cycling in the country despite the topographical and financial challenges. Thus, the government donated around 600 bikes to its employees (Gerth-Niculescu, 2019).

Overall, although the improvements achieved by the Addis Ababa government seem limited when compared to the European countries, however, their efforts are organized, targeted, and in line with the Ethiopian society culture. Addis Ababa is going on the right track in achieving its goals and further efforts to provide its people a healthier life.

Madrid, Spain

The car-free city initiative in Madrid is not new. Madrid joined the movement that started late in the 1980s to claim cities back for pedestrians through limiting spaces for private cars (Medina, 2018). Madrid is one of the 17 Regions in Spain, with a traditional dense morphology, a population of more than 6 million inhabitants and an area of 8.028,5 km^2. The car-free city movement in Madrid was supported by strong transportation infrastructure. In Madrid Region, there are seven public transport modes: Metro (underground) system, Madrid city urban buses system, Suburban railway system, Metropolitan and regional buses, Interurban buses, Urban buses in other municipalities, Light rail systems, and Big bus interchanges. The Public Transport Authority of the Madrid Region- (Consorcio Regional de Transportes (CRTM))- created in May 1985 by law as an autonomous agency of the Madrid Regional Government- has developed an Integration Model of the Public Transport System, and an interchange plan to encourage people using public transportation and to improve the quality of the service. The integration model has many objectives, including establishing an integrated Fare System for the whole Public Transport Network, and the planning of transport services and defining the coordinate operating programs for all Transport Modes. The interchange plan aims to facilitate the combined use of different means of transportation through connecting the urban bus network to the circular metro line at the main interchanges nodes, in which the metro is the core of the public transport system in Madrid (Velasco, 2016), figure 3. The interchange plan has already reduced journey time and improved the quality of the service.

Also, Madrid Public Transport Authority or Consorcio Regional de Transportes de Madrid (CRTM) has developed in late 2013 Madrid ambitious plan that is called "*2020 Sustainable Mobility Urban Plan*". The 2020 Strategic Mobility Urban Plan (SMUP) of the Madrid Region is covering 179 municipalities. It aims not only to reduce urban congestion through having an efficient, safe, and sustainable integrated public transport but also to improve the quality of life of the citizens through encouraging them to minimize private car dependency. The analysis of SMUP evolution until 2017 shows that measures of the plan are focusing more on public transport, and less on social culture aspects (Mozos-Blanco, et al., 2018)

To encourage people to travel further with less car use, Madrid was the latest city to start the public bike share scheme. However, it was the first European city to use electric bikes. The advantage of the electric bike is that it has an electric motor that provides the push on the hilly urban landscape in Madrid, which would

Figure 3. Buses interchange plan- urban bus network connecting to the circular metro line, (Velasco, 2016)

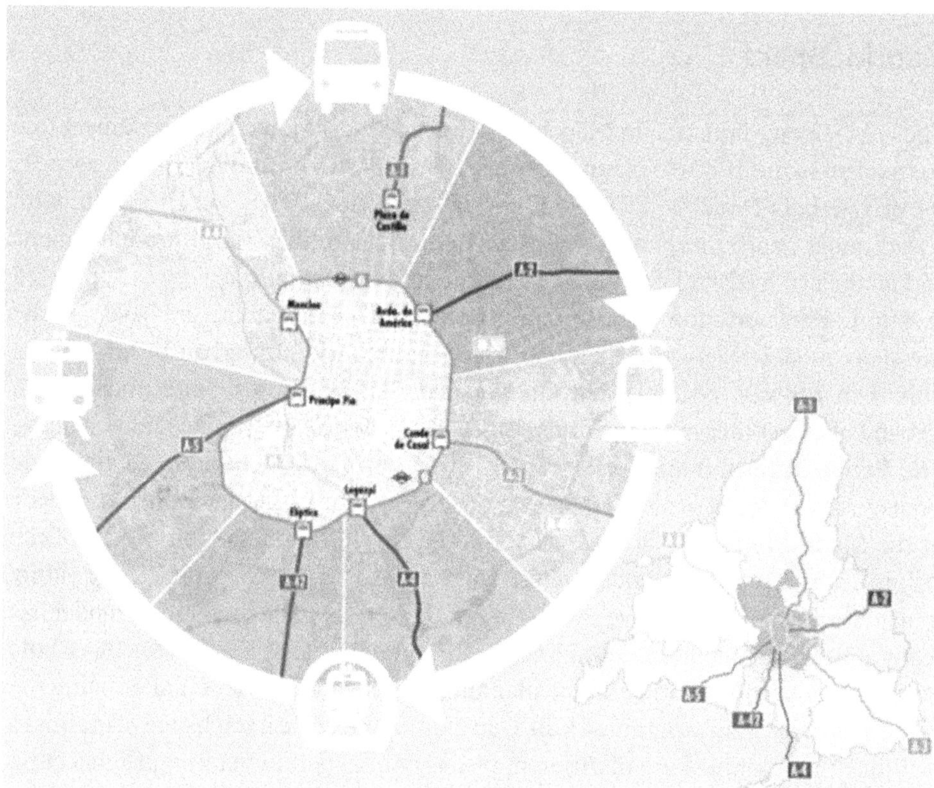

encourage people to get out of their cars (Walker, 2014). Despite the efforts invested in sustainable mobility plans, the quality of cycling mobility and pedestrian initiatives and implementation needs to be improved (Mozos-Blanco, et al., 2018).

Despite the early start in the car-free city movement to reduce greenhouse gas emissions and improve air quality, Madrid has oriented its efforts to invest in planning public transport infrastructure, and has not spent enough effort to involve and engage the community in decision making. Thus, and similarly to Paris, and Mexico City, Madrid struggled with the strong opposition from local business owners and citizens towards any traffic restrictions. For example, Madrid City set up its first pedestrian-only zone (Residential Priority Zones) in a historic, densely packed neighborhood. This action aims to exclude the downtown streets from being "through-routes" across the city and to reduce the traffic passing through its neighborhoods (Garfield, 2018). However, there was not enough support to extend these traffic restrictions to other areas to the extent that ten years later, there was a strong opposition towards any traffic restrictions from local business owners, transportation professionals, and citizens (Medina, 2018).

In line with the restricted policy, the Spanish capital confirmed that in November 2018, Madrid downtown streets are not for drivers, and planned for gradual implementation. Madrid set a mobility plan that aims to gradually pedestrianize the city center to become completely car-free by 2020 and increase space for pedestrians by nearly 22,000 m^2. This was carried out by keeping cars out of 500 acres (2.02 km) in the city center where the city urban planner redesign 24 streets of the busiest in the city to be for walking rather than driving. The objective was to keep cars out of passing through neighborhoods and to allow only cars that belong to local residents, zero-emissions delivery vehicles, taxis, and public transit, and any other vehicles entering this zone will be automatically ticketed. This would potentially cut down on car traffic, car usage from 29%-23% daily, and air pollution in the city. To facilitate implementation, two months were given as an allowance period to warn cars driving through downtown and breaking the new rules about the fine that they will be charged in the future. In the third month, an automated system with cameras took place, and they started enforcing the rule and charging people who breach the rule (Medina, 2018).

The integrated model of the Public Transport Network and its connectivity in Madrid is the key of success to achieve the sustainable urban mobility plan. However, more efforts are needed to improve the cycling mobility and pedestrian spaces to minimize the private car dependency and empower citizens of a better quality of life. An additional improvement is also needed in the area of community involvement.

Oslo, Norway

What distinguishes Oslo's experience in transforming into a car-free city is that it is one of the earliest experiences in Europe that started in the 1970s. The long and early process that Oslo passed through has prepared the city to overcome the challenges, evolve effective solutions, and to develop a clear vision that is based on putting people first and engage them in the decision making. From the beginning, Oslo adopted a gradual implementation method to achieve its goals of freeing Oslo streets from cars, eliminating the environmental pollution, and engaging citizens smoothly in the entire process.

Oslo has started the car-ban plans as early as the 1970s through pedestrianizing some streets in the city center. In the 1980s, the capital invested heavily in public transportation. Furthermore, in 2015, the city government in Oslo planned for a more significant car ban transformation and announced the city center to be car-free by 2019 as a part of the 'slash' greenhouse gas emissions plan (Cathcart-Keays, 2017; Bliss, 2018). In order to transform into the car-free center, changes started in the city center in 2017 through planning to create more space for city life and build a minimum of 60km of bicycle lanes, whereas buses and trams continue to serve central Oslo with more new lines, more frequent departures and lower cost of tickets. Oslo started the implementation gradually through banning parking and replacing over 700 parking spots with bike lanes and small parks, streets were filled with people, and the only access to the city center was limited to emergency vehicles, few spots for disabled drivers, and limited hours for delivery cars. Additionally, the capital invested heavily in public transportation and replaced more than 35 miles of roads that have the capacity of about 350,000 cars with bike lanes (Peters, 2019; Bliss, 2018).

Additionally, Oslo runs a City Bike project called Urban Sharing, which is a local bike-share system that has quickly grown and is operated by the Urban Infrastructure Partner (UIP), a technology-driven operator of shared infrastructure. The City Bike project is a technology platform for forward-thinking efficient and resilient mobility solutions that cope with the needs of cities as they grow and evolve. Oslo City Bike that was launched in Spring 2016 is considered one of the most efficiently-run bike-sharing schemes in the world that locates and unlocks bikes through a user-friendly App (Hendrix, 2017). Expanding the biking infrastructure improves the bike-share experience and would encourage people to switch to bike commuting. This would result in mitigating congestion on public transportation and would, in turn, encourage car users to use public transportation (Shemkus, 2019).

Oslo has taken different actions towards the car emissions problems as the transport sector accounts for 65% of greenhouse gas emissions (Ruter#, 2018). Complementing the efforts made previously and continuity to achieve the city's vision, Oslo Started with the Fossil Free 2020 plan and planned to have only buses that run on clean and renewable energy and to replace the current diesel buses with electric ones. Oslo also has further developments up to 2025 to ensure that most of the city is zero-emission by 2030 (C40, 2019). Furthermore, Oslo plans to have only electric cars in the city by 2030, and by 2050 to be fossil-fuel-free (Cathcart-Keays, 2017; Bliss, 2018).

Oslo aims to reduce pollution and improve air quality, improve the quality of life, provide safer streets for pedestrians and cyclists, and dedicate the spaces that will be available when stop using private cars to public spaces and sidewalks (Garfield, 2015(a)). Along with environmental benefits that would result from slashing transportation emissions, the core of Oslo's plan is putting people first, providing more public space, and encouraging using it (Shemkus, 2019).

Oslo experience is an evidence that the gradual implementation of policies and actions, and involving the community in the process from the beginning results in a smooth and successful transformation into a car-free city, while avoiding public resistance and dramatic collisions with the community.

Copenhagen- Denmark

Similarly to Oslo, Copenhagen represents one of the early practices of the car-free city. It is a strong example of managing and implementing car-free practice over the long-term (Tønnesen, et al., 2016). The car-free activities started at the historical center in 1962, when the government began a process of pedestrianization for its central core and the central street network (Wright, 2005). Additionally, parking has been removed from a total of 18 squares to give space to more city life. Today the city has about 3,000 parking spaces; however, the city government classifies the parking areas into zones, where the red zone is the most expensive one. Middelalderbyen, the historic central area, is located in the red parking zone, where residents have to buy a yearly resident license. However, it is not desirable for all residents not only because it is costly, but also because the use of the residents' license is limited to specific places at specific days and times where they have to buy parking tickets like everyone else. Therefore, the available parking zones becomes a more suitable option for visitors and businesspeople (Maiken, 2016).

In the early 1970s, 'Car-Free Sundays' were introduced in Copenhagen, and there were demands that all of Copenhagen became car-free. Nowadays, an annual weekly event of car-free days in September since 2002 has been held in the area of Copenhagen center "Middelalderbyen". The main objective of these car-free days was to raise awareness about the benefits of the alternatives to individual car traffic. The events spanned from one day to three days in a row. During the events, cars were banished from the city center in the period between 06:00 am to 08:00 pm due to the fact that peak commuting to and from the area is in this period. However, exceptions were given to residents and businesses in the area, which means that they were not affected by the restrictions during the event. Moreover, deliveries of goods were allowed between 04:00 am to 11:00 am (Maiken, 2016).

The city began the transformation process gradually with Strøget Street, which is the longest pedestrian shopping street in the center, and even in Europe, then it transformed many other streets such as Vimmelskaflet, Vesterbrogade, Ostergade, and many others. Along Strøget Street, there are many local streets with mix-used buildings used by both pedestrians and vehicles; however, the speed is limited. In the central area, more than 20 different strategies for walkable streets and alleys have been woven into a pedestrian network restricted motor vehicle traffic (Wang, 2011). However, the transformation of the main street into a pedestrian street caused much public debate "No cars means no customers, and no customers means no business," said tradesmen. Other voices claimed that "there is no tradition for outdoor public life in Scandinavia." However, in no time, Strøget proved to be a huge success as a pedestrian street, in both popular and commercial terms, when the culture of walking spreads among citizens. Since 2002, more streets and squares have been pedestrianized. Gradually a cohesive network of pedestrian streets was developed for people. It is easy to walk around in Copenhagen from one end of the city to the other, and today foot traffic represents about 80% of the movements in the inner city (Maiken, 2016). Although the public debate, the number of people who engage in recreational activities on the streets and squares of the city center on a summer day has increased during summer's day, and the area brings year-round benefits to the merchants (Wright, 2005). This is because people have had time to change their commuting culture and traffic habits and patterns (Gehl Architects, 2019).

In addition to walking, Copenhagen encourages the use of bikes as well. Concerns about air pollution, climate change, and the need for people to get enough exercise in addition to Denmark's heavy taxes on petrol and automobiles are all factors that encouraged Copenhagen to shift to biking. In Denmark, people cycle in all types of weather and at all times of the day. At rush hour, bike lanes in Copenhagen can be as crowded as car traffic in other parts of the world as over half of Copenhagen's population bikes to work every day. The city provides wide cycle paths and cycle bridges to increase safety. A cycle superhighway creates better conditions for cyclists, and connects work, study, and residential areas and run near stations, which makes it attractive to combine cycling with public transportation (denmark.dk, 2019). Today, Copenhagen boasts more than 200 miles of bike lanes and has one of the lowest percentages of car ownership in Europe. The first of 28 planned routes opened in 2014, and 11 more are completed by the end of 2018. The city has also pledged to become completely carbon-neutral by 2025 (Tønnesen, et al., 2016). Copenhagen might be the only city providing free use of bicycles (Wang, 2011). Many projects for developing bikes to facilitate mobility and foster sustainability were initiated, for example, a super pedestrian project for intelligent electric vehicles, which produces smart, responsive hybrid e-bikes that function as mobile sensing units. It captures the energy dissipated while cycling and braking and saves it when a boost is needed. It also maps pollution levels, traffic congestion, and road conditions in real-time. The bike can be controlled through smartphones. The wheel's sensing unit can capture the rider's effort level and information about the surrounding environment, including road conditions, carbon monoxide, NOx, noise, ambient temperature, and relative humidity. People can access this data through the phone or the web and use it to plan healthier bike routes (The Copenhagen wheel, 2019).

The long-term planning and gradual implementation were the success keys that helped Copenhagen in its transformation into a car-free city. The essence of this success can be concluded in two points: 1) car owners had time to get accustomed to the idea that it is much easier to bicycle or use public transport as more sustainable commuting alternatives, and 2) residents have developed a new commuting culture smoothly. Therefore, Copenhagen is considered as one of the most successful experiences in transforming gradually into a car-free city and culture.

CHALLENGES AND APPROACHES FOR SOLUTIONS

Reviewing the nine case studies of leading cities in their leading initiatives to transform into car-free cities have shown that the risk of pollution stimulates the importance of shifting towards the car-free city paradigm as the main motive. The pollution ratios have exceeded the acceptable threshold and are posing a high health risk. This chapter presents some leading global practices of transforming into car-free cities around the world and discusses the problems that some cities have solved while implementing and other problems that appeared due to the approach of implementing. This chapter aims to identify those challenges and to identify the success measures and recommended strategies for efficient car-free implementation.

The following part synthesizes the major challenges that faced the selected cities while transforming into car-free. The challenges include:

1. **Public resistance**: Public resistance and collisional confrontation is a result of resorting to quicker and more severe solutions because of the urgent need for immediate action to deal with the high risk of pollution. Not involving the public with the implemented policy has a negative impact, in which people would resist any policy when they get fined either because they did not know about the banning time or because they have to use their private car during the banning time as in Paris. People may not resist, but they can circumvent the regulations as in Mexico.

In any transformation or changing behavior or habits, people tend to be anxious to change what they used to do. Therefore, the public should be involved in making future urban development decisions. This has proven to be effective with several cities, in which some held information campaigns to provide knowledge about expected results and to explain the policy straightforwardly with details, as in London. Moreover, it helped the authority in other cities to modify the decision after being taken as in Oslo and Copenhagen, where business owners objected, and they were given exceptional access to some areas in specific times for delivery.

2. **Insufficient Public transport infrastructure:** lack of reliable public transport that is insufficient to accommodate the city's population is a very critical factor when considering transformation into a car-free city. It results in increasing the dependency on private cars, as in Johannesburg, where congestion increased, and in Addis-Ababa, where the insufficient transportation infrastructure is not encouraging people to use it.

Providing well-established public transportation alternatives supported by a robust infrastructure connecting all parts of the city minimizes the private car dependency as in Copenhagen, Bologna, Oslo, and Madrid.

3. **Insufficient cycling and pedestrians' network:** Lack and insufficient pedestrian spaces and network, and cycling network influence negatively the dependence on private cars. The challenge of cycling and pedestrian networks has three sides: lack of networks, improper selection of the zones selected to be pedestrianized, and/or lack of walkability culture. Some cities had insufficient pedestrians and cycling networks or that those networks were not connected to public transport networks. Other cities had pedestrianized zones where people have business in and can't afford not driving through. And, other cities had the challenge that people are not using the pedestrians and cycling mobility infrastructure as walking and cycling are not part of their community culture, which resulted in more congestion and useless network and pedestrian zones.
4. **Lack of humanized urban planning**: The current urban planning design in most of the cities around the world is based on giving priority to cars over pedestrians. This kind of planning is very challenging for and hindering the implementation of the car-free initiative. One of the challenges is the centralization of new extended sustainable commuting services in some areas and depriving others. This means that the public transportation networks do not reach a more significant fraction of residents, who would not have any other option but to use private cars for commuting.
5. **Topography**: Topography has been a challenge for some transport modes. For example, when it comes to using bicycles to commute, it might be difficult for people to bike regularly on elevated and hilly lands. It was a challenge that faced some cities when implementing car-free. In Madrid, this challenge was overcome by using the electric bike with an electric motor to provide push on the hilly urban landscape. In Addis Ababa, the hilly land still a challenge for decision-makers that need an affordable solution.
6. **Lack of financial funding**: Lack of funds poses a significant challenge, especially in developing countries, where it would affect promoting mobility services and infrastructure. In Addis Ababa, the lack of funds affected the quality and quantity of the transportation system to fulfill the essential demand; hence, Ethiopian people restored to more affordable transportation modes like skating.

CONCLUSION

To conclude, the idea of the car-free city was not wholly accepted, and the possibility of achieving was doubted. However, some cities in different parts of the world have attempted, and others are still attempting to shift to car-free in varying degrees by setting appropriate policies and committing to them. All these attempts of positive transformation and emancipation of private fuel vehicles were so challenging that they needed great efforts to overcome; however, they were indeed inspiring for many countries.

After reviewing the above practices and identifying the challenges that different cities have faced, it can be concluded that the successful practice of transforming into the car-free city is based on three main pillars: conscious planning and sufficient and robust infrastructure, efficient and resilient policies, and enlightened and cooperative community members. Those three pillars are the recommended actions for successful implementation and can be described below as Planning, Policies, and People (3 Ps):

1. **Planning: Conscious planning and strong and sufficient infrastructure:** planning should facilitate people's mobility and commuting services to dispense of private car use. In order to do this, well-established public transportation infrastructure and effective urban planning must be in action. Both are important factors that helped cities in their transformation into car-free cities. To encourage people to reduce dependency on private cars, a well-developed and integrated transportation system that reaches all parts of the city is essential. The establishment of a strong infrastructure and the improvement of existing public transport services require sustained and robust funding. Moreover, investing in public transportation while considering the supply and demand encourages people to reduce car-dependency.

From a planning perspective, effective urban planning can play a significant role in encouraging the sustainable practices of commuting and improving the efficiency of the implementation. Encouraging walkability can be achieved through creating public transportation stations within a five-minute walk, a clear and organized street networks that link different places of activity, and a safe and attractive environment with an adjacent mix of uses and activities that provide an enjoyable walking environment (Priyank P. Patel, et al.,2016).

On the other hand, the decision-makers in each city have to assure the plan decentralization of the transportation system. Instead of accumulating the whole new activities, systems, infrastructure, strategies, etc. in a specific area. The decentralization contributes to making all parts of the city more accessible and more likely to be within walking distance for residents.

The frequency of public transportation in relation to the population and demand is another important factor. To manage transport and congestion and to improve the traffic flow of all modes, both: 'supply-side' or 'demand-side' measures should be considered, whereby supply-side measures aim to increase the capacity of the transport system and demand-side measures aim to reduce the level of car demand (Enoch, 2016).

2. **Polices: Efficient and resilient policies:** as one of the main pillars in the success of car-free city implementation, it was found that the diversity and flexibility of policies, introduce different alternate solutions (strategies and actions) for different challenges. Reviewing those strategies and actions has shown that each city used a variety of strategies that can be graded from decisive strategies to flexible ones. However, encouraging and rewarding policies have been proven to be more acceptable, hence, more successful for implementation. Therefore, policies should be designed based on incentives rather than penalties to avoid public resistance. Decisive strategies such as paying a fine may not be desirable for people and sometimes reflect weakness and inefficiency of the strategy. Thus, it is important to provide people with a proper alternative before enforcing them to specific actions that may obstacle their life routine. Encouraging strategies and using incentives such as lowering the price of public transportation, free bikes, car-sharing service, and public transit pass, or making public transport more attractive, affordable, regular, and extensive can have a positive impact on people.

The car industry is an important party and should be involved in the process of transformation as well. They will get impacted by the decisive governmental decisions to ban sales of diesel and gas cars. Therefore, governments should involve them in the development of policies and decision making. This would significantly contribute to orienting the community travel behavior towards more sustainable alternatives through investing in clean and eco-mobility alternatives.

Also, the cities' experiences in transforming into car-free cities showed that as earlier the cities started the transforming process, the more efficient decisions taken by the government. The long-term plans for the transforming process allow the government and authorities to modify as early as possible the implemented strategies upon the consequences of the decisions taken. In turn, more suitable policies, strategies, and action plans are developed and are more applicable and acceptable by the community.

3. **People: Enlightened and cooperative community members:** the core concept of successful car-free city transformation is based on putting people

first. Public participation is an important factor in promoting the car-free concept. People should implement car-free policies willingly and to support it. In order to achieve this, people's awareness in this regard should be raised, and the other 2 Ps (policies and planning) should serve people to improve the quality of their lives, in which they should feel the real benefits of the car-free transformation in terms of health, less congestion, more outdoor spaces, less noise, easy mobility, safe walking, etc. This would encourage people to interact and participate more positively in the transformation process.

People's awareness can be raised using different means, including involving people in the decision-making process to explain and assure how the new plans fulfill their needs. Sharing ideas in a public dialogue between citizens and other parties, such as the municipal technicians, could significantly develop the whole process of transforming into car-free cities. Discussing the applied policies and regulations with people open their minds to accept the idea of changing their commuting behavior and reducing car use. Also, the weekly/monthly initiative of the car-free day allows public participation by letting people experience social and environmental benefits, which in turn would increase the efficiency of implementation. This will ensure that new legislation would be accepted and applied.

Another important factor that reduces public resistance is involving different parties of the community in the process of discussing the policies and implementing the strategies. For example, the cooperation between the government and the non-governmental associations and car industry. The role of local organizations also has shown efficiency in this regard, in which they provide funds and develop new initiatives.

Also, being a partner with an international community such as C40 organization would ensure that the city would have continuous efforts to reduce pollution and encourage car-free practice.

Allowing the public to be involved in fulfilling the vision during the decision-making process and before making any decisions would increase the efficiency of the implementation of the car-free city plan. This can be achieved through utilizing smart governance and using advanced technology to coordinate among the different municipalities and stakeholders (Doheim, et al., 2019). Most of the mentioned cities applied the concept of smart governance by using different tools to achieve car-free cities' goals. For example, establishing a technological transport system that integrates traffic monitoring and rule enforcement, or surveying people on proposed policies. Based on this, better communications between people, different parties, and the government is assured; hence, better mobility plans can be applied.

The above 3 Ps should be balanced and incorporated together for a successful car-free implementation process. The people are the core of the process, whereas the quality of their lives is the main target. Planning is the tool to achieve this target, and the policy is to assure the effectiveness of the process and the continuity on the right track. Although all precedent practices are mainly built on the three pillars discussed above, each experience was distinctive and had its uniqueness and identity due to the differences in cultural, social, geographical, and political aspects.

Futuristic short and long-run plans will put the city on track of continuous achievements. In this regard, it is essential to keep measuring, recording, and evaluating the transforming results. In order to ensure the continued success of Car-Free Cities, there should be an evaluation of such success. The evaluation of the measures, strategies, and policies used help in solving and avoiding any problems in the future. Benchmarking against the good practice (international/ regional/ national) may help to identify the level of success.

REFERENCES

C40. (2019). Celebrating the best performing cities in C40 regions in 2018. *C40 Blog*. Retrieved from https://www.c40.org/blog_posts/participation-standards-2018

Abubaker, I. (2014). Friday Fun: Community building through skateboarding in Addis Ababa. *The City Fix*. Retrieved from https://thecityfix.com/blog/friday-fun-community-building-through-skateboarding-addis-ababa-ethiopia-iman-abubaker/

Abubaker, I. (2019). Reclaiming the Streets: Addis Ababa, Other African Cities Launch Car-Free Days. *The City Fix*. Retrieved from https://thecityfix.com/blog/reclaiming-streets-addis-ababa-african-cities-launch-car-free-days-iman-abubaker/

Adbelhamid, M. M., Elfakharany, M. M., & Elfakharany, A. M. (2018). Reducing Private Cars Dependency to Achieve Sustainable Urban Environment in Congested Cities. In *7th International Conference on Modern Research in* (pp. 1-13). Munich, Germany: Academic Press.

Amoly, E., Dadvand, P., Forns, J., López-Vicente, M., Basagaña, X., Julvez, J., ... Sunyer, J. (2014). Green and blue spaces and behavioral development in Barcelona schoolchildren: The BREATHE project. *Environmental Health Perspectives*, *122*(12), 1351–1358.

Asthana, A., & Taylor, M. (2017). Britain to ban sale of all diesel and petrol cars and vans from 2040. *The Guardian*. Retrieved from https://www.theguardian.com/politics/2017/jul/25/britain-to-ban-sale-of-all-diesel-and-petrol-cars-and-vans-from-2040

Avila-Forcada, S., & Medina-Martínez, I. (2019). *Commuting Trends in Mexico City (2007-2017). Urban Mobility in Mexico City*. Working Paper. FLACSO.

Badger, E. (2016). Paris just banned all cars made before 1997. *The Washington Post*. Retrieved from https://www.washingtonpost.com/news/wonk/wp/2016/07/02/paris-just-banned-all-cars-made-before-1997/?fbclid=IwAR2ZqVtH03PEilOgVyVteBton3GIoZgH3c8WGPGbSUwjPXkkO-JEQjMLuKQ&noredirect=on

Baynes, C. (2017). Thousands of cars banned from roads as Mexico City chokes in longest pollution alert for two decades. *Independent*. Retrieved from https://www.independent.co.uk/news/world/americas/cars-banned-mexico-city-longest-pollution-alert-smog-a7746756.html

BBC. (2018). Thousands walk on Ethiopia Car Free Day. *BBC News*. Retrieved from https://www.bbc.com/news/world-africa-46499036

Bernardi, S., & Rupi, F. (2015). An analysis of bicycle travel speed and disturbances on off-street and on-street facilities. *Transportation Research Procedia*, *5*, 82–94. doi:10.1016/j.trpro.2015.01.004

Bliss, L. (2018). The War on Cars, Norwegian Edition. *City Lab*. Retrieved from https://www.citylab.com/transportation/2018/05/oslos-race-to-become-a-major-bike-haven/559358/

Boffey, D. (2020). European commuters still choose cars and congestion over public transport. *The Guardian*. Retrieved from https://www.theguardian.com/world/2020/mar/03/european-commuters-still-prefer-cars-to-public-transport?fbclid=IwAR28I-p95vCiT2h_h3bS64koAJjURcDSnIaHVQ6ORRPXTbU6OZbyKbT1wMw

Bongardt, D., Creutzig, F., Hüging, H., Sakamoto, K., Bakker, S., Gota, S., & Böhler-Baedeker, S. (2013). *Low-carbon land transport: policy handbook*. Routledge. doi:10.4324/9780203071229

Buehler, R., Pucher, J., Gerike, R., & Götschi, T. (2017). Reducing car dependence in the heart of Europe: Lessons from Germany, Austria, and Switzerland. *Transport Reviews*, *37*(1), 4–28. doi:10.1080/01441647.2016.1177799

Bulteau, J., Feuillet, T., & Dantan, S. (2019). Carpooling and car sharing for commuting in the Paris region: A comprehensive exploration of the individual and contextual correlates of their uses. *Travel Behaviour & Society*, *16*, 77–87. doi:10.1016/j.tbs.2019.04.007

Cathcart-Keays, A. (2017). Oslo's car ban sounded simple enough. Then the backlash began. *The Guardian*. Retrieved from https://www.theguardian.com/cities/2017/jun/13/oslo-ban-cars-backlash-parking

Chrisafis, A., & Vaughan, A. (2017). France to ban sales of petrol and diesel cars by 2040. *The Guardian*. Retrieved from https://www.theguardian.com/business/2017/jul/06/france-ban-petrol-diesel-cars-2040-emmanuel-macron-volvo

CIVITAS. (2013). *Exploring innovative urban mobility solutions*. Retrieved from https://civitas.eu/mobility-solutions

CIVITAS. (2019). *Implementing sustainable mobility*. Retrieved from https://civitas.eu/measure/mobility-managers

Coffey, H. (2018). Paris to ban cars in city centre one Sunday a month. *Independent*. Retrieved from https://www.independent.co.uk/travel/news-and-advice/paris-car-free-sundays-city-centre-france-pedestrian-a8566991.html

Crerar, P. (2017). Radical plan to pedestrianise Oxford Street by Christmas 2018. *City Hall Editor*. Retrieved from https://www.standard.co.uk/news/transport/unveiled-radical-plan-to-pedestrianise-oxford-street-by-next-christmas-a3677081.html

Deloitte City Mobility Index. (2018). *Johannesburg Global City Mobility*. Retrieved from https://www2.deloitte.com/content/dam/insights/us/articles/4331_Deloitte-City-Mobility-Index/Johannesburg_GlobalCityMobility_WEB.pdf

denmark.dk. (2019). *A nation of cyclists*. Retrieved from https://denmark.dk/people-and-culture/biking

Di Bartolo, C. (2015). *Incentivising a shift to more cycling in Bologna (Italy)*. Retrieved from https://www.eltis.org/discover/case-studies/incentivising-shift-more-cycling-bologna-italy

Dieleman, H. (2013). Mexico-City, sustainability and culture: A plea for hybrid sustainabilities for a baroque and labyrinthine city. *City. Cultura e Scuola*, *4*(3), 163–172. doi:10.1016/j.ccs.2013.05.001

Dusart, A. (2015). Paris leads the way in electro-mobility. *OECD Observer*. Retrieved from https://oecdobserver.org/news/fullstory.php/aid/5305/Paris_leads_the_way_in_electro-mobility.html

Eltis. (2014). Bologna: the most mobility friendly city in Italy. *The Urban Mobility Observatory*. Retrieved from https://www.eltis.org/discover/news/bologna-most-mobility-friendly-city-italy-0

Eskeland, G. S., & Feyzioglu, T. (1997). Rationing can backfire: The "day without a car" in Mexico City. *The World Bank Economic Review, 11*(3), 383–408. doi:10.1093/wber/11.3.383

Garfield, L. (2015a). *Oslo just declared that it will become the first major city to ban cars*. Retrieved from. https://www.businessinsider.com/oslo-bans-cars-from-its-city-center-2015-10

Garfield, L. (2015b). Johannesburg tried to get people to give up driving for a month — here's what happened. *Business Insider*. Retrieved from https://www.businessinsider.com/johannesburgs-experiment-with-no-cars-2015-12

Garfield, L. (2018). *13 cities that are starting to ban cars*. Retrieved from https://www.businessinsider.com/cities-going-car-free-ban-2017-8

Gehl Architects. (2019). Public Spaces in Copenhagen, A guide to the public spaces in Copenhagen. *Urban Quality Consultants*. Retrieved from https://www.akershus.no/file/c112a084c81172d57c8dba94e41113cc/091217_Cph_Guide%20(2).pdf

Gerth-Niculescu, M. (2019). *A car-free day in Addis Ababa*. Retrieved from https://www.dw.com/en/a-car-free-day-in-addis-ababa/g-48715793

Gilroy, P. (2000). Driving while black. In D. Miller (Ed.), *Car Cultures* (pp. 81–104). Berg Publishers.

Gössling, S. (2013). Urban transport transitions: Copenhagen, city of cyclists. *Journal of Transport Geography, 33*, 196–206. doi:10.1016/j.jtrangeo.2013.10.013

Graham, G. (2014). London will follow Paris and ban diesel cars, campaigners warn. *The Telegraph*. Retrieved from https://www.telegraph.co.uk/news/earth/environment/11280067/London-will-follow-Paris-and-ban-diesel-cars-campaigners-warn.html

Griffiths, S. (2014). The most polluted street in the world is in London: Oxford Street has the highest levels of nitrogen dioxide, claims expert. *Daily Mail*. Retrieved from https://www.dailymail.co.uk/sciencetech/article-2688686/The-polluted-street-world-LONDON-Oxford-Street-highest-levels-nitrogen-dioxide-claims-expert.html

Guerra, E. (2014). The built environment and car use in Mexico City: Is the relationship changing over time? *Journal of Planning Education and Research, 34*(4), 394–408. doi:10.1177/0739456X14545170

Hendrix, A. (2017). *Oslo: A small city with big ideas*. Retrieved from https://medium.com/urbansharing/oslo-a-small-city-with-big-ideas-dc3c22b807a9

Joburg. (2015). *EcoMobility World Festival 2015*. Retrieved from https://ecomobility.org/about-the-ecomobility-world-festival-2015/

Lerner, W., & Audenhove, V. F. (2012). The future of urban mobility: Towards networked, multimodal cities in 2050. *Public Transport International*, *61*(2), 14–18.

Linda, S. T. (2003). Can public transport compete with the private car? *IATSS Research*, *27*(2), 27–35. doi:10.1016/S0386-1112(14)60141-2

Maiken, T. (2016). *Copenhagen – After the Car? An examination of existing car-free projects and a future development of a car-free environment in Middelalderen (Master's Thesis)*. Aalborg University. Retrieved from https://projekter.aau.dk/projekter/files/239504819/COPENHAGEN_____AFTER_THE_CAR.pdf

Maota, R. (2015). Sandton goes car-free for a month. *Brand South Africa*. Retrieved from https://www.brandsouthafrica.com/governance/developmentnews/sandton-goes-car-free-for-a-month

Medina, M. Á. (2018). Madrid takes historic step to becoming a car-free city center-New measures are in effect to curb air pollution and increase space given over to pedestrians. *EL PAÍS*. Retrieved from https://english.elpais.com/elpais/2018/11/30/inenglish/1543565577_207058.html

Memmott, M. (2014). Paris Ends 'Odd Or Even' Limit On Cars After One Day. *NPR*. Retrieved from https://www.npr.org/sections/thetwo-way/2014/03/17/290849704/to-clear-its-air-paris-tests-odd-or-even-driving-days?fbclid=IwAR1KU2LRTqHeAGBVGqH-GK9ZbYQVhkBe29KL83Y7pRQySitTwQAoJ-dbYqM

Mendez, G. (2014). Mexico City's car-free Sundays are catalyst for change Integrating sustainable mobility into the everyday. *The City Fix*. Retrieved from https://thecityfix.com/blog/beyond-move-mexico-city-integrating-sustainable-mobility-ecobici-biking-gisela-mendez/

Morris, D., Enoch, M., Pitfield, D., & Ison, S. (2009). Car-free development through UK community travel plans. *Proceedings of the Institution of Civil Engineers. Urban Design and Planning*, *162*(1), 19–27. doi:10.1680/udap.2009.162.1.19

Mozos-Blanco, M. Á., Pozo-Menéndez, E., Arce-Ruiz, R., & Baucells-Aletà, N. (2018). The way to sustainable mobility. *A comparative analysis of sustainable mobility plans in Spain, 45*.

Najdovski, C. (2017). Roads and Mobility-Parisian Policy. *Direction De La Voirie Et Des Déplacements*. Retrieved from https://www.eltis.org/sites/default/files/sump_conference_2017_ops2_1_najdovski.pdf

Nieuwenhuijsen, M. J., & Khreis, H. (2016). Car free cities: Pathway to healthy urban living. *Environment International, 94*, 251–262. doi:10.1016/j.envint.2016.05.032 PMID:27276440

O'Sullivan, F. (2016). More Trees, Fewer Cars for the Central Squares of Paris. *City Lab*. Retrieved from https://www.citylab.com/solutions/2016/04/paris-public-square-plan/476463/

O'Sullivan, F. (2017). How to Pedestrianize a Vital Urban Street. *City Lab*. Retrieved from https://www.citylab.com/transportation/2017/11/london-oxford-street-pedestrian-plan-buses-car-ban/545171/

Oranga, B. (2015). Addis Ababa on the frontier of sustainable transport for African cities. *The City Fix*. Retrieved from https://thecityfix.com/blog/friday-fun-addis-ababa-ethiopia-light-rail-urban-africa-public-transport-beryl-oranga/

Osborne, S. (2018). Paris considers making public transport free to reduce pollution. *Independent*. Retrieved from https://www.independent.co.uk/news/world/europe/paris-public-transport-free-pollution-anne-hidalgo-cars-a8269581.html

Peters, A. (2018). The entire city of Paris will be car-free for a day. *Fast Company*. Retrieved from https://www.fastcompany.com/90236541/the-entire-city-of-paris-will-be-car-free-for-a-day

Peters, A. (2019). What happened when Oslo decided to make its downtown basically car-free? *Fast Company*. Retrieved from https://www.fastcompany.com/90294948/what-happened-when-oslo-decided-to-make-its-downtown-basically-car-free

Porta, S., Strano, E., Iacoviello, V., Messora, R., Latora, V., Cardillo, A., Wang, F., & Scellato, S. (2009). Street centrality and densities of retail and services in Bologna, Italy. *Environment and Planning. B, Planning & Design, 36*(3), 450–465. doi:10.1068/b34098

Richard, M. G. (2010). What is Car Sharing and How Does it Work? *Treehugger*. Retrieved from https://www.treehugger.com/cars/what-is-car-sharing-and-how-does-it-work.html

Rodriguez. C. (2015). Paris Will Stop All Traffic, Literally, For One Day in September. *Forbes*. Retrieved from https://www.forbes.com/sites/ceciliarodriguez/2015/08/19/french-car-ban-makes-history-for-one-day-in-september-there-will-be-no-traffic-in-paris/?fbclid=IwAR37zXakGjdupAUTAVCafFyBDvZGSGZIWyF2QuMe9bYkoO0Nl8ZQefsvuok#6020bad47793

Ruter#. (2018). *Emission Free public transport in Oslo and Akershus - Ruter. Version 10* (English translation of original document in Norwegian). Retrieved from https://ruter.no/contentassets/4fceaf4f35f54c0bbc0298805bc29241/report-zero-emission-english.pdf

Santos, G., Behrendt, H., Maconi, L., Shirvani, T., & Teytelboym, A. (2010a). Part I: Externalities and economic policies in road transport. *Research in Transportation Economics*, *28*(1), 2–45. doi:10.1016/j.retrec.2009.11.002

Santos, G., Behrendt, H., & Teytelboym, A. (2010b). Part II: Policy instruments for sustainable road transport. *Research in Transportation Economics*, *28*(1), 46–91. doi:10.1016/j.retrec.2010.03.002

Shemkus, S. (2019). In Norway, Oslo leaders are creating a 'car-free city center.' Could Boston be next? *Energy News*. Retrieved from https://energynews.us/2019/07/15/northeast/in-norway-oslo-leaders-are-creating-a-car-free-city-center-could-boston-be-next/?fbclid=IwAR0S7tEIMp7vtybaudBq_1c_duF26IkHklrpc5DYDSxu0a_hPHe3mLu1_F8

Taylor, M. (2017). Four select committees launch joint inquiry into UK air pollution crisis. *The Guardian*. Retrieved from https://www.theguardian.com/environment/2017/mar/20/four-select-committees-launch-joint-inquiry-uk-air-pollution-crisis

The Copenhagen Wheel. (2019). Retrieved from http://senseable.mit.edu/copenhagenwheel/

Thøgersen, J. (2009). Promoting public transport as a subscription service: Effects of a free month travel card. *Transport Policy*, *16*(6), 335–343. doi:10.1016/j.tranpol.2009.10.008

Tønnesen, A., Meyer, S. F., Skartland, E. G., & Sundfør, H. B. (2016). *European cities with car-free city centres*. Institute of Transport Economics.

Topp, H., & Pharoah, T. (1994). Car-free city centres. *Transportation*, *21*(3), 231–247. doi:10.1007/BF01099212

Velasco, A. (2016). Integration of the Public Transport System in Madrid Region. *Consorcio Regional de Transportes de Madrid (CRTM)*. Retrieved from https://ebsf2.eu/sites/default/files/8.1%20EBSF_2%20Madrid%20Demo%20Context.CRTM.pdf

Wahlgren, L., & Schantz, P. (2014). Exploring bikeability in a suburban metropolitan area using the Active Commuting Route Environment Scale (ACRES). *International Journal of Environmental Research and Public Health, 11*(8), 8276–8300. doi:10.3390/ijerph110808276 PMID:25153462

Walker, A. (2014). *Why Don't More Cities Have E-Bike Shares?* Retrieved from https://gizmodo.com/why-dont-more-cities-have-e-bike-shares-1595348781

Wang, G. (2011). *Public Walking Space. A study of Commercial Pedestrian Streets in Copenhagen, Stockholm and Nanjing* (Diploma thesis). Karlskrona, Sweden.

WHO. (2018). Global status report on road safety: Summary. *WHO Press*. Retrieved from https://www.who.int/violence_injury_prevention/road_safety_status/2018/English-Summary-GSRRS2018.pdf

WHO/UNEP. (1992). Urban air pollution in megacities of the World. World Health Organization. *United Nations Environment Programme*. Retrieved from https://apps.who.int/iris/handle/10665/39902

Wright, L. (2005). *Car free Development, Sustainable Transport. A source book for Policy Makers in developing Countries, Module 3e*. Retrieved from https://www.worldcarfree.net/resources/freesources/carfree_dev.pdf

Zarrilli, D. (2019). *Zero Emission Vehicles*. Retrieved 2019, Retrieved from C40Cities: https://www.c40.org/networks/zero-emission-vehicles

Zhu, Y., & Diao, M. (2016). The impacts of urban mass rapid transit lines on the density and mobility of high-income households: A case study of Singapore. *Transport Policy, 51*, 70–80. doi:10.1016/j.tranpol.2016.03.013

Compilation of References

Abubaker, I. (2014). Friday Fun: Community building through skateboarding in Addis Ababa. *The City Fix*. Retrieved from https://thecityfix.com/blog/friday-fun-community-building-through-skateboarding-addis-ababa-ethiopia-iman-abubaker/

Abubaker, I. (2019). Reclaiming the Streets: Addis Ababa, Other African Cities Launch Car-Free Days. *The City Fix*. Retrieved from https://thecityfix.com/blog/reclaiming-streets-addis-ababa-african-cities-launch-car-free-days-iman-abubaker/

Abusaada, H., & Elshater, A. (2020). Effects of Urban Atmospheres on Changing Attitudes of Crowded Public Places: An Action Plan. *Int. Journal of Community Well-being*, 1-51.

Abusaada, H., Vellguth, C., & Elshater, A. (Eds.). (2019). *Handbook of Research on Digital research Methods and Architectural Tools in Urban Planning and Design*. IGI Global. doi:10.4018/978-1-5225-9238-9

Acedo, A., Painho, M., & Casteleyn, S. (2017). Place and City: Operationalizing Sense of Place and Social Capital in the Urban Context. *Transactions in GIS*, *21*(3), 503–520. doi:10.1111/tgis.12282

Adbelhamid, M. M., Elfakharany, M. M., & Elfakharany, A. M. (2018). Reducing Private Cars Dependency to Achieve Sustainable Urban Environment in Congested Cities. In *7th International Conference on Modern Research in* (pp. 1-13). Munich, Germany: Academic Press.

Adel, H., Salheen, M., & Mahmoud, R. A. (2016). Crime in relation to urban design. Case study: The Greater Cairo Region. *Ain Shams Engineering Journal*, *7*(3), 925–938. doi:10.1016/j.asej.2015.08.009

Agyeman, J., & McLaren, D. (2017). Sharing cities. *Environment*, *59*(3), 22–27. doi:10.1080/00139157.2017.1301168

Ahern, J. (2003). Greenways in the USA: theory, trends and prospects. *Ecological Networks and Greenways: Concept, Design, Implementation*, 34–55.

Ahern, J. (1995). Greenways as a planning strategy. *Landscape and Urban Planning*, *33*(1–3), 131–155. doi:10.1016/0169-2046(95)02039-V

Ahram Online. (2010). *Egypt vehicles up 13.5 percent in 2010*. Ahram Online.

Alexander, J. C. (2002). On the Social Construction of Moral Universals: The 'Holocaust' from War Crime to Trauma Drama. *European Journal of Social Theory*, *5*(1), 5–85. doi:10.1177/1368431002005001001

Amin, A. (2004). Regions unbound: towards a new politics of place. *Geografiska Annaler, B, 36*, 33-44.

Amoly, E., Dadvand, P., Forns, J., López-Vicente, M., Basagaña, X., Julvez, J., ... Sunyer, J. (2014). Green and blue spaces and behavioral development in Barcelona schoolchildren: The BREATHE project. *Environmental Health Perspectives*, *122*(12), 1351–1358.

Arisha, A. R., & El-Moneim, N. M. (2019). Space Syntax Beyond Cairo Street Markets: An Application of a Theory on Socio-Spatial Contemporary Practices. In H. Abusaada, C. Vellguth, & A. Elshater (Eds.), *Handbook of Research on Digital Research Methods and Architectural Tools in Urban Planning and Design* (pp. 250–281). IGI Global. doi:10.4018/978-1-5225-9238-9.ch012

ARTISTS Consortium. (2004). *Arterial streets towards sustainability. A Guide for European Cities*. Lund University.

Asthana, A., & Taylor, M. (2017). Britain to ban sale of all diesel and petrol cars and vans from 2040. *The Guardian*. Retrieved from https://www.theguardian.com/politics/2017/jul/25/britain-to-ban-sale-of-all-diesel-and-petrol-cars-and-vans-from-2040

Aurbach, L. (2009). *Towards a functional classification replacement*. PedShed.net.

Austin, G. (2014). *Green Infrastructure for Landscape Planning: Integrating human and natural systems*. Routledge. doi:10.4324/9781315856780

Avila-Forcada, S., & Medina-Martínez, I. (2019). *Commuting Trends in Mexico City (2007-2017). Urban Mobility in Mexico City*. Working Paper. FLACSO.

Axhausen, K. W., & Gärling, T. (1992). Activity-based approaches to travel analysis: Conceptual frameworks, models, and research problems. *Transport Reviews*, *12*(4), 323–341. doi:10.1080/01441649208716826

Badger, E. (2016). Paris just banned all cars made before 1997. *The Washington Post*. Retrieved from https://www.washingtonpost.com/news/wonk/wp/2016/07/02/paris-just-banned-all-cars-made-before-1997/?fbclid=IwAR2ZqVtH03PEilOgVyVteBton3GIoZgH3c8WGPGbSUwjPXkkO-JEQjMLuKQ&noredirect=on

Bahrainy, H., & Khosravi, H. (2013). The Impact of Urban Design Features and Qualities on Walkability and Health in Under-Construction Environments: The Case of Hashtgerd New Town in Iran. *Cities (London, England)*, *31*, 17–28. doi:10.1016/j.cities.2013.01.001

Bakker, L. (2018). Security Blurs and Citizenship: Consequenses in Indonesia. In E. G. Tessa Diphoorn (Ed.), *Security Blurs: The Politics of Plural Security Provision*. Routledge. doi:10.4324/9781351127387-4

Barker, J. (2009b). Negara Beling: Street level authority in an Indonesian slum. In G. v. K. a. J. Barker (Ed.), State of authority: The state in society in Indonesia (pp. 47–72). Ithaca: Cornell University Press.

Barker, J. (2009). Introduction. *City & Society*, *21*(2), 155–162. doi:10.1111/j.1548-744X.2009.01019.x

Baumgardner, K. (2015). *Beyond Google's cute car*. Retrieved September 10, 2019, from http://offcite.org/wp-content/uploads/sites/3/2015/10/Baumgardner_Autonomous_Cars.pdf

Baynes, C. (2017). Thousands of cars banned from roads as Mexico City chokes in longest pollution alert for two decades. *Independent*. Retrieved from https://www.independent.co.uk/news/world/americas/cars-banned-mexico-city-longest-pollution-alert-smog-a7746756.html

BBC. (2018). Thousands walk on Ethiopia Car Free Day. *BBC News*. Retrieved from https://www.bbc.com/news/world-africa-46499036

Belanger, L. (2018). *'Put them near a park' and other things this coworking founder finds companies miss when picking an office space*. Retrieved September 1, 2019, from https://www.entrepreneur.com/article/316938

Bélanger, P. (2016). *Landscape as infrastructure : a base primer* (1st ed.). Routledge. doi:10.4324/9781315629155

Benedict, M. A. E. T., & Mcmahon, J. D. (2002). *Green Infrastructure: Smart Conservation for the 21st Century*. The Conservation Fund. Retrieved from www.conservationfund.org

Benedict, M. A., & McMahon, E. T. (2006). *Green Infrastructure: Linking Landscapes and Communities*. Island Press.

Benjamin, W. (2002). *The Arcades Project*. Harvard University Press.

Bently, I., Alcock, A., Murrain, P., McGlynn, S., & Smith, G. (1985). *Responsive environments: A manual for designers*. Elsevier.

Bereitschaft, B. (2018). Walk Score® versus Residents' Perceptions of Walkability in Omaha, NE. *Journal of Urbanism*, *11*(4), 412–435. doi:10.1080/17549175.2018.1484795

Bernardi, S., & Rupi, F. (2015). An analysis of bicycle travel speed and disturbances on off-street and on-street facilities. *Transportation Research Procedia*, *5*, 82–94. doi:10.1016/j.trpro.2015.01.004

Bishop, R., Philips, J., & Yeo, W. W. (2003). Perpetuating cities: Excepting globalization and the Southeast Asia supplement. In J. P. W. Y. R. Bishop (Ed.), *Postcolonial Urbanism: Southeast Asian Cities and Global Processes* (pp. 1–36). Routledge.

Bliss, L. (2018). The War on Cars, Norwegian Edition. *City Lab*. Retrieved from https://www.citylab.com/transportation/2018/05/oslos-race-to-become-a-major-bike-haven/559358/

Boffey, D. (2020). European commuters still choose cars and congestion over public transport. *The Guardian.* Retrieved from https://www.theguardian.com/world/2020/mar/03/european-commuters-still-prefer-cars-to-public-transport?fbclid=IwAR28I-p95vCiT2h_h3bS64koAJjURcDSnIaHVQ6ORRPXTbU6OZbyKbT1wMw

Bongardt, D., Creutzig, F., Hüging, H., Sakamoto, K., Bakker, S., Gota, S., & Böhler Baedeker, S. (2013). *Low-carbon land transport: policy handbook.* Routledge. doi:10.4324/9780203071229

Borsi, S. (1993). *Roma di Benedetto XIV: La Pianta di Giovan Battista Nolli, 1748.* Officina Edizione.

Bower, B. (1963). MacKaye, Benton, The New Exploration. *Natural Resources Journal, 3*(2), 371. https://digitalrepository.unm.edu/nrj/vol3/iss2/14

Boyer, B. (2017). Are designers letting autonomous vehicles shape future cities? *Medium.* Retrieved September 9, 2019, from https://medium.com/dashmarshall/are-designers-letting-autonomous-vehicles-shape-future-cities-153c4b42a999

Brown, H. (2014). *Next Generation Infrastructure, Principles for Post Industrial Public Works.* Island Press. doi:10.5822/978-1-61091-202-0

Brown, H., & Stigge, B. (2017). *Infrastructural ecologies : alternative development models for emerging economies.* The MIT Press. doi:10.7551/mitpress/10558.001.0001

Bryman, A. (2016). *Social research methods* (5th ed.). OUP Oxford.

Buehler, R., Pucher, J., Gerike, R., & Götschi, T. (2017). Reducing car dependence in the heart of Europe: Lessons from Germany, Austria, and Switzerland. *Transport Reviews, 37*(1), 4–28. doi:10.1080/01441647.2016.1177799

Bueno, J. A., Tsihrintzisb, A., & Alvareza, L. (1995). South Florida greenways: a conceptual framework for the ecological reconnectivity of the region. *Landscape and Urban Planning, 33*, 249. Retrieved from https://ac-els-cdn-com.ezproxy.nottingham.ac.uk/0169204694020217/1-s2.0-0169204694020217-main.pdf?_tid=248f8fc4-c759-449a-90c1-3d142a6deb34&acdnat=1540031359_b766fac58a088ae00d5000a83a40ac41

Bulteau, J., Feuillet, T., & Dantan, S. (2019). Carpooling and car sharing for commuting in the Paris region: A comprehensive exploration of the individual and contextual correlates of their uses. *Travel Behaviour & Society, 16*, 77–87. doi:10.1016/j.tbs.2019.04.007

Burgen, S. (2019). *Barcelona's car-free 'superblocks' could save hundreds of lives.* Retrieved September 15, 2019, from https://www.theguardian.com/cities/2019/sep/10/barcelonas-car-free-superblocks-could-save-hundreds-of lives?fbclid= IwAR0rb9bGTNUMhUZKkqIOE9ndAIZgT9TCfgeg VREP1qkum6GMsManIANpnY4

Burns, L. D., & Golob, T. F. (1976). The role of accessibility in basic transportation choice behavior. *Transportation, 5*(2), 175–198. doi:10.1007/BF00167272

C40. (2019). Celebrating the best performing cities in C40 regions in 2018. *C40 Blog*. Retrieved from https://www.c40.org/blog_posts/participation-standards-2018

CABE. (2002). Paving the way: How we achieve clean, safe, and attractive streets. CABE (Commission for Architecture and the Built Environment) Space Publications.

CABE. (2008). Civilized streets. CABE (Commission for Architecture and the Built Environment) Space Publications.

Cairns, S., Sloman, L., Newson, C., Anable, J., Kirkbride, A., & Goodwin, P. (2008). Smarter choices: Assessing the potential to achieve traffic reduction using "soft measures". *Transport Reviews*, *28*(5), 593–618. doi:10.1080/01441640801892504

CAPMAS. (2014). *Income, Expenditure & Consumption survey*. CAPMAS.

Carlson, R. (2006). *A concrete introduction to real analysis*. CRC Press. doi:10.1201/b15894

Carmona, M., Tirsdell, S., Heath, T., & Oc, T. (2010). *Public Places Urban Spaces: The Dimensions of Urban Design* (2nd ed.). Elsevier Ltd.

Carson, R. (2000). *Silent spring*. Penguin.

Cathcart-Keays, A. (2017). Oslo's car ban sounded simple enough. Then the backlash began. *The Guardian*. Retrieved from https://www.theguardian.com/cities/2017/jun/13/oslo-ban-cars-backlash-parking

Cats, O., Susilo, Y. O., & Reimal, T. (2017). The prospects of fare-free public transport: Evidence from Tallinn. *Transportation*, *44*(5), 1083–1104. doi:10.100711116-016-9695-5

CE 261. (2017, September). Retrieved from http://athena.ecs.csus.edu/: http://athena.ecs.csus.edu/~yaoz/CE261_class3.html

Cervero, R., Guerra, E., & Al, S. (2017). *Beyond mobility : planning cities for people and places*. Island Press. doi:10.5822/978-1-61091-835-0

Chandler, D. (2015). *Stockport: Bricolage Town* (MA Thesis). Manchester School of Architecture.

Chatterjee, K., Chng, S., Clark, B., Davis, A., De Vos, J., Ettema, D., ... Reardon, L. (2019). Commuting and wellbeing: A critical overview of the literature with implications for policy and future research. *Transport Reviews*, •••, 1–30.

Chrisafis, A., & Vaughan, A. (2017). France to ban sales of petrol and diesel cars by 2040. *The Guardian*. Retrieved from https://www.theguardian.com/business/2017/jul/06/france-ban-petrol-diesel-cars-2040-emmanuel-macron-volvo

Christa, B., Taylor, M., Joe, H., & Bettencourt, L. M. (2015). *The Topology of Cities*. Santa Fe Institute.

CIVITAS. (2013). *Exploring innovative urban mobility solutions*. Retrieved from https://civitas.eu/mobility-solutions

CIVITAS. (2019). *Implementing sustainable mobility*. Retrieved from https://civitas.eu/measure/mobility-managers

CIVITAS. (2019). Making European cities greener Towards clean and smart mobility. doi:10.2840/616004

Claude Comtois, B. S. (2006). *The Geography of Transport Systems*. Routledge.

Clewell, A., Rieger, J., & Munro, J. (2005). Guidelines for Developing and Managing Ecological. *Ecological Restoration*, *1*(December), 1–16. doi:10.1098/rspb.2013.2236

Coffey, H. (2018). Paris to ban cars in city centre one Sunday a month. *Independent*. Retrieved from https://www.independent.co.uk/travel/news-and-advice/paris-car-free-sundays-city-centre-france-pedestrian-a8566991.html

Cohen, S. A., & Gössling, S. (2015). A darker side of hypermobility. Environment and Planning A. *Economy and Space*, *47*(8), 166–1679.

Corner, J. (2006). Terra Fluxus. In C. Waldheim (Ed.), *Landscape Urbanism Reader* (pp. 21–34). Princeton Architectural Press.

Cox, C., & Hart, A. (2016). *How autonomous vehicles could relieve or worsen traffic congestion*. Retrieved October 1, 2019, from https://www.here.com/sites/g/files/odxslz166/files/2018-12/HERE_How_autonomous_vehicles_could_relieve_or_worsen_traffic_congestion_white_paper.pdf

Crawshaw, P. (2009). *The Future of Greenways in Sydney*. Retrieved from https://www.be.unsw.edu.au/sites/default/files/upload/pdf/schools_and_engagement/resources/_notes/5A2_41.pdf

Crayton, T. J., & Meier, B. M. (2017). Autonomous vehicles: Developing a public health research agenda to frame the future of transportation policy. *Journal of Transport & Health*, *6*, 245–252. doi:10.1016/j.jth.2017.04.004

Crerar, P. (2017). Radical plan to pedestrianise Oxford Street by Christmas 2018. *City Hall Editor*. Retrieved from https://www.standard.co.uk/news/transport/unveiled-radical-plan-to-pedestrianise-oxford-street-by-next-christmas-a3677081.html

Creswell, J. W., & Clark, V. L. (2011). *Designing and conducting mixed methods research*. Sage.

Croci, E. (2016). Urban road pricing: A comparative study on the experiences of London, Stockholm and Milan. *Transportation Research Procedia*, *14*, 253–262. doi:10.1016/j.trpro.2016.05.062

Davis, R., Duany, A., & Plater-Zyberk, E. (2002). *The Lexicon of New Urbanism*. Duany Plater-Zyberk & Company. Retrieved from https://www.dpz.com/uploads/Books/Lexicon-2014.pdf

de Certeau, M. (1984). *The Practice of Everyday Life* (S. Rendall, Trans.). University of California Press.

De Vos, J. (2019). Analysing the effect of trip satisfaction on satisfaction with the leisure activity at the destination of the trip, in relationship with life satisfaction. *Transportation, 46*(3), 623–645. doi:10.100711116-017-9812-0

Deloitte City Mobility Index. (2018). *Johannesburg Global City Mobility*. Retrieved from https://www2.deloitte.com/content/dam/insights/us/articles/4331_Deloitte-City-Mobility-Index/Johannesburg_GlobalCityMobility_WEB.pdf

denmark.dk. (2019). *A nation of cyclists*. Retrieved from https://denmark.dk/people-and-culture/biking

Department for Transport. (2007). *Manual for Streets 1*. Thomas Telford Publishing.

DfT. (2007). *Manual for Streets. D. f. Transport*. Thomas Telford Publishing.

DfT. (2011). *Local Transport Note 1/11 - Shared Space*. TSO (The Stationery Office). Retrieved from https://assets.publishing.service.gov.uk/government/uploads/system/uploads/attachment_data/file/732739/ltn-1-11.pdf

Di Bartolo, C. (2015). *Incentivising a shift to more cycling in Bologna (Italy)*. Retrieved from https://www.eltis.org/discover/case-studies/incentivising-shift-more-cycling-bologna-italy

Dieleman, H. (2013). Mexico-City, sustainability and culture: A plea for hybrid sustainabilities for a baroque and labyrinthine city. *City. Cultura e Scuola, 4*(3), 163–172. doi:10.1016/j.ccs.2013.05.001

Diener, E., Emmons, R. A., Larsen, R. J., & Griffin, S. (1985). The satisfaction with life scale. *Journal of Personality Assessment, 49*(1), 71–75. doi:10.120715327752jpa4901_13 PMID:16367493

Diener, E., Suh, E. M., Lucas, R. E., & Smith, H. L. (1999). Subjective wellbeing: Three decades of progress. *Psychological Bulletin, 125*(2), 276–302. doi:10.1037/0033-2909.125.2.276

Dobrowolska, A., & Dobrowolski, J. (2006). *Heliopolis: Rebirth of the City of the Sun*. American Univesity in Cairo Press.

Dolan, P., Peasgood, T., & White, M. (2008). Do we really know what makes us happy: A review of the economic literature on the factors associated with subjective wellbeing. *Journal of Economic Psychology, 29*(1), 94–122. doi:10.1016/j.joep.2007.09.001

Dora, C., Hosking, J., Mudu, P., & Fletcher, E. R. (2011). *Sustainable Transport: A Sourcebook for Policy-makers in Developing Cities*. Retrieved from https://www.who.int/hia/green_economy/giz_transport.pdf?ua=1

Douglass, M. (2008). Civil society for itself and in the public sphere: Comparative research on globalization, cities and civic space in Pacific Asia. In K. C. H. G. L. O. M. Douglass (Ed.), *Globalization, The City and Civil Society in Pacific Asia* (pp. 27–49). Routledge.

Duany, A., & Plater-Zyberk, E. (1999). *The lexicon of the new urbanism*. Duany Plater-Zyberk & Company.

Duncan, D. T., Aldstadt, J., Whalen, J., Melly, S. J., & Gortmaker, S. L. (2011). Validation of Walk Score® for Estimating Neighborhood Walkability: An Analysis of Four US Metropolitan Areas. *International Journal of Environmental Research and Public Health*, *8*(11), 4160–4179. doi:10.3390/ijerph8114160 PMID:22163200

Dusart, A. (2015). Paris leads the way in electro-mobility. *OECD Observer*. Retrieved from https://oecdobserver.org/news/fullstory.php/aid/5305/Paris_leads_the_way_in_electro-mobility.html

Edwards, P., & Tsouros, A. (2006). Active living, health and local leadership. In *Promoting physical activity and active living in urban environments: The role of local governments* (pp. 1–5). WHO Regional Office for Europe.

EGWA. (2000). *Greenways | Asociación Europea de Vías Verdes*. Retrieved January 20, 2019, from http://www.aevv-egwa.org/greenways/#

Elshater, A. (2012). The New Urbanism Principles vs. Urban Design Dimensions towards Behavioral Performance in Egyptian Neighborhood. In Future Communities: Socio-Cultural & Environmental Challenges (pp. 826-843). Giza: AicE-Bs-ASIA Pacific International Conference on Environment-Behavior Studies.

Elshater, A. (2016). The Ten-Minute Neighbourhood Is[not] a Basic Planning Unit for Happiness in Egypt. *Archnet-IJAR*, *10*(1), 344–357. doi:10.26687/archnet-ijar.v10i1.878

Elshater, A. (2019). Food consumption in the everyday life of liveable cities: Design implications for conviviality. *Journal of Urbanism*, 1–29.

Elshater, A., Abusaada, H., & Afifi, S. (2019). What makes livable cities of today alike? Revisiting the criterion of singularity through two case studies. *Cities (London, England)*, *92*, 273–291. doi:10.1016/j.cities.2019.04.008

Eltis. (2014). Bologna: the most mobility friendly city in Italy. *The Urban Mobility Observatory*. Retrieved from https://www.eltis.org/discover/news/bologna-most-mobility-friendly-city-italy-0

Eskeland, G. S., & Feyzioglu, T. (1997). Rationing can backfire: The "day without a car" in Mexico City. *The World Bank Economic Review*, *11*(3), 383–408. doi:10.1093/wber/11.3.383

Ettema, D., Gärling, T., Friman, M., & Olsson, L. E. (2016). Travel mode use, travel mode shift and subjective wellbeing: Overview of theories, empirical findings and policy implications. In D. Wang & S. He (Eds.), *Mobility, sociability and wellbeing of urban living* (pp. 129–150). Springer. doi:10.1007/978-3-662-48184-4_7

Ettema, D., Gärling, T., Olsson, L. E., & Friman, M. (2010). Out-of-home activities, daily travel, and subjective wellbeing. *Transportation Research Part A, Policy and Practice*, *44*(9), 723–732. doi:10.1016/j.tra.2010.07.005

Ettema, D., & Smajic, I. (2015). Walking, places and wellbeing. *The Geographical Journal*, *181*(2), 102–109. doi:10.1111/geoj.12065

European Commission. (2015). European Accessibility Act. Employment, Social affairs, & Inclusion. Downloaded from: https://ec.europa.eu/social/main.jsp?catId=1202&langId=en&moreDocuments=yes

Eusuf, M. A., Mohit, M. A., Eusuf, M. M. R. S., & Ibrahim, M. (2014). Impact of Outdoor Environment to the Quality of Life. *Procedia: Social and Behavioral Sciences*, *153*, 639–654. doi:10.1016/j.sbspro.2014.10.096

Ewing, R., & Cervero, R. (2010). Travel and the Built Environment: A Meta-Analysis. *Journal of the American Planning Association*, *76*(3), 265–294. doi:10.1080/01944361003766766

Ewing, R., & Handy, S. (2009). Measuring the Unmeasurable: Urban design qualities related to walkability. *Journal of Urban Design*, *14*(1), 65–84. doi:10.1080/13574800802451155

Fabos, J. G. (1995). Introduction and overview: The greenway movement, uses and potentials of greenways. *Landscape and Urban Planning*, *33*(1–3), 1–13. doi:10.1016/0169-2046(95)02035-R

Fábos, J. G. (2004). Greenway planning in the United States: Its origins and recent case studies. J.G. Fábos. *Landscape and Urban Planning*, *68*(2-3), 321–342. doi:10.1016/j.landurbplan.2003.07.003

Farini Orleans, E., & Nijhuis, S. (2013). Flowscapes : exploring landscape infrastructures. Francisco de Vitoria University, Architecture Department.

Farr, D. (2012). *Sustainable Urbanism: Urban Design with Nature*. John Wiley & Sons. Retrieved from https://books.google.co.uk/books?id=meuuBgAAQBAJ&printsec=frontcover&source=gbs_ge_summary_r&cad=0#v=onepage&q&f=false

Flink, C. A., & Searns, R. M. (1993). *Greenways: A Guide to Planning, Design, and Development*. The Conservation Fund.

Forsyth, A. (2015). What is a Walkable Place? The Walkability Debate in Urban Design. *URBAN DESIGN International*, *20*(4), 274–292. doi:10.1057/udi.2015.22

Foucault, M. (1967). *Of Other Spaces*. Academic Press.

Friedmann, J. (2007). Reflections on Place and Place-making in the Cities of China. *International Journal of Urban and Regional Research*, *31*(2), 257–279. doi:10.1111/j.1468-2427.2007.00726.x

Friedman, T. (2005). *The World is Flat: A Brief History of the Twenty-First Century*. Farrar Straus and Giroux.

Friman, M., Fujii, S., Ettema, D., Gärling, T., & Olsson, L. E. (2013). Psychometric analysis of the satisfaction with travel scale. *Transportation Research Part A, Policy and Practice*, *48*, 132–145. doi:10.1016/j.tra.2012.10.012

Friman, M., Huck, J., & Olsson, L. E. (2017). Transtheoretical Model of Change during Travel Behavior Interventions: An Integrative Review. *International Journal of Environmental Research and Public Health*, *14*(6), 581. doi:10.3390/ijerph14060581 PMID:28556810

Friman, M., Olsson, L. E., & Ettema, D. (Eds.). (2018). *Quality of Life and Daily Travel. Springer.* Elsevier. doi:10.1007/978-3-319-76623-2

Fujii, S., Gärling, T., Jakobsson, C., & Jou, R. C. (2004). A cross-country study of fairness and infringement on freedom as determinants of car owners' acceptance of road pricing. *Transportation, 31*(3), 285–295. doi:10.1023/B:PORT.0000025395.17250.49

Furman, A. (2017). Exploring affordances of the street. *International Journal of Sustainable Development and Planning, 12*(3), 606-615.

Fyhri, A., Heinen, E., Fearnley, N., & Sundfør, H. B. (2017). A push to cycling—Exploring the e-bike's role in overcoming barriers to bicycle use with a survey and an intervention study. *International Journal of Sustainable Transportation, 11*(9), 681–695. doi:10.1080/15568318.2017.1302526

Garfield, L. (2015a). *Oslo just declared that it will become the first major city to ban cars.* Retrieved from. https://www.businessinsider.com/oslo-bans-cars-from-its-city-center-2015-10

Garfield, L. (2015b). Johannesburg tried to get people to give up driving for a month — here's what happened. *Business Insider*. Retrieved from https://www.businessinsider.com/johannesburgs-experiment-with-no-cars-2015-12

Garfield, L. (2018). *13 cities that are starting to ban cars.* Retrieved from https://www.businessinsider.com/cities-going-car-free-ban-2017-8

Gärling, T., Eek, D., Loukopoulos, P., Fujii, S., Johansson-Stenman, O., Kitamura, R., Pendyala, R., & Vilhelmson, B. (2002). A conceptual analysis of the impact of travel demand management on private car use. *Transport Policy, 9*(1), 59–70. doi:10.1016/S0967-070X(01)00035-X

Gärling, T., & Friman, M. (2012). A Behavioral Perspective on Voluntary Reduction of Private Car Use. In B. van Wee (Ed.), *Keep moving, towards sustainable mobility* (pp. 109–134). Eleven International Publishing.

Gärling, T., & Fujii, S. (2002). Structural equation modeling of determinants of implementation intentions. *Scandinavian Journal of Psychology, 43*, 1–8. doi:10.1111/1467-9450.00263 PMID:11885756

Gärling, T., & Schuitema, G. (2007). Travel demand management targeting reduced private car use: Effectiveness, public acceptability and political feasibility. *The Journal of Social Issues, 63*(1), 139–153. doi:10.1111/j.1540-4560.2007.00500.x

Garrett, B. (2015, Aug. 4). The privatisation of cities' public spaces is escalating. It is time to take a stand. *The Guardian*. Retrieved September 14, 2019, from https://eprints.soton.ac.uk/380084/1/pops-privately-owned-public-space-cities-direct-action

Gavanas, N. (2019). Autonomous road vehicles: Challenges for urban planning in European cities. *Urban Science, 3*(2), 61. doi:10.3390/urbansci3020061

Gehl Architects. (2019). Public Spaces in Copenhagen, A guide to the public spaces in Copenhagen. *Urban Quality Consultants*. Retrieved from https://www.akershus.no/file/c112a084c81172d57c8dba94e41113cc/091217_Cph_Guide%20(2).pdf

Gehl, J. (2010). *Cities for people*. Island Press.

Gehl, J. (2011). *Life Between Buildings*. Island Press / Center for Resource Economics.

Gehl, J. (2011). *Life between buildings: using public space* (6th ed.). Island Press.

Gehl, J., & Svarre, B. (2013). *How to study public life*. Island Press.

Gerth-Niculescu, M. (2019). *A car-free day in Addis Ababa*. Retrieved from https://www.dw.com/en/a-car-free-day-in-addis-ababa/g-48715793

Gibson, J. J. (1977). The theory of affordances. Hilldale.

Gilroy, P. (2000). Driving while black. In D. Miller (Ed.), *Car Cultures* (pp. 81–104). Berg Publishers.

Girling, C. L., & Kellett, R. (2005). *Skinny streets and green neighborhoods: design for environment and community*. Island Press.

Goldenberg, S. (2016). Masdar's zero-carbon dream could become world's first green ghost town. *The Guardian*.

Gössling, S. (2013). Urban transport transitions: Copenhagen, city of cyclists. *Journal of Transport Geography*, *33*, 196–206. doi:10.1016/j.jtrangeo.2013.10.013

Graham, G. (2014). London will follow Paris and ban diesel cars, campaigners warn. *The Telegraph*. Retrieved from https://www.telegraph.co.uk/news/earth/environment/11280067/London-will-follow-Paris-and-ban-diesel-cars-campaigners-warn.html

Griffiths, S. (2014). The most polluted street in the world is in London: Oxford Street has the highest levels of nitrogen dioxide, claims expert. *Daily Mail*. Retrieved from https://www.dailymail.co.uk/sciencetech/article-2688686/The-polluted-street-world-LONDON-Oxford-Street-highest-levels-nitrogen-dioxide-claims-expert.html

Griffiths, P. E., & West, C. (2015). A balanced intervention ladder: Promoting autonomy through public health action. *Public Health*, *129*(8), 1092–1098. doi:10.1016/j.puhe.2015.08.007 PMID:26330372

Guerra, E. (2014). The built environment and car use in Mexico City: Is the relationship changing over time? *Journal of Planning Education and Research*, *34*(4), 394–408. doi:10.1177/0739456X14545170

Guide-Dogs. (2020). *Streets Ahead*. Retrieved 01/01, 2020, from https://www.guidedogs.org.uk/how-you-can-help/campaigning/our-current-campaigns/streets-ahead#shared-surfaces

Habermas, J. (1990). *The Structural Transformation of the Public Sphere: an Inquiry into a Category of Bourgeois Society*. MIT Press.

Hall, S. (2012). *City, Street and Citizen: The Measure of the Ordinary*. Routledge. doi:10.4324/9780203118597

Hamilton-Baillie, B. (2001). *Home Zones - Reconciling People, Places and Transport Study Tour of Denmark, Germany, Holland and Sweden - July to August 2000*. Academic Press.

Hamilton-Baillie, B. (2008). Shared Space: Reconciling People, Places and Traffic. *Built Environment, 34*(2), 161-181.

Hamilton-Baillie, B. (2004). Urban design: Why don't we do it in the road? Modifying traffic behavior through legible urban design. *Journal of Urban Technology, 11*(1), 43–62. doi:10.1080/1063073042000341970

Handy, S., Cao, X., & Mokhtarian, P. (2005). Correlation or Causality between the Built Environment and Travel Behavior? Evidence from Northern California. *Transportation Research Part D, Transport and Environment, 10*(6), 427–444. doi:10.1016/j.trd.2005.05.002

Han, H., & Hawken, S. (2018). Introduction: Innovation and identity in next-generation smart cities. City, Culture. *Society, 12*, 1–4.

Haq, S., & Berhie, G. (2018). Space syntax investigation of Lubbock, a grid-like American city and some insights into isotropic layouts. *Journal of Urban Design, 23*(1), 5–22. doi:10.1080/13574809.2017.1369872

Harvey, D. (1973). *Social Justice and the City*. Johns Hopkins University Press.

Harvey, D. (1989). *The Condition of Postmodernity: An Enquiry into the Origins of Cultural Change*. Blackwell Publisher.

Harvey, D. (2003). *The New Imperialism*. Oxford University Press.

Harvey, D. (2012). *Rebel Cities: From The Right To The City To The Urban Revolution*. Verso.

Hatherley, O. (2012). Rebel cities: from right to the city to the urban revolution by David Harvey– review. *The Guardian*, 12.

Heathcote, E. (2017, May 5). *How architects have tried to create 'streets in the sky*. Retrieved from https://www.ft.com/content/8f2bab62-2b32-11e7-bc4b-5528796fe35c

Heinen, E., Van Wee, B., & Maat, K. (2010). Commuting by Bicycle: An Overview of the Literature. Transport Reviews: A Transnational Transdisciplinary Journal, 30, 59-96.

Hellmund, P. C., & Smith, D. S. (2006). *Designing Greenways: Sustainable Landscape for Nature and People*. Washington, DC: Island Press.

Hendrix, A. (2017). *Oslo: A small city with big ideas*. Retrieved from https://medium.com/urbansharing/oslo-a-small-city-with-big-ideas-dc3c22b807a9

Hess, P. M., & Milroy, B. M. (2006). *Making Toronto's streets*. The Centre for Urban Health Initiatives, University of Toronto.

Hilburg, J. (2018). New York City launches interactive map of its privately-owned public spaces. *The Architect's Newspaper*. Retrieved October 20, 2019, from https://archpaper.com/2018/09/new-york-city- privately-owned-public-spaces-pops-map/

Hillier, B. (1996). *Space is the machine: A configurational theory of architecture*. Cambridge University Press.

Hillier, B. (2015). *Space Is the Machine: A Configurational Theory of Architecture*. Space Syntax.

Hillier, B., & Hanson, J. (1989). *The Social Logic of Space*. Cambridge University Press.

Howard, E. (1902). *Garden Cities of Tomorrow*. Swan Sonnenschein & Co., Ltd. Retrieved from https://archive.org/details/gardencitiestom00howagoog/page/n31

Howard, E. (2007). *Garden Cities of To-Morrow*. Routledge.

Hung, Y.-Y., Aquino, G., Waldheim, C., Bélanger, P., Czerniak, J., & Geuze, A. … Robinson, A. (2013). Landscape Infrastructure: Case Studies by SWA (2nd ed.). Birkhäuser.

Imam, K. Z. E. A. (2006). Role of urban greenway systems in planning residential communities: A case study from Egypt. *Landscape and Urban Planning*, *76*(1–4), 192–209. doi:10.1016/j.landurbplan.2004.09.032

Ingvardson, J. B., & Nielsen, O. A. (2018). Effects of new bus and rail rapid transit systems–an international review. *Transport Reviews*, *38*(1), 96–116. doi:10.1080/01441647.2017.1301594

Ison, S., & Mulley, C. (2014). Parking: issues and policies. In M. C. J. Bliemer, C. Mulley, & C. J. Moutou (Eds.), *Handbook on transport and urban planning in the developed world*. Emerald.

Jabbari, M., Fonseca, F., & Ramos, R. (2018). Combining multi-criteria and space syntax analysis to assess a pedestrian network: The case of Oporto. *Journal of Urban Design*, *23*(1), 23–41. do i:10.1080/13574809.2017.1343087

Jacobs, A. B. (1993). *Great streets*. MIT Press.

Jacobs, J. (1961). *Death and life of great American cities*. Random House.

Jacobs, J. (1961). *The Death and Life of Great American Cities*. Random House.

Jacobs, J. (1961). *The death and life of Great American cities*. Vintage Books.

Jakobsson Bergstad, C., Gamble, A., Hagman, O., Polk, M., Gärling, T., Ettema, D., Friman, M., & Olsson, L. E. (2012). Influences on subjective wellbeing of affect associated with routine out-of-home activities. *Applied Research in Quality of Life*, *7*, 49–62. doi:10.100711482-011-9143-9

Jakobsson Bergstad, C., Gamble, A., Hagman, O., Polk, M., Gärling, T., & Olsson, L. E. (2011). Affective-Symbolic and Instrumental-Independence Psychological Motives Mediating Effects of Socio-Demographic Variables on Daily Car Use. *Journal of Transport Geography*, *19*(1), 33–38. doi:10.1016/j.jtrangeo.2009.11.006

Jeffrey, K., & Newman, P. (1999). *Sustainability and cities: Overcoming automobile dependence*. Island Press.

Jessop, B., Brenner, N., & Jones, M. (2008). Theorizing sociospatial relations. *Environment and Planning. D, Society & Space*, *26*(3), 389–401. doi:10.1068/d9107

Jin, X. (2010). *Modelling the influence of neighbourhood design on daily trip patterns in urban neighbourhoods*. Retrieved from Memorial University Research Repository: https://research.library.mun.ca/9553/

Joburg. (2015). *EcoMobility World Festival 2015*. Retrieved from https://ecomobility.org/about-the-ecomobility-world-festival-2015/

Jolma Architects. (2018). *How autonomous vehicles are influencing urban design*. Retrieved October 20, 2019, from https://land8.com/how-autonomous-vehicles-are-influencing-urban-design/

Jones, M. (2009). Phase space: Geography, relational thinking, and beyond. *Human Geographies*, *33*, 802–822.

Jones, P., & Di Guardo, G. (2019). Analysing pedestrian and vehicle interaction at courtesy crossings. In *The 17th Annual Transport Practitioners' Meeting*. PTRC.

Kahneman, D., Diener, E., & Schwartz, N. (1999). *Well-Being: The Foundations of Hedonic Psychology*. Russell Sage.

Kahneman, D., Slovic, P., & Tversky, A. (Eds.). (1982). *Judgment under uncertainty: Heuristics and biases*. Cambridge University Press. doi:10.1017/CBO9780511809477

Kang, C. D., & Cervero, R. (2009). From elevated freeway to Urban Greenway: Land value impacts of the CGC project in Seoul, Korea. pp.2772–2791. *Urban Studies (Edinburgh, Scotland)*. Advance online publication. doi:10.1177/0042098009345166

Kaparias, I., Bell, M. G. H., Biagioli, T., Bellezza, L., & Mount, B. (2015). Behavioural analysis of interactions between pedestrians and vehicles in street designs with elements of shared space. *Transportation Research Part F: Traffic Psychology and Behaviour*, *30*, 115–127. doi:10.1016/j.trf.2015.02.009

Kaparias, I., Bell, M. G. H., Dong, W., Sastrawinata, A., Singh, A., Wang, X., & Mount, B. (2013). Analysis of pedestrian-vehicle traffic conflicts in street designs with elements of shared space. *Transportation Research Record: Journal of the Transportation Research Board*, *2393*(1), 21–30. doi:10.3141/2393-03

Kariminia, S., & Sh Ahmad, S. (2013). Dependence of Visitors' Thermal Sensations on Built Environments at an Urban Square. *Procedia: Social and Behavioral Sciences*, *85*, 523–534. doi:10.1016/j.sbspro.2013.08.381

Kennedy, C., Miller, E., Shalaby, A., MacLean, H., & Coleman, J. (2005). The four pillars of sustainable urban transportation. *Transport Reviews*, *25*(4), 393–414. doi:10.1080/01441640500115835

Kherdeen, R. M. (2016). Masdar City: Oriental City of the Twenty-First Century. *Academia (Caracas)*.

Kilian, T. (1998). Public and private, power and space. In A. Light & J. Smith (Eds.), The production of public space. Lanham, MD: Rowman and Littlefield.

Kim, Y., & Penn, A. (2004). Linking the Spatial Syntax of Cognitive Maps to the Spatial Syntax of the Environment. *Environment and Behavior*, *36*(4), 483–504. doi:10.1177/0013916503261384

Kitamura, R., Fujii, S., & Pas, E. I. (1997). Time-use data, analysis and modeling: Toward the next generation of transportation planning methodologies. *Transport Policy*, *4*(4), 225–235. doi:10.1016/S0967-070X(97)00018-8

Kitchin, R., Lauriault, T. P., & McArdle, G. (2016). Smart cities and the politics of urban data. In S. Marvin, A. Luque-Ayala, & C. Mcfarlane (Eds.), *Smart urbanism: Utopian vision or false dawn* (pp. 16–34). Routledge.

Koohsari, M. J., Takemi Sugiyama, A. T., & Owen, N. (2014). Associations of Leisure-Time Sitting in Cars With Neighborhood Walkability. *Journal of Physical Activity & Health*, *11*(6), 1129–1132. doi:10.1123/jpah.2012-0385 PMID:25368941

Koschinsky, J., Talen, E., Alfonzo, M., & Lee, S. (2017). How Walkable Is Walker's Paradise? *Environment and Planning. B, Planning & Design*, *44*(2), 343–363.

Kota Bandung, B. P. S. (2017). *Indeks Kebahagiaan 2017* [Bandung Happyness Indeks]. Retrieved from https://www.bps.go.id/publication/2017/12/05/1f99cefd596c449b93405fcd/indeks-kebahagiaan-2017.html

Kota Bandung, B. P. S. (2018). *Kota Bandung dalam angka* [Bandung in numbers]. Retrieved from https://bandungkota.bps.go.id/

Kullmann, K. (2011). Thin parks/thick edges: Towards a linear park typology for (post)infrastructural sites. *Journal of Landscape Architecture*, *6*(2), 70–81. doi:10.1080/18626033.2011.9723456

Kullmann, K. (2013). Green-Networks: Integrating Alternative Circulation Systems into Post-industrial Cities. *Journal of Urban Design*, *18*(1), 36–58. doi:10.1080/13574809.2012.739545

Kunstler, J. H. (1994). *The geography of nowhere: The rise and decline of America's man-made landscape*. Simon and Schuster.

Kunto, H. (1984). *Wajah Bandoeng Tempo Doeloe*. Granesia.

Kusno, A. (2000). *Behind the Postcolonial: Architecture, Urban Space and Political Cultures in Indonesia*. Routledge.

La Rocca, R. A. (2010). Soft Mobility and Urban Transformation. *TeMaLab Journal of Land Use. Mobility and Environment, 3*, 85–90. doi:10.6092/1970-9870/125

Lättman, K., Friman, M., & Olsson, L. E. (2016). Perceived accessibility of public transport as a potential indicator of social inclusion. *Social Inclusion (Lisboa), 4*(3), 36–45. doi:10.17645i.v4i3.481

Lättman, K., Olsson, L. E., & Friman, M. (2018). A new approach to accessibility–Examining perceived accessibility in contrast to objectively measured accessibility in daily travel. *Research in Transportation Economics, 69*, 501–511. doi:10.1016/j.retrec.2018.06.002

Lättman, K., Olsson, L. E., & Friman, M. (2020). Restricted car-use and perceived accessibility. *Transportation Research Part D, Transport and Environment, 78*, 102213. doi:10.1016/j.trd.2019.102213

Lättman, K., Olsson, L. E., Friman, M., & Fujii, S. (2019). Perceived Accessibility, Satisfaction with Daily Travel, and Life Satisfaction among the Elderly. *International Journal of Environmental Research and Public Health, 16*(22), 4498. doi:10.3390/ijerph16224498 PMID:31739648

Leary-Owhin, M. E. (2016). *Exploring the production of urban space: Differential space in three post-industrial cities*. Policy Press.

Lefebvre, H. (1991). *The Production of Space* (D. Nicholshon-Smith, Trans.). Blackwell.

Lefebvre, H., & Nicholson-Smith, D. (1991). *The production of space* (Vol. 142). Blackwell.

Lerner, W., & Audenhove, V. F. (2012). The future of urban mobility: Towards networked, multimodal cities in 2050. *Public Transport International, 61*(2), 14–18.

Lewis, P. H. (1996). *Tomorrow by design: a regional design process for sustainability*. Academic Press.

Liang, C., & Ng, E. (2012). Outdoor thermal comfort and outdoor activities: A review of research in the past decade. *Cities (London, England), 29*(2), 118–125. doi:10.1016/j.cities.2011.08.006

Linda, S. T. (2003). Can public transport compete with the private car? *IATSS Research, 27*(2), 27–35. doi:10.1016/S0386-1112(14)60141-2

Linehan, J., Grossa, M., & Finnb, J. (1995). Greenway planning : Developing a landscape ecological network approach. *Landscape and Urban Planning, 33*(1–3), 179–193. doi:10.1016/0169-2046(94)02017-A

Litman, T. (2017). *Autonomous vehicle implementation predictions*. Victoria Transport Policy Institute.

Little, C. E. (1995). *Greenways for America*. Johns Hopkins University Press.

Lomholt, I. (2011). *Madrid Rio: Spanish Landscape Design*. Retrieved December 25, 2018, from e-architect website: https://www.e-architect.co.uk/madrid/madrid-rio-landscape

Loos, F., & van Vliet, M. (2014). Greenway Design: Context, Identity, and Sustainability. In *Landscape Record* (pp. 120–123). Design media Pub Ltd. Retrieved from https://www.ribabookshops.com/item/landscape-record-greenway-design-no.-6-2014/86565/

Loukaitou-Sideris, A. (1993). *Retrofit of Urban Corridors: Land Use Policies and Design Guidelines for Transit-Friendly Environment*. Academic Press.

Lubell, S. (2016). Here's how self-driving cars will transform your city. *Wired Magazine*. Retrieved October 25, 2019, from https://www.wired.com/2016/10/heres-self-driving-cars-will-transform-city/

Luberoff, D. (2016). *Reimagining and reconfiguring New York City's streets*. Transforming Urban Transportation.

Lucas, K. (2012). Transport and social exclusion: Where are we now? *Transport Policy*, *20*, 105–113. doi:10.1016/j.tranpol.2012.01.013

Lynch, K. (1960). *The image of the city*. MIT Press.

Lynch, K. (1960). *The Image of the City*. MIT Press.

Lynch, K. (1981). *Good City Form*. The MIT Press.

Madanipour, A. (2003). *Public and private spaces of the city*. Routledge. doi:10.4324/9780203402856

Maiken, T. (2016). *Copenhagen – After the Car? An examination of existing car-free projects and a future development of a car-free environment in Middelalderen (Master's Thesis)*. Aalborg University. Retrieved from https://projekter.aau.dk/projekter/files/239504819/COPENHAGEN_____AFTER_THE_CAR.pdf

Manzo, L. C. (2005). For better or worse: Exploring multiple dimensions of place meaning. *Journal of Environmental Psychology*, *25*(1), 67–86. doi:10.1016/j.jenvp.2005.01.002

Maota, R. (2015). Sandton goes car-free for a month. *Brand South Africa*. Retrieved from https://www.brandsouthafrica.com/governance/developmentnews/sandton-goes-car-free-for-a-month

Marshall, S. (2009). *Cities design and evolution*. Routledge.

Marshall, S., Jones, P., Svensson, Å., Hydén, C., Draskoczy, M., Papaioannou, P., Thomsen, J., & Boujenko, N. (2004). *ARTISTS - Arterial streets for people: guidance for planners and decision-makers when reconstructing arterial streets*. European Commission Fifth Framework Program.

Marston, S. A., Jones, J. P. III, & Woodward, K. (2005). Human geography without scale. *Transactions of the Institute of British Geographers*, *30*(4), 416–432. doi:10.1111/j.1475-5661.2005.00180.x

Martin, A., Goryakin, Y., & Suhrcke, M. (2014). Does active commuting improve psychological wellbeing? Longitudinal evidence from eighteen waves of the British Household Panel Survey. *Preventive Medicine*, *69*, 296–303. doi:10.1016/j.ypmed.2014.08.023 PMID:25152507

Martin, D. G. (2003). Place-Framing' as Place-Making: Constituting a Neighborhood for Organizing and Activism. *Annals of the Association of American Geographers*, *93*(3), 730–750. doi:10.1111/1467-8306.9303011

Massey, D. (1999) *Power-Geometries and the Politics of Space-Time.* Hettner Lecture 1998. Heidelberg, University of Heidelberg-Institute of Geography.

Massey, D. (1995). Places and their pasts. *History Workshop Journal*, *39*(1), 182–192. doi:10.1093/hwj/39.1.182

Masterplan, U. (2017). *UCLan Masterplan.* Retrieved 02/03, 2020, from https://uclanmasterplan.co.uk/

Mathur, S., Menon, S. B., & Kacker, P. (2013). Greenways for a Greater Good- An Indian Perspective. *Proceedings of Fabos Conference on Landscape and Greenway Planning.* Retrieved from https://works.bepress.com/cgi/viewcontent.cgi?article=1001&context=mark_lindhult

McFarland, M. (2017). *Cities warm up to designated Uber, Lyft pick-up spots.* Retrieved October 20, 2019, from https://money.cnn.com/2017/11/16/technology/uber-lyft-designated- pickup-spots/index.html

McGee, T. (2002). Jalan, Jalan: Invading, destroying and reconstructing the Southeast Asian city. *Bijdragen tot de Taal-, Land- en Volkenkunde*, *158*(4), 637–652.

McMahon, D. M. (2008). The pursuit of happiness in history. In M. Eid & R. J. Larsen (Eds.), *The science of subjective wellbeing* (pp. 80–93). Guildford Press.

Medina, M. Á. (2018). Madrid takes historic step to becoming a car-free city center- New measures are in effect to curb air pollution and increase space given over to pedestrians. *EL PAÍS.* Retrieved from https://english.elpais.com/elpais/2018/11/30/inenglish/1543565577_207058.html

Memmott, M. (2014). Paris Ends 'Odd Or Even' Limit On Cars After One Day. *NPR.* Retrieved from https://www.npr.org/sections/thetwo-way/2014/03/17/290849704/to-clear-its-air-paris-tests-odd-or-even-driving-days?fbclid=IwAR1KU2LRTqHeAGBVGqH-GK9ZbYQVhkBe29KL83Y7pRQySitTwQAoJ-dbYqM

Mendez, G. (2014). Mexico City's car-free Sundays are catalyst for change Integrating sustainable mobility into the everyday. *The City Fix.* Retrieved from https://thecityfix.com/blog/beyond-move-mexico-city-integrating-sustainable-mobility-ecobici-biking-gisela-mendez/

Mitchell, W. (1995). *City of bits: space, place, and the infobahn.* The MIT Press.

Mohamed, A. A. (2019). Space Syntax Approach for Articulating Space and Social Life. In H. Abusaada, C. Vellguth, & A. Elshater (Eds.), *Handbook of Research on Digital Research Methods and Architectural Tools in Urban Planning and Design* (pp. 223–249). IGI Global. doi:10.4018/978-1-5225-9238-9.ch011

Molotch, H., Freudenburg, W., & Paulsen, K. E. (2000). History Repeats Itself, But How? City Character, Urban Tradition, and the Accomplishment of Place. *American Sociological Review*, *65*(6), 791–823. doi:10.2307/2657514

Morris, D., Enoch, M., Pitfield, D., & Ison, S. (2009). Car-free development through UK community travel plans. *Proceedings of the Institution of Civil Engineers. Urban Design and Planning*, *162*(1), 19–27. doi:10.1680/udap.2009.162.1.19

Mossop, E. (2006). Landscapes of infrastructure. In C. Waldheim (Ed.), *Landscape Urbanism Reader* (Vol. 1, pp. 163–178). Princeton Architectural Press., doi:10.1017/CBO9781107415324.004

Mostafavi, M. (2015). Why Ecological Urbanism? Why Now? In *Ecological Urbanism* (pp. 12–45). Lars Muller Publishers.

Mostafavi, M., & Najle, C. (2003). *Landscape urbanism : A manual for the machinic landscape*. Architectural Association. Retrieved from https://copac.jisc.ac.uk/id/24979355?style=html

Moughtin, C. (1991). The European city street. Part 1: Paths and places. *The Town Planning Review*, *62*(1), 51–77.

Moughtin, C. (2003). *Urban design: street and square*. Architectural Press.

Mozos-Blanco, M. Á., Pozo-Menéndez, E., Arce-Ruiz, R., & Baucells-Aletà, N. (2018). The way to sustainable mobility. *A comparative analysis of sustainable mobility plans in Spain, 45*.

Mumford, L. (1961). *The city in history: Its origins, its transformations, and its prospects* (Vol. 67). Houghton Mifflin Harcourt.

Mumford, L. (1963). *The highway and the city*. Harcourt.

Najdovski, C. (2017). Roads and Mobility-Parisian Policy. *Direction De La Voirie Et Des Déplacements*. Retrieved from https://www.eltis.org/sites/default/files/sump_conference_2017_ops2_1_najdovski.pdf

Nas, P. P. (2003). The streets of Jakarta: Fear, trust and amnesia in urban development. In P. N. R. Jaffe (Ed.), *Framing Indonesian Realities* (pp. 275–294). KITLV Press.

Nassauer, J. I. O. P., & Opdam, P. (2008). Design in science: Extending the landscape ecology paradigm. *Landscape Ecology*, *23*(6), 633–644. doi:10.100710980-008-9226-7

National Association of City Transportation Officials. (2016). *Global Street Design Guide*. Island Press.

Ndubisi, F., DeMeo, T., & Ditto, N. D. (1995). Environmentally sensitive areas: A template for developing greenway corridors. *Landscape and Urban Planning*, *33*(1–3), 159–177. doi:10.1016/0169-2046(94)02016-9

Newman, P., & Kenworthy, J. (2015). *The end of automobile dependence: How cities are moving beyond car based planning*. Island Press., doi:10.5822/978-1-61091-613-4

Newman, P., & Kenworthy, J. R. (1999). *Sustainability and cities : overcoming automobile dependence*. Island Press.

NFBUK. (n.d.). *Shared Spaces*. Retrieved 04/03, 2020, from https://www.nfbuk.org/campaign/shared-spaces/

Nieuwenhuijsen, M. J., Bastiaanssen, J., Sersli, S., Waygood, E. O. D., & Khreis, H. (2018). Implementing car free cities: rationale, requirements, barriers and facilitators. In M. Nieuwenhuijsen & H. Khreis (Eds.), *Integrating Human Health into Urban and Transport Planning*. Springer.

Nieuwenhuijsen, M. J., & Khreis, H. (2016). Car free cities: Pathway to healthy urban living. *Environment International*, *94*, 251–262. doi:10.1016/j.envint.2016.05.032 PMID:27276440

Nijhuis, S., & Jauslin, D. (2015). Urban landscape infrastructures Designing. In *Flowscapes: designing infrastructure as landscape* (pp. 13–34). TU Delft. Retrieved from https://books.bk.tudelft.nl/index.php/press/catalog/view/isbn.9789461864727/435/88-1

Norberg-Schulz, C. (1980). Genius loci: towards a phenomenology of architecture. Rizzoli, International Publications, Inc.

Norman, D. (2013). *The design of everyday things*. New York: Basic Books.

O'Sullivan, F. (2016). More Trees, Fewer Cars for the Central Squares of Paris. *City Lab*. Retrieved from https://www.citylab.com/solutions/2016/04/paris-public-square-plan/476463/

O'Sullivan, F. (2017). How to Pedestrianize a Vital Urban Street. *City Lab*. Retrieved from https://www.citylab.com/transportation/2017/11/london-oxford-street-pedestrian-plan-buses-car-ban/545171/

Olsson, L. E., Gärling, T., Ettema, D., Friman, M., & Fujii, S. (2013). Happiness and Satisfaction with Work Commute. *Social Indicators Research*, *111*(1), 255–263. doi:10.100711205-012-0003-2 PMID:23378683

Oranga, B. (2015). Addis Ababa on the frontier of sustainable transport for African cities. *The City Fix*. Retrieved from https://thecityfix.com/blog/friday-fun-addis-ababa-ethiopia-light-rail-urban-africa-public-transport-beryl-oranga/

Osborne, S. (2018). Paris considers making public transport free to reduce pollution. *Independent*. Retrieved from https://www.independent.co.uk/news/world/europe/paris-public-transport-free-pollution-anne-hidalgo-cars-a8269581.html

Paneerchelvam, P. T., Maruthaveeran, S., Maulan, S., & Shureen, S. F. (2020). The use and associated constraints of urban greenway from a socioecological perspective: A systematic review. *Urban Forestry & Urban Greening, 47*. Advance online publication. doi:10.1016/j.ufug.2019.126508

Pasalar, C., & Hallowell, G. (2018). *Toward human-centered smart cities: understanding emerging technologies and their effect on the urban experience*. Paper presented at AMPS, ARCHITECTURE_MPS, University of Arizona.

Pasalar, C., & Hallowell, G. D. (2019). Sharing economies, technologies, and the changing nature of urban public space in medium-sized cities. *Terr@ Plural, 13*(3), 418-433.

Pasalar, C., & Hallowell, G. (2020). Toward human-centered smart cities: Understanding emerging technologies and their effect on the urban experience. In J. Bean, S. Dickinson, & A. Ida (Eds.), *Critical Practices in Architecture: The Unexamined* (pp. 317–339). Cambridge Scholars Publishing.

Peck, E., & Maldonado, C. (2017). How Airbnb is pushing locals out of New Orleans' coolest neighborhoods. *Huffpost*. Retrieved December 30, 2019, from https://www.huffpost.com/entry/airbnb-new-orleans-housing _n_59f33054e4b03cd20b811699

Peters, A. (2018). The entire city of Paris will be car-free for a day. *Fast Company*. Retrieved from https://www.fastcompany.com/90236541/the-entire-city-of-paris-will-be-car-free-for-a-day

Peters, A. (2019). What happened when Oslo decided to make its downtown basically car-free? *Fast Company*. Retrieved from https://www.fastcompany.com/90294948/what-happened-when-oslo-decided-to-make-its-downtown-basically-car-free

Plumer, B. (2016). *Cars take up way too much space in cities. New technology could change that*. Retrieved October 20, 2019, from https://www.vox.com/a/new-economy-future/cars-cities-technologies

Polyák, L. (2017). *A bundle of rights and obligations: Privately owned public spaces*. Retrieved October 20, 2019, from A https://cooperativecity.org/2017/11/01/privately-owned-public-spaces/

Pope, R., & Phillips, K. (1995). *University of Central Lancashire A History of the Development of the Institution since 1828*. University of Central Lancashire.

Porta, S., Strano, E., Iacoviello, V., Messora, R., Latora, V., Cardillo, A., Wang, F., & Scellato, S. (2009). Street centrality and densities of retail and services in Bologna, Italy. *Environment and Planning. B, Planning & Design, 36*(3), 450–465. doi:10.1068/b34098

Pourjafar, M., & Moradi, A. (2015). Explaining Design Dimensions of Ecological Greenways. *Open Journal of Ecology, 5*(03), 66–79. doi:10.4236/oje.2015.53007

Prasetyo, F. (2014). *Car Free Day: Contestation of 'Third Space' as a Phenomenon in the Production of Urban Public Space in Bandung* (Unpublished Thesis). ITB-School of Architecture, Planning and Policy Development-Department of Regional and City Planning, Bandung, Indonesia.

Prasetyo, F. (2017). Car Free Day: Transformasi Ruang dan Globalisasi Urbanisme Kontemporer di Bandung. *Jurnal Pemikiran Sosiologi -UGM, 4*(1), 1-24.

Prasetyo, F., & Iverson, S. M. (2013). *Art, activism and the 'Creative Kampong': A case study from Dago Pojok, Bandung, Indonesia*. Paper presented at the PlanInk International Conference, Malang, Indonesia.

Prasetyo, F., & Sean, M. Iverson (2014). *Playing Under The Fly Over: A Collaborative Creative Community in Bandung*. Paper presented at the 5th Arte Polis International Conference and Workshop, Bandung, Indonesia.

Project for Public Spaces, Inc. (2008). *Streets as places: using streets to rebuild communities*. Author.

Rapoport, A. (1991). *Pedestrian street use: culture and perception in public streets for public use*. Columbia University Press.

Redman, L., Friman, M., Gärling, T., & Hartig, T. (2013). Quality attributes of public transport that attract car users: A research review. *Transport Policy*, *25*, 119–127. doi:10.1016/j.tranpol.2012.11.005

Refaat, A. M. (2019). Space Syntax Theory and Its Contribution to Urban Design. In H. Abusaada, C. Vellguth, & A. Elshater (Eds.), *Handbook of Research on Digital Research Methods and Architectural Tools in Urban Planning and Design* (pp. 203–222). IGI Global. doi:10.4018/978-1-5225-9238-9.ch010

Richard, M. G. (2010). What is Car Sharing and How Does it Work? *Treehugger*. Retrieved from https://www.treehugger.com/cars/what-is-car-sharing-and-how-does-it-work.html

Richter, J., Friman, M., & Gärling, T. (2010). Review of Implementations of Soft Transport Policy Measures. Transportation. *Theory & Application*, *2*, 5–18.

Rodriguez. C. (2015). Paris Will Stop All Traffic, Literally, For One Day in September. *Forbes*. Retrieved from https://www.forbes.com/sites/ceciliarodriguez/2015/08/19/french-car-ban-makes-history-for-one-day-in-september-there-will-be-no-traffic-in-paris/?fbclid=IwAR37zXakGjdupAUTAVCafFyBDvZGSGZIWyF2QuMe9bYko00Nl8ZQefsvuok#6020bad47793

Rogers, R. (1997). *Cities for a Small Planet*. Faber and Faber.

Rojas-Rueda, D., Nieuwenhuijsen, M., & Khreis, H. (2017). Autonomous vehicles and public health: Literature Review. *Journal of Transport & Health*, *5*, S13. doi:10.1016/j.jth.2017.05.292

Ruter#. (2018). *Emission Free public transport in Oslo and Akershus - Ruter. Version 10* (English translation of original document in Norwegian). Retrieved from https://ruter.no/contentassets/4fceaf4f35f54c0bbc0298805bc29241/report-zero-emission-english.pdf

Ryan, K.-L., Flink, C. A., Lagerwey, P., Balmori, D., & Searns, R. M. (1993). *Trails for the Twenty-First Century: Planning, Design, and Management for Multi-Use Trails*. Rails-to-Trails Conservancy.

Saar, M. a. H. P. (2009). *The Dimensions of Place Meanings Living* Rev (Vol. 3). Academic Press.

Salici, A. (2013). Greenways as a Sustainable Urban Planning Strategy. In M. Ozyavuz (Ed.), *Advances in Landscape Architecture* (p. 645). InTech. doi:10.5772/55757

Santos, G., Behrendt, H., Maconi, L., Shirvani, T., & Teytelboym, A. (2010a). Part I: Externalities and economic policies in road transport. *Research in Transportation Economics*, *28*(1), 2–45. doi:10.1016/j.retrec.2009.11.002

Santos, G., Behrendt, H., & Teytelboym, A. (2010b). Part II: Policy instruments for sustainable road transport. *Research in Transportation Economics*, *28*(1), 46–91. doi:10.1016/j.retrec.2010.03.002

Schiller, P. L., & Kenworthy, J. R. (2017). An introduction to sustainable transportation: Policy, planning and implementation. In *An Introduction to Sustainable Transportation* (2nd ed.). Policy, Planning and Implementation. doi:10.4324/9781315644486

Schimmack, U. (2008). The structure of subjective wellbeing. In M. Eid & R. J. Larsen (Eds.), *The science of subjective wellbeing* (pp. 97–123). Guilford Press.

Searns, R. M. (1995). The evolution of greenways as an adaptive urban landscape form. *Landscape and Urban Planning*, *33*(1–3), 65–80. doi:10.1016/0169-2046(94)02014-7

Sennett, R. (1977). *The fall of public man*. Knopf.

SFMTA - San Francisco Municipal Transportation Agency. (2016). *City of San Francisco: Meeting the smart city challenge*. Retrieved October 20, 2019, from https://www.sfmta.com/sites/default/files/projects/2016/SF%20Smart%20City%20Challenge_Final.pdf

Sharma, A. (2015). Urban greenways: Operationalizing design syntax and integrating mathematics and science in design. *Frontiers of Architectural Research*, *4*(1), 24–34. doi:10.1016/j.foar.2014.11.002

Shemkus, S. (2019). In Norway, Oslo leaders are creating a 'car-free city center.' Could Boston be next? *Energy News*. Retrieved from https://energynews.us/2019/07/15/northeast/in-norway-oslo-leaders-are-creating-a-car-free-city-center-could-boston-be-next/?fbclid=IwAR0S7tEIMp7vtybaudBq_1c_duF26IkHklrpc5DYDSxu0a_hPHe3mLu1_F8

Simsekoglu, Ö., & Klöckner, C. (2019). Factors related to the intention to buy an e-bike: A survey study from Norway. *Transportation Research Part F: Traffic Psychology and Behaviour*, *60*, 573–581. doi:10.1016/j.trf.2018.11.008

Singleton, P. A. (2019a). Validating the Satisfaction with Travel Scale as a measure of hedonic subjective wellbeing for commuting in a US city. *Transportation Research Part F: Traffic Psychology and Behaviour*, *60*, 399–414. doi:10.1016/j.trf.2018.10.029

Singleton, P. A. (2019b). Walking (and cycling) to wellbeing: Modal and other determinants of subjective wellbeing during the commute. *Travel Behaviour & Society*, *16*, 249–261. doi:10.1016/j.tbs.2018.02.005

Siregar, A. (1990). *Bandung, the architecture of a city in development: urban analysis of a regional capital as a contribution to the present debate on Indonesian urbanity and architectural identity* (Unpublished doctoral dissertation). Katholieke Universiteit Leuven, Leuven, Belgium.

Skinner, R., & Bidwell, N. (2016). *Making better places: Autonomous vehicles and future opportunities*. WSP Parsons Brinckerhoff Engineering Services.

Smaldone, D., Harris, C., & Sanyal, N. (2005). An exploration of place as a process: The case of Jackson Hole, WY. *Journal of Environmental Psychology*, *25*(4), 397–414. doi:10.1016/j.jenvp.2005.12.003

Smith, N. (1984). *Uneven Development*. Blackwell.

Soja, E. (1989). *Postmodern Geographies: The Reassertion of Space in Critical Social Theory*. Verso Press.

Soja, E. (1996). *Thirdspace: Journey to Los Angeles and Orher Real-and-Imagined Places*. Blackwell Publishers Inc.

Southworth, M. (2005). Designing the walkable city. *Journal of Urban Planning and Development*, *131*(4), 246–257. doi:10.1061/(ASCE)0733-9488(2005)131:4(246)

Speck, J. (2012). *Walkable city : how downtown can save America, one step at a time*. North Point Press.

Stanley, J. K., Hensher, D. A., Stanley, J. R., & Vella-Brodrick, D. (2011). Mobility, social exclusion and wellbeing: Exploring the links. *Transportation Research Part A, Policy and Practice*, *45*(8), 789–801. doi:10.1016/j.tra.2011.06.007

Steg, L. (2005). Car use: Lust and must. Instrumental, symbolic and affective motives for car use. *Transportation Research Part A, Policy and Practice*, *39*(2-3), 147–162. doi:10.1016/j.tra.2004.07.001

Surico, J. (2018). *What Happens When a City Bans Non-Resident Drivers? Citylab.com*. Retrieved October 20, 2019, from https://www.citylab.com/transportation/2018/04/the-small-town-that-took-on-waze/558215/

Suzuki, H., Murakami, J., Hong, Y.-H., & Tamayose, B. (2015). Financing Transit-Oriented Development with Land Values: Adapting Land Value Capture in Developing Countries. doi:10.1596/978-1-4648-0149-5

Suzuki, H., Cervero, R., & Iuchi, K. (2013). *Transforming cities with transit : transit and land-use integration for sustainable urban development*. World Bank. doi:10.1596/978-0-8213-9745-9

Taniguchi, A., Suzuki, H., & Fujii, S. (2007). Mobility management in Japan: Its development and meta-analysis of travel feedback programs. *Transportation Research Record: Journal of the Transportation Research Board*, *2021*(1), 100–109. doi:10.3141/2021-12

Tarigan, A. K. M., Sagala, S., Samsura, D. A. A., Fiisabiilillah, D. F., Simarmata, H. A., & Nababan, M. (2015). Bandung City, Indonesia. *Cities (London, England)*, *50*, 100–110. doi:10.1016/j.cities.2015.09.005

Taylor, M. (2017). Four select committees launch joint inquiry into UK air pollution crisis. *The Guardian*. Retrieved from https://www.theguardian.com/environment/2017/mar/20/four-select-committees-launch-joint-inquiry-uk-air-pollution-crisis

Taylor, S., Giang, C., Chau, P., & Aumann, P. (2017). *Cycling aspects of austroads guides* (3rd ed.). Austroads Ltd. Retrieved from www.austroads.com.au

Teipelke, R. (2016, April 13). *An Implementation Critique of Jan Gehl's "Cities for People"*. Retrieved from https://placemanagementandbranding.wordpress.com/2016/04/13/an-implementation-critique-of-jan-gehls-cities-for-people/

Thaler, R. H., & Sunstein, C. R. (2009). *Nudge: Improving decisions about health, wealth, and happiness*. Penguin.

The Behavioral Risk Factor Surveillance System. (2003). Walking trends among U.S. adults. *American Journal of Preventive Medicine*, 95–100. PMID:12880875

The Conversation. (2017). *Here's how driverless vehicles will utterly transform how our cities look*. Retrieved October 20, 2019, from https://futurism.com/heres-how-driverless-vehicles-will-utterly-transform-how-our-cities-look

The Copenhagen Wheel. (2019). Retrieved from http://senseable.mit.edu/copenhagenwheel/

Thøgersen, J. (2009). Promoting public transport as a subscription service: Effects of a free month travel card. *Transport Policy*, *16*(6), 335–343. doi:10.1016/j.tranpol.2009.10.008

Thompson, C. (2018). *The vehicle of the future has two wheels, handlebars, and is a bike*. Retrieved October 20, 2019, from https://www.wired.com/story/vehicle-future-bike/

Tibbalds, F. (1992). *Place matter most in making people-friendly towns: improving the public environment in towns and cities*. Taylor & Francis.

TomTom. (2019). Retrieved from TomTom: https://www.tomtom.com/en_gb/traffic-index/

Tønnesen, A., Meyer, S. F., Skartland, E. G., & Sundfør, H. B. (2016). *European cities with car-free city centres*. Institute of Transport Economics.

Topp, H., & Pharoah, T. (1994). Car-free city centres. *Transportation*, *21*(3), 231–247. doi:10.1007/BF01099212

Tuan, Y.-F. (1974). Space and Place: Humanistic Perspective. *Progress in Geography*, *6*, 211–252.

Tuan, Y.-F. (2001). *Space and Place: The Perspective Experience*. University of Minnesotta Press.

Turner, T. (1995). Greenways, blueways, skyways and other ways to a better London. *Landscape and Urban Planning*, *33*(1–3), 269–282. doi:10.1016/0169-2046(94)02022-8

Turner, T. (2006). Greenway planning in Britain: Recent work and future plans. *Landscape and Urban Planning*, *76*(1–4), 240–251. doi:10.1016/j.landurbplan.2004.09.035

Turner, T. H. D. (1984). Landscape planning: The need to train specialists. *Landscape Planning*, *11*(1), 73–79. doi:10.1016/0304-3924(84)90019-4

U.S. Department of Transportation. (2015). *Smart city challenge: lessons for building cities of the future.* Retrieved September 20, 2019, from https://www.transportation.gov/policy-initiatives/smartcity/smart-city-challenge-lessons-building-cities-future.

UNDESA. (2018). *The 2018 revision of the world urbanization prospects.* Retrieved October 10, 2019, from: https://www.un.org/development/desa/en

UN-Habitat. (2013). *Planning and Design for Sustainable Urban Mobility: Global Report on Human Settlements 2013. United Nations Human Settlements Programme (UN-Habitat).* Routledge. Retrieved from www.unhabitat.org

United Nations Advisory Committee of Local Authorities, UNACLA Secretariat. (2013). *Transport and mobility snapshots.* United Nations Advisory Committee of Local Authorities (UNACLA) and United Nations Human Settlements Programme (UN-Habitat). Retrieved from www.unhabitat.org

Urban Planet. (2018). Urban Planet: Knowledge towards Sustainable Cities. Cambridge: Cambridge University Press.

Urry, J. (2017). The Little Book of Car Free Cities. Imagination Lancaster, Lancaster University.

Van Der Velde, R., & De Wit, S. (2015). Late twentieth century green infrastructures in Paris. In *Flowscapes: designing infrastructure as landscape* (p. 208). Research in Urbanism Series., doi:10.7480/rius.3.838

Van der Vlugt, A. L., Curl, A., & Wittowsky, D. (2019). What about the people? Developing measures of perceived accessibility from case studies in Germany and the UK. Applied Mobilities, 1-21.

Velasco, A. (2016). Integration of the Public Transport System in Madrid Region. *Consorcio Regional de Transportes de Madrid (CRTM).* Retrieved from https://ebsf2.eu/sites/default/files/8.1%20EBSF_2%20Madrid%20Demo%20Context.CRTM.pdf

Vella-Brodrick, D. A., & Stanley, J. (2013). The significance of transport mobility in predicting wellbeing. *Transport Policy*, *29*, 236–242. doi:10.1016/j.tranpol.2013.06.005

Vorkinn, M. R., & Riese, H. (2011). Environmental Concern in a Local Context: The Significance of Place Attachment. *Environment and Behavior*, *33*(2), 249–263. doi:10.1177/00139160121972972

Voskuil, R. P. G. A. (2007). *Bandoeng, Beeld van een stad.* Planologi-ITB and Jagaddhita.

Waber, B., Magnolfi, J., & Lindsay, G. (2014). Workspaces that move people. *Harvard Business Review*, *92*(10), 68–77. PMID:25509577

Wachsmuth, D., Chaney, D., Kerrigan, D., Shillolo, A., & Basalaev-Binder, R. (2018). *The high cost of short-term rentals in New York City. A Report from the Urban Politics and Governance Research Group School of Urban Planning*. McGill University.

Wadud, Z., Mackenzie, D., & Leiby, P. (2016). Help or hindrance? the travel, energy, and carbon impacts of highly automated vehicles. *Transportation Research Part A, Policy and Practice*, *86*, 1–18. doi:10.1016/j.tra.2015.12.001

Wahlgren, L., & Schantz, P. (2014). Exploring bikeability in a suburban metropolitan area using the Active Commuting Route Environment Scale (ACRES). *International Journal of Environmental Research and Public Health*, *11*(8), 8276–8300. doi:10.3390/ijerph110808276 PMID:25153462

Waldheim, C. (2006). Landscape as Urbanism. In C. Waldheim (Ed.), *The Landscape Urbanism Reader - Review* (pp. 35–53). doi:10.1021/bi0013905

Waldheim, C. (2016). *Landscape as urbanism: a general theory*. Princeton University Press.

Walk Score. (2015). *Walk Score®*. Retrieved 2018, from Walk Score Methodology: https://www.walkscore.com/methodology.shtml

Walker, A. (2014). *Why Don't More Cities Have E-Bike Shares?* Retrieved from https://gizmodo.com/why-dont-more-cities-have-e-bike-shares-1595348781

Walker, J., & Johnson, C. (2016). *Peak car ownership: the market opportunity of electric automated mobility services*. Rocky Mountain Institute.

Walmsley, A. (1995). Greenways and the making of urban form. *Landscape and Urban Planning*, *33*(1–3), 81–127. doi:10.1016/0169-2046(95)02015-L

Walmsley, A. (2006). Greenways: Multiplying and diversifying in the 21st century. *Landscape and Urban Planning*, *76*(1–4), 252–290. doi:10.1016/j.landurbplan.2004.09.036

Wang, G. (2011). *Public Walking Space. A study of Commercial Pedestrian Streets in Copenhagen, Stockholm and Nanjing* (Diploma thesis). Karlskrona, Sweden.

Wang, L. (2014). *How the Cheonggyecheon river urban design restored the green heart of Seoul*. Retrieved November 1, 2019, from https://inhabitat.com/how-the-cheonggyecheon-river-urban-design-restored-the-green-heart-of-seoul/

Waygood, E. O. D., Friman, M., Taniguchi, A., & Olsson, L. E. (2019). Children's life satisfaction and travel satisfaction: Evidence from Canada, Japan, and Sweden. *Travel Behaviour & Society*, *16*, 214–223. doi:10.1016/j.tbs.2018.04.004

Waygood, E. O., Friman, M., Olsson, L. E., & Mitra, R. (Eds.). (2019). *Transport and Children's Wellbeing*. Elsevier.

Weinberger, R., & Sweet, M. N. (2012). Integrating walkability into planning practice. *Transportation Research Record: Journal of the Transportation Research Board*, *2322*(1), 20–30. doi:10.3141/2322-03

Westman, J., Olsson, L. E., Gärling, T., & Friman, M. (2017). Children's travel to school: Satisfaction, current mood, and cognitive performance. *Transportation*, *44*(6), 1365–1382. doi:10.100711116-016-9705-7

WHO Regional Office for Europe. (2017). *Urban green space interventions and health: A review of impacts and effectiveness*. Retrieved from http://www.euro.who.int/pubrequest

WHO. (2018). Global status report on road safety: Summary. *WHO Press*. Retrieved from https://www.who.int/violence_injury_prevention/road_safety_status/2018/English-Summary-GSRRS2018.pdf

WHO/UNEP. (1992). Urban air pollution in megacities of the World. World Health Organization. *United Nations Environment Programme*. Retrieved from https://apps.who.int/iris/handle/10665/39902

Whyte, W. H. (1980). *The Social Life of small urban spaces*. The Conservation Foundation.

Women and Equality Committee. (2017). *Building for Equality: Disability and the Built Environment- Ninth Report of Session 2016–17. Online, House of Commons*. Retrieved 03/03, 2020, from https://publications.parliament.uk/pa/cm201617/cmselect/cmwomeq/631/631.pdf

World Bank. (2012). *Who needs smart cities for sustainable development?* Retrieved October 20, 2019, from https://www.worldbank.org/en/news/feature/2012/03/20/who-needs-smart-cities-for-sustainable-development

Wright, L. (2005). *Carfree Development, Sustainable Transport. A source book for Policy Makers in developing Countries, Module 3e*. Retrieved from https://www.worldcarfree.net/resources/freesources/carfree_dev.pdf

Write, L. (2005). *Car Free Development*. GTZ.

Xiongbing, J., & Grammenos, F. (2013, May 21). *Taking the Guesswork out of Designing for Walkability*. Retrieved from Planetizen: https://www.planetizen.com/node/63248

Young, R., & Lieberknecht, K. (2016). From smart cities to wise cities: Ecological wisdom as a basis for sustainable urban development. *Journal of Environmental Planning and Management*, 1–18.

Zarrilli, D. (2019). *Zero Emission Vehicles*. Retrieved 2019, Retrieved from C40Cities: https://www.c40.org/networks/zero-emission-vehicles

Zhu, Y., & Diao, M. (2016). The impacts of urban mass rapid transit lines on the density and mobility of high-income households: A case study of Singapore. *Transport Policy*, *51*, 70–80. doi:10.1016/j.tranpol.2016.03.013

Zube, E. H. (1995). Greenways and the US National Park System. *Landscape and Urban Planning*, *33*, 17–25. Retrieved from https://ac-els-cdn-com.ezproxy.nottingham.ac.uk/0169204694020114/1-s2.0-0169204694020114-main.pdf?_tid=c31cb0ae-9269-4e2a-a1fb-bb8c3179fc57&acdnat=1540157014_6f8cf8cc5000a1c3683b83f7750c397a

About the Contributors

Rahma Doheim is an assistant professor in Architecture and the Built Environment. She received her Ph.D. in the Built Environment from the University of Ulster, UK. And, she holds a Master's in Safety management from the University of West Virginia, USA. Dr. Doheim's research interest falls under two research areas: urban ecology and architectural pedagogy. Her recent research focuses on Futur cities and has published recently a book chapter entitled: 'Smart city vision and practices across the Kingdom of Saudi Arabia—a review' Elsevier. She has a long track record of publications participating in international and Scopus indexed conferences, publishing in peer-reviewed journals, and writing several book chapters. Dr. Doheim is a registered architect in Egypt, and a member at the Charted Institute of Building Services Engineering (CIBSE), UK. She has taught and practiced architecture in the USA, the Middle East, and the UK. She was a Senior Designer with BONIAN Architects Firm, Egypt, and an Architect at the Herberger Centre for Design Excellence, Arizona State University, USA.

Alshimaa Aboelmakarem Farag is a researcher, educator, and urban designer, who has a special passion for creating smart cities that are inclusive, livable, healthy, and people-oriented. Her research interest lies in the Ecological Psychology that concerns human behaviors and their responsive reactions to the built environment, especially in the Middle East and Northern Africa Region (MENA). Also, she has a special interest in Urban Sustainability; therefore, she got the certificate (LEED AP ND) and (LEED AP Home). She is an Assistant Professor at Zagazig University, Egypt, in which she got her Ph.D. in Urban Design. She holds a Master degree from Cairo University, Egypt. Dr. Alshimaa is a member of the Editorial Review Board for the International Journal of Urban Planning and Smart Cities, in addition to participating in reviewing research works for several indexed journals, books, and conferences. She has taught and practiced Urban Design and Architecture in Egypt and Saudi Arabia, where she recently got the Queen Effat Award for Excellence in Teaching for the Academic Year 2018-2019.

Ehab Kamel is a senior lecturer and active researcher in architecture and urban design, with over twenty years of experience in both academia and practice. He is a registered architect (in Egypt and MENA countries) and an urban designer. Since 1998, he has worked on several urban and architectural design projects in the Middle East, North Africa, Asia, and Europe, leading design teams through several projects and winning architectural and urban design competitions. He taught architecture at top universities across three different continents. Just after completing his PhD at The University of Nottingham (UK), Ehab moved to China to establish the first international architecture course at the University of Nottingham's campus in Ningbo, China, before he started his work at UCLan in 2014. Ehab studies the relation between city cultures, and users' perception and wellbeing, which investigates how design for culturally rich sites learn from the past to create sustainable environments and people-focused public spaces. Ehab's work engages the study of urban spaces and architecture within, and in relation to, cultural heritage contexts. His research concerns the understanding of heritage as an asset to promote better sustainable urban development and city wellbeing.

* * *

Mohamed Abedo worked as a teaching assistant at the Architecture and Urban Design program at German University in Cairo. He is a graduate of Urban Planning and Design Department, Ain Shams University in 2014. Abedo is also a postgraduate student in the same department he graduated from. His interested is in studying people perception of cities and how the urban form affects it. He participated in several workshops and competitions addressing that issue. Abedo had a previous professional experience working as a landscape architect participating in local and international projects. Abedo also is involved in education development and interaction among students; he was the president and one of the cofounders of the first architectural students' community in Ain Shams University so-called FEDA Society. His role in this community was managing and organising student-to-student workshops, events and local and national student competitions.

Nora Osama Ahmed is an associate researcher at Architecture and Housing Research Institute, at the Housing and Building National Research Center (HBRC) in Cairo, Egypt, graduated from the Faculty of Regional and Urban Planning (FRUP) at Cairo University. Her Master's thesis in Livable Streets Guidelines received the Award for Excellence in Graduate Studies (Master of Science) from the Faculty of Engineering, Cairo University, as a Ph.D. candidate she focuses her postgraduate degree on creating urban livability by reorienting city design towards pedestrians, cyclists, and transport users. Her research interest focuses on the livability of cities and neighborhoods, with a particular focus on streets. She is concerned with understanding transport planning from a social sciences perspective, she addressed the issue of mobility in the human scale and the design of cities around the needs of people instead of cars through participating in international conferences and workshops as well as publishing in scientific journals and conference proceedings.

Frans Ari Prasetyo is an independent researcher and photographer. He is with the Ethnography Lab, University of Toronto. He works in Indonesia on urban politics with various grassroots communities, underground scenes and collectives, urban poor and the urban marginalized, and has published several articles-photos and chapters on these issues.

Samaa Badawi is an Associate Professor of Architecture and urban design at Mansoura University, Egypt. She received her Ph.D. in urban design in 2013. Her Ph.D. focuses on the role of form-based code in achieving sustainable urban communities in Egypt. Dr. Samaa was awarded the university prize for the best Ph.D. in 2014 and the Queen Effat Award for excellence in Teaching in 2016. She has several published papers in the field of urban design. She works as a reviewer for national and international journals. Her research interest focuses on sustainable urban movements.

Jim Dyson is a practising registered architect and studio tutor at universities in Manchester, Preston and Hong Kong. With over 30 years at the 'sharp end', Jim brings a wealth of professional experience to students of this practical art. He is also developing a body of work in contemplation of the urban condition, with regular art gallery exhibits and research into typological studies. All these activities have in common through pedagogy, activism, architecture, and its representation, the ambition to question our current environmental governance and enrich studio projects with good civic sense, appropriate to a program which values this contextual approach.

Abeer Elshater, PhD, is a professor of Urban Morphology at Ain Shams University (ASU), Cairo, Egypt, teaching and supervising multidisciplinary topics in the ideology of urban design. Her area of interest is in urban design discipline. She has worked on some international research projects with international universities. Elshater is an author and co-author of published books as well as a number of scientific manuscripts in international journals and conference proceedings. Elshater is an editor and reviewer in several indexed journals. Recently, she acts as an ambassador of the Regional Studies Association, United Kingdom.

Margareta Friman is a Professor in Psychology. Pro-Vice Chancellor at Karlstad University.

George Hallowell is a faculty member of the North Carolina State University's College of Design, in the School of Architecture. He is a registered architect and has worked in the profession for 30 years. He has a Ph.D. in Design, a Master of Architecture degree, and B.A. in Urban Geography. His research focuses on urban and suburban morphology, change, and the concept of structural inertia relating to the formal and spatial characteristics of older urban neighborhoods. Other research and publications have concentrated on acoustics, environmental justice, healthy and smart city design, and sharing technologies and their relationship to the design and use of urban public spaces. He has presented at numerous architecture and urban design conferences and has peer-reviewed articles or proceedings published by ISUF, EDRA, ACSA, AAG, ARCC, Spaces and Flows, Landscape and Urban Planning, International Journal of Architectural Research, Terra@Plural Journal, and the Journal of Space Syntax.

Christopher D. Ives is an interdisciplinary researcher with expertise in urban social-ecological systems and the human dimensions of environmental management. He joined the University of Nottingham in 2016 after holding research positions at universities in Australia and Germany. Chris is interested in understanding how people relate to and interact with the natural world, and how social and environmental knowledge can be used to inform action for sustainability.

Sarmada Madhulika Kone is an independent researcher. She received her Bachelors degree in Architecture from School of Planning and Architecture, India in 2014 and Masters degree in Urban Development from Maulana Azad National Institute of Technology, India in 2017. Since 2017, she has been with Sri Venkateshwara College of Architecture, where she is an assistant professor. Her current research interests include urban management, participatory planning and urban cognition. 'A Planning Model of Cognitive Cities' is her recent publication in the book 'Spatial Planning in Big Data Revolution'.

Yanhua Lu is a PhD in Design student at North Carolina State University. She has a Master of Landscape Architecture degree from University of Massachusetts Amherst and Bachelor of Landscape Architecture degree from Beijing Forestry University. With a background in Landscape Architecture and Urban Design, Lu's research focuses on walkability in smart cities, with an emphasis on technologies that change human travel behavior and promote walking in terms of public health and wellbeing.

Celen Pasalar, Ph.D., is an Associate Professor of Landscape Architecture at College of Design, North Carolina State University in United States. She has a Ph.D. in Design, a Master of Science in Urban Design, and Bachelor of Architecture degree. She is a registered architect in Northern Cyprus. Her research focuses on human-environment relationship, sustainability, and the design of smart, healthy and resilient communities. Her current research explores the impact of sharing technologies on human experiences and the design of urban public spaces and overall cities. She worked on numerous award-winning design projects promoting economic and sustainable development in urban and rural communities that also help strengthen human health. Dr. Pasalar is also a recipient of 2020 Council of Educators in Landscape Architecture - Excellence in Service Learning award. She was the co-chair of the Environmental Design Research Association's 47th annual international conference. She actively shares her research and scholarship at interdisciplinary conferences. She has numerous peer-reviewed chapters, articles, proceedings, and abstracts published by Cambridge Scholars Publishing, Comportements, Journal of Informatics, International Journal of Architectural Research, Terr@Plural Journal, Ekistics, and EDRA.

Nicole Porter (PhD, MArch, BPD, PGCHE, FHEA) is an Associate Professor of Architecture and Landscape at the University of Nottingham, UK. Her work focuses on the many values that humans associate with nature – ecological, philosophical, psychological, symbolic, spiritual, economic – and interprets how these values are embodied in contemporary landscape design and policy. She is engaged in a range of research and teaching projects, including cross-cultural comparisons of biophilic design, blue-green infrastructure, national park design, place-making and spirituality, and integrating mindfulness and landscape for wellbeing. She is the author of Landscape and Branding: the promotion and production of place (Routledge, 2016).

Amal Ramadan is a registered architect in Egypt and the Middle East. She had her BSc in Architectural Engineering from Ain Shams University, Cairo, Egypt in 1998, where she had practiced architecture at one of Egypt's largest architecture and engineering consultancy bureaux for ten years. Amal completed an MSc in Property Management (with distinction) in 2010 at Nottingham Trent University, in the United Kingdom. At the time of finalizing this Book, Amal is completing her final stage of minor corrections, post passing the oral examination for her Ph.D. from The University of Nottingham, conducted between China and the UK. Amal's research concerns users' wellbeing, through developing psychologically supportive design strategies and measures, where her Ph.D. bridges across architecture, psychology, and medicine disciplines. Currently, Amal works within a mental-health support team at the UK's National Health Service (NHS).

Mohamed Salheen obtained his BSc in Urban Planning and Design in 1993 from Ain Shams University, Cairo. He was appointed as teaching staff at the department and later received a PhD scholarship to obtain his PhD in Urban Design from Edinburgh College of Art, UK in 1997 with a thesis on "Comprehensive Analysis Approaches in Urban Settings". From 2001 until 2014 he acted as an assistant and associate professor at Ain Shams University teaching and supervising multidisciplinary topics. Developing a clear research line, he was then promoted as the first Professor in Egypt of "Integrated Planning and Design". To develop this research field Salheen had to develop various research lines at different levels ranging from Architectural Design to Regional Planning with diverse topics ranging from spatial analysis to socioeconomic and cultural aspects, yet all focused on the intermediate gaps and Integrated solutions.

Robert Turner is a Senior Project Manager working client-side in the construction industry with over fifteen years' experience of delivering major capital projects in the north of England. He is currently leading on developing the University of Central Lancashire's £57m flagship Student Centre building & University Squares public realm scheme in Preston, being delivered as part of a wider £200m Campus Masterplan to create enhanced connectivity both across campus and between the University Quarter and City Centre. He has previously been the principal lead on developing and delivering major capital investment projects across the educational estate in Liverpool, England. Bob has a successful record of applying project management methodologies to a range of capital schemes with significant experience of working across multi-disciplinary teams focused on the procurement and operation of new building accommodation and public realm spaces, leading on the planning, development, delivery and operation of new educational facilities and urban spaces. To date, Bob has delivered £230million worth of projects and approximately 800,000sq feet of internal building space with a keen focus on sustainability and Social Value. This is Bob's first foray into formal writing following his recent interest and experiences in understanding how accessibility issues can both enhance and hinder the design outcomes of integrated public realm spaces.

Abdulrahman A. Zawawi (BLA, MLA) is a lecturer at the Department of Landscape Architecture, King Abdulaziz University who is on academic leave to pursue a PhD degree. Currently, Mr Zawawi is a PhD in Architecture (Social Science) student at the University of Nottingham, Department of Architecture and Built Environment. He is interested in research about overcoming automobile dependence and the transition towards sustainable urban mobility. With an educational and a professional background in the field of landscape architecture, Mr Zawawi's current research focuses on greenways, walkability, non-motorized transport, and public open spaces.

Index

I

K

L

M

N

O

P

S

T

U

W

Y

www.ingramcontent.com/pod-product-compliance
Lightning Source LLC
LaVergne TN
LVHW081338300926
849161LV00008B/2003